Glycopolymer Code

Synthesis of Glycopolymers and Their Applications

RSC Polymer Chemistry Series

Editor-in-Chief:
Professor Ben Zhong Tang, *The Hong Kong University of Science and Technology, Hong Kong, China*

Series Editors:
Professor Alaa S. Abd-El-Aziz, *University of Prince Edward Island, Canada*
Professor Stephen Craig, *Duke University, USA*
Professor Jianhua Dong, *National Natural Science Foundation of China, China*
Professor Toshio Masuda, *Shanghai University, China*
Professor Dr Christoph Weder, *University of Fribourg, Switzerland*

Titles in the Series:

1: Renewable Resources for Functional Polymers and Biomaterials
2: Molecular Design and Applications of Photofunctional Polymers and Materials
3: Functional Polymers for Nanomedicine
4: Fundamentals of Controlled/Living Radical Polymerization
5: Healable Polymer Systems
6: Thiol-X Chemistries in Polymer and Materials Science
7: Natural Rubber Materials: Volume 1: Blends and IPNs
8: Natural Rubber Materials: Volume 2: Composites and Nanocomposites
9: Conjugated Polymers: A Practical Guide to Synthesis
10: Polymeric Materials with Antimicrobial Activity: From Synthesis to Applications
11: Phosphorus-Based Polymers: From Synthesis to Applications
12: Poly(lactic acid) Science and Technology: Processing, Properties, Additives and Applications
13: Cationic Polymers in Regenerative Medicine
14: Electrospinning: Principles, Practice and Possibilities
15: Glycopolymer Code: Synthesis of Glycopolymers and Their Applications

How to obtain future titles on publication:
A standing order plan is available for this series. A standing order will bring delivery of each new volume immediately on publication.

For further information please contact:
Book Sales Department, Royal Society of Chemistry, Thomas Graham House, Science Park, Milton Road, Cambridge, CB4 0WF, UK
Telephone: +44 (0)1223 420066, Fax: +44 (0)1223 420247
Email: booksales@rsc.org
Visit our website at www.rsc.org/books

Glycopolymer Code
Synthesis of Glycopolymers and Their Applications

Edited by

C. Remzi Becer
Queen Mary University of London, UK
Email: r.becer@qmul.ac.uk

and

Laura Hartmann
MPI of Colloids and Interfaces, Potsdam-Golm, Germany
Email: laura.hartmann@hhu.de

ROYAL SOCIETY OF CHEMISTRY

THE QUEEN'S AWARDS FOR ENTERPRISE: INTERNATIONAL TRADE 2013

RSC Polymer Chemistry Series No. 15

Print ISBN: 978-1-84973-978-8
PDF eISBN: 978-1-78262-266-6
ISSN: 2044-0790

A catalogue record for this book is available from the British Library

Published by The Royal Society of Chemistry,
Thomas Graham House, Science Park, Milton Road,
Cambridge CB4 0WF, UK

Registered Charity Number 207890

For further information see our web site at www.rsc.org

Preface

Sugars have long been recognized for their role in biology and medicine and their synthesis and modifications are a constant challenge for organic chemists. However, in the world of polymer chemistry and material sciences, sugars have often been neglected. While peptides and oligonucleotides – the other two biomacromolecules – have inspired tremendous efforts in creating biohybrid or bioinspired polymers and materials, the sugars have been mainly used as a renewable resource for starting materials not taking into account their biological properties. Sugars mediate a number of biological events such as inflammation and infection *via* their interaction with protein receptors. However, the interaction of a single sugar ligand is very weak and Nature uses the so-called glycocluster or multivalency effect to have several sugar ligands interact simultaneously and create a strong binding event. Based on this concept, polymer chemists have synthesized glycopolymers presenting a large number of sugar ligands along a polymeric backbone, thus creating high-affinity ligands. Although the concept is fairly simple, the covalent attachment of sugar ligands to a polymeric backbone, the synthesis can be challenging, including different strategies for the conjugation of sugar ligands and the variation of the polymer chains. Tremendous progress has been made in recent years in synthesizing glycopolymers and glycomaterials. The ability of sugars to introduce biofunctionality into synthetic materials has been recognized and shown for a number of applications ranging from drug design to biosensors. However, the field is still in its infancy and many of the synthesized glycopolymers and -materials have not yet been studied for their potential properties and applications.

The chapters in this book are concerned with different classes of glycopolymers and glycomaterials and specifically focus on the different synthetic strategies that have been developed over the last few years. In the first chapter,

RSC Polymer Chemistry Series No. 15
Glycopolymer Code: Synthesis of Glycopolymers and Their Applications
Edited by C. Remzi Becer and Laura Hartmann

Published by the Royal Society of Chemistry, www.rsc.org

Lindhorst focuses on the lectin structures that are the lock of Glycopolymer Code. We believe that it is crucial to understand the locking mechanism first in order to attempt to crack the code. The following chapter is concerned with the preparation of glycopolymers, where Ting and Stenzel provide an extremely detailed account of state-of-the-art synthesis techniques for glycopolymers. In the third chapter, Krannig and Schlaad provide insights into glycopolypeptides. This chapter is followed by Dondoni and Marra's excellent introduction to glycocalixarenes and their molecular recognition. In Chapter 5, Voit and her colleagues focus on the dendritic architectures of synthetic glycan structures and their use in brain disease therapy. In Chapter 6, Chen and her colleagues discuss the glycomaterials that grow larger as we look at the self-assembly of glycopolymers and their formation of vesicles and hydrogels. In the last three chapters, the focus is concentrated on the applications of glycopolymers. In Chapter 7, Narain and his colleagues discuss glyconanoparticles and their biomedical applications. Following on, Chapter 8, by Fernández-García and Muñoz-Bonilla, then outlines the great potential for such hybrid glycomaterials in various biomedical applications. In the final chapter, Miura and Seto discuss recent literature examples on the use of glycopolymers in biosensing applications. We believe that all these various applications provide insights into Glycopolymer Code that require a more systematic approach, similar to glycomics, in order to crack Nature's sugar code and create the Glycopolymer Code.

Overall, we hope that this book will inspire interested research students and academics alike and support their education, teaching and research and thus further promote the field of glycopolymers and glycomaterials.

We are very grateful to all authors of the chapters in this book who kindly agreed to support this project and we would like to thank them for their enthusiasm and excellent work.

Contents

RSC Polymer Chemistry Series No. 15
Glycopolymer Code: Synthesis of Glycopolymers and Their Applications
Edited by C. Remzi Becer and Laura Hartmann

Published by the Royal Society of Chemistry, www.rsc.org

CHAPTER 1

Small Molecule Ligands for Bacterial Lectins: Letters of an Antiadhesive Glycopolymer Code

THISBE K. LINDHORST*[a]

[a]Otto Diels Institute of Organic Chemistry, Christiana Albertina University of Kiel, Otto-Hahn-Platz 3–4, 24118 Kiel, Germany
*E-mail: tklind@oc.uni-kiel.de

1.1 Introduction

This chapter discusses how glycopolymers might function in the context of microbial adhesion. This is an important topic as attachment of viruses and bacteria to surfaces is a global problem and for host organisms it has fundamental implications for their vitality. This was considered when the human microbiome project was launched in 2008. Consequently, the human microbiome project is dedicated to research into how changes of microbial colonization influence human health and disease.[1]

It has turned out that microbial colonization of the body is largely associated with the glycoconjugate decoration of the host cells, named the 'glycocalyx.' The glycocalyx of a cell is an extracellular compartment comprising a huge variety of different glycoconjugates. Strikingly, it forms an anchoring

RSC Polymer Chemistry Series No. 15
Glycopolymer Code: Synthesis of Glycopolymers and Their Applications
Edited by C. Remzi Becer and Laura Hartmann

Published by the Royal Society of Chemistry, www.rsc.org

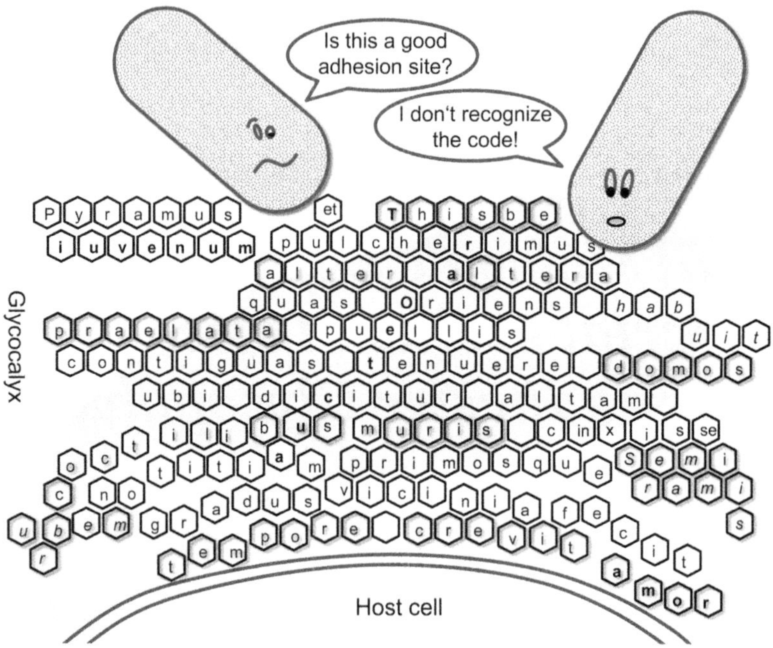

Figure 1.1 'Deciphering the glycocode.'[4] Cartoon to exemplify that carbohydrate-specific adhesion of, *e.g.*, a bacterial cell to the glycocalyx of a host cell might be looked at as reading a code.

platform for invading microbes. It has been asked how carbohydrate recognition has evolved among microbes,[2] how it is regulated and how it develops during the lifetime of an organism, in other words, how binding to cell surface carbohydrates is being 'spelled' (Figure 1.1). It has been suggested that the oligo- and polysaccharide structures that are expressed on cell surfaces function in the sense of a 'glycocode,'[3] thus paralleling the biology of carbohydrates with the alphabet of a language, in order to decipher its meaning.[4] Of course, it is sensible to consider the diversity of carbohydrate structures as a biologically meaningful concert corresponding to the whole of molecular interactions. Glycopolymers can be regarded as a means to interrogate a putative carbohydrate alphabet and, moreover, as a powerful tool to prevent microbial colonization of surfaces.

1.2 Lectin-Mediated Bacterial Adhesion

To colonize cell surfaces of the host, bacteria, for example, have to accomplish a process of adhesion in order to withstand natural defence mechanisms and mechanical shear stress. Stable adhesion can lead to the formation of bacterial biofilms, which is accompanied by vital advantages for the microbial colonies[5] but disadvantages for the host. Finally, adhesion apparently is a prerequisite for bacterial infections that constitute a major global health

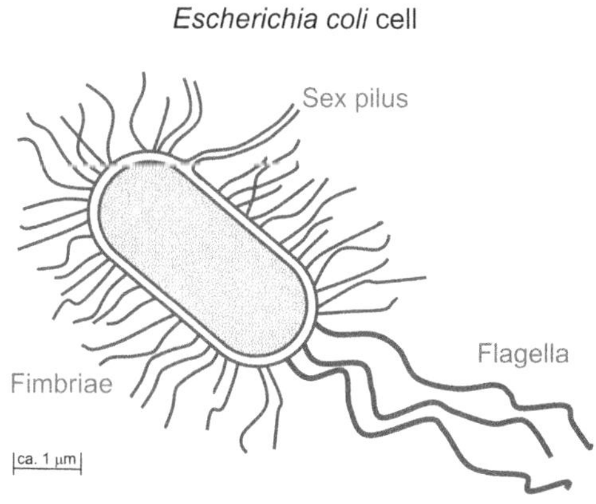

Figure 1.2 A majority of bacterial cells, such as *E. coli*, are equipped with three types of hair-like protein appendages, named pili, fimbriae and flagella. Fimbriae serve as adhesive organelles, mediating adhesion to the glycocalyx of host cells. *E. coli* cells are covered with several hundred copies of fimbriae of different carbohydrate specificity.

problem, in particular in developing countries. Bacterial infections are especially dangerous for newborns and young children,[6] with the most common serious neonatal infections involving bacteremia, meningitis and respiratory tract infections. Key pathogens in these infections are *Escherichia coli*, *Klebsiella* sp., *Staphylococcus aureus* and *Streptococcus pyogenes*.[7]

One important mechanism of bacterial adhesion is based on molecular interactions between cell surface carbohydrates of the host and specialized carbohydrate-specific bacterial proteins called adhesins or lectins. Lectins were first described at the end of the 19th century,[8] when it was shown that plant lectins have the ability to agglutinate erythrocytes blood group specifically. As we know today, this is a result of a multivalent carbohydrate–lectin interaction. In 1954, Boyd and Shapleigh proposed the term lectin 'for these and other antibody-like substances' with blood group-specific agglutination properties.[9] In the 1990s, Lis and Sharon[10] suggested that 'lectin' should be used as a general name for all proteins of non-immune origin that possess the ability to agglutinate erythrocytes and other cell types. Early classification of lectins relied on their carbohydrate specificity. However, today lectins are grouped on the basis of their structural features and especially the relatedness of their carbohydrate binding sites, which are often called 'carbohydrate recognition domains,' or CRDs.[11–13]

It is common knowledge today that lectins are ubiquitously spread in Nature, comprising many different functions in different organisms.[14] Also, many bacteria, in particular those of the Enterobacteriaceae family, have the ability to agglutinate erythrocytes by their own lectins. This haemagglutination activity of bacteria is almost always associated with the presence of multiple filamentous protein appendages projecting from the surface of the bacteria.[15] These are called fimbriae (from the Latin word for 'thread') and also, less correctly, pili (from the Latin word for 'hair') (Figure 1.2). Whereas pili are involved in gene transfer between bacteria ('sex pili') and flagellae have the role of sensory organelles used for moving, fimbriae serve as adhesive organelles. Fimbriae contain lectin subunits, which mediate carbohydrate-specific adhesion to cell surfaces (and also cell agglutination). Thus, bacteria utilize the sugar decoration of cells – the glycocalyx – to colonize the cell surface, wherever cells are in contact with the outside environment, as for example in the case of epithelial cells.

1.3 Carbohydrate Specificity of Type 1 Fimbriae

Type 1 fimbriae are particularly efficient adhesion tools of bacteria to mediate the colonization of various biotic and abiotic surfaces. They are uniformly distributed on the bacterial cell surface with their length varying between 0.1 and 2 μm and a width of ~7 nm. Since the 1970s, numerous studies have been carried out to elucidate the carbohydrate specificities of bacterial adhesion, in particular of type 1 fimbriae-mediated adhesion of *E. coli.*[6] A key finding of this research was that the type 1 fimbrial lectin, called FimH, requires α-D-mannose and α-D-mannosides for binding. The other anomer, namely β-mannosides, cannot be complexed within the carbohydrate binding site. This knowledge suggested that type 1 fimbriated bacteria can adhere to tissues expressing glycoproteins of the high-mannose type, exposing multiple terminal α-D-mannosyl units.[16] For example, urinary tract infections are caused by uropathogenic *E. coli* (UPEC). Type 1 fimbriae are present in at least 90% of all known UPEC strains, where they are important pathogenicity factors.[6,15] Today, it is known that bacterial adhesion to the surface of urothelial cells is mediated by FimH binding to oligomannoside residues of the glycoprotein uroplakin Ia. This interaction is a prerequisite for bacterial invasion.[17] Consequently, much effort has been invested in the development of potent inhibitors of type 1 fimbriae-mediated bacterial adhesion in order to prevent bacterial adhesion to mucosa and thus treat bacterial infection in an approach that has been called antiadhesion therapy.[18,19]

In this context, a second feature of type 1 fimbriae-mediated bacterial adhesion that was discovered already quite early is important.[16] It was found that α-D-mannosides with an aromatic aglycone moiety exhibit an improved affinity to the bacterial lectin and an enhanced potency as inhibitors of type 1 fimbriae-mediated bacterial adhesion to surfaces. Today, this finding is well understood based on the X-ray studies of the structure of the type 1 fimbrial lectin FimH that have been published since 1999.[20–23] Structural biology has shown that the entrance of the carbohydrate binding site of FimH is flanked by two tyrosine residues, Y48 and Y137, which make π–π interactions with an aromatic aglycone of an α-D-mannoside ligand that is complexed within the cavity of the FimH carbohydrate binding site (Figure 1.3).

This and other structural features of the bacterial lectin FimH have been described elsewhere[6,24–26] and are not further detailed in this account. Similarly, the biosynthesis of type 1 fimbriae has been elucidated and reviewed.[27,28] Briefly, the fimbrial appendage is assembled in the outer membrane of

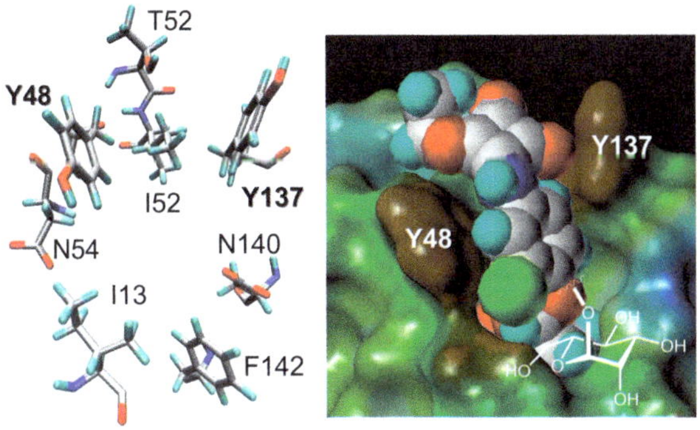

Figure 1.3 Left: spatial orientation of the amino acid residues at the entrance of the carbohydrate binding site of the bacterial lectin FimH as revealed by crystallography. The tyrosine residues Y48 and Y137 form a so-called 'tyrosine gate' that mediates the comparatively high affinity of mannosides with an aromatic aglycone by formation of π–π interactions. Right: the FimH carbohydrate binding site depicted as a Connolly surface, complexed with the mannoside **2** (*cf.* Figure 1.5). Mannoside **2** is shown as a CPK model. The large chloro substituent of **2** pointing towards the observer fills a depression at the ridge of the carbohydrate binding site, thereby improving affinity. The ring structure in white explains how α-D-mannoside ligands are located within the FimH cleft, the α-glycosidic bond sticking out of the binding site.

Gram-negative bacteria in a process called the chaperone–usher pathway. To be able to judge the potential value of especially glycopolymers as inhibitors of type 1 fimbriae-mediated bacterial adhesion, it is important to know that FimH is a two-domain protein, terminating every type 1 fimbrial rod (Figure 1.4). The so-called pilin domain of FimH, $FimH_P$, is required to anchor the protein at the fimbrial tip, comprising also the subunits FimF and FimG. The lectin domain $FimH_L$, on the other hand, accommodates the α-D-mannoside-specific carbohydrate-binding site. Both FimH domains are interconnected by a hinge region, permitting allosteric regulation of the carbohydrate binding site (see below).[29,30]

1.4 Tailor-Made FimH Antagonists

Based on the information obtained in structural biology studies, molecular modelling was employed to design tailor-made ligands of FimH and FimH antagonists. Synthesis and testing of these non-natural α-D-mannosides have only recently led to a revival of the idea of an antiadhesion therapy against microbial infection.[31] In the 1990s, Lindhorst *et al.* introduced the idea of using multivalent α-D-mannoside clusters to inhibit effectively type 1 fimbriae-mediated bacterial adhesion in a potential therapeutic context.[32] Later, molecular docking studies led to promising new monovalent mannosides as potent FimH antagonists.[33] Many groups have added various

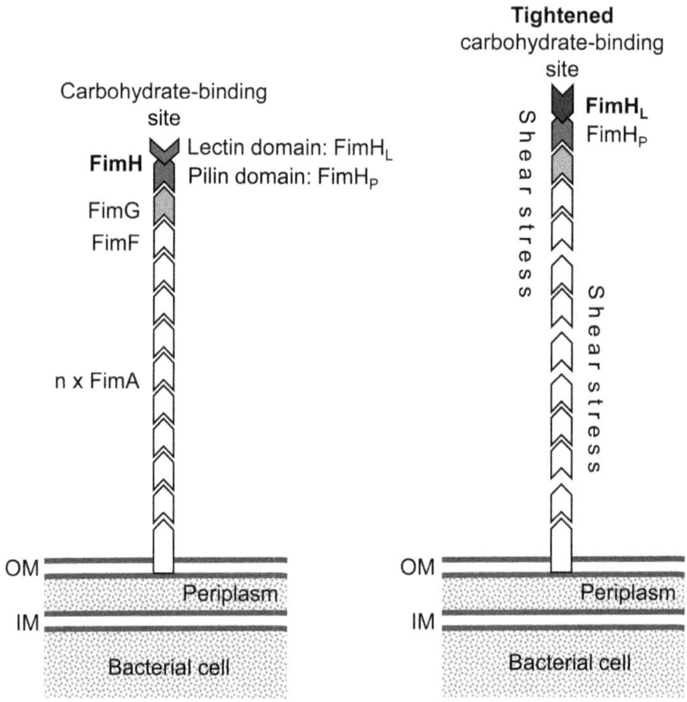

Figure 1.4 Left: type 1 fimbriae are projecting from the outer membrane (OM) of Gram-negative bacteria where they are assembled in a process named the chaperone–usher pathway. The fimbrial rod is a right-handed helical structure composed of numerous FimA subunits, terminated by the fimbrial tip comprising FimF, FimG and the lectin FimH, the last mediating α-D-mannose-specific adhesion. FimH is a two-domain protein composed of a pilin and a lectin domain. Right: when shear stress is exerted, type 1 fimbriae can be elongated along the FimA helix and, in addition, FimH undergoes an allosteric rearrangement in which the lectin domain is stretched and mannose binding is enhanced concurrently. This allosteric process leading to enhanced ligand binding as a result of mechanical stress is called the 'catch bond' mechanism.

high-affinity ligands of FimH to the arsenal of synthetic FimH antagonists with even nanomolar affinities, which is particularly potent in the context of lectin binding.[34] The literature on the design and testing of FimH antagonists and inhibitors of type 1 fimbriae-mediated bacterial adhesion was reviewed in 2011.[35] Since then, Ernst's group has complemented the collection of promising FimH antagonists with new mannosides that were shown to be non-toxic and active *in vivo*. The best representatives have even entered clinical studies concerning their value in the treatment of cystitis.[36–39]

Five key representatives of highly potent FimH ligands are illustrated in Figure 1.5. Heptyl α-D-mannoside (**1**) was found to possess a 440-times greater power as inhibitor of type 1 fimbriae-mediated bacterial adhesion than methyl α-D-mannoside (not shown).[40] The squaric acid derivative **2** has

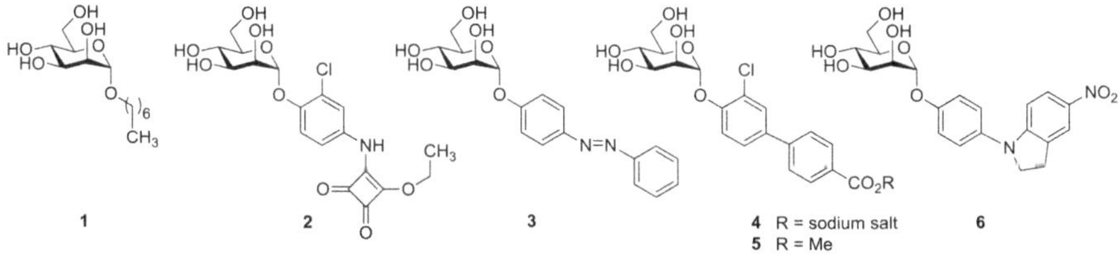

Figure 1.5 Structures of five potent representatives of small molecule FimH antagonists.

approximately 10 times the inhibitory potency of its analogue *o*-chloro-*p*-nitrophenyl α-D-mannoside (not shown).[33] The photosensitive azobenzene mannoside **3** has almost the same inhibitory power as *p*-nitrophenyl α-D-mannoside (not shown) but has little value as an inhibitor of bacterial adhesion *in vivo* owing to its limited water solubility.[41,42] The respective mannobioside on the other hand, is beneficial as a water-soluble and photosensitive FimH ligand.[41]

The biphenyl mannosides **4** and **5** are the result of recent systematic lead optimization.[37,43] Ernst and co-workers determined the K_D values of **1** and **4**, among others, by an in-solution affinity assay as K_D (**1**) = 5.6 ± 1.6 nM and K_D (**4**) = 0.71 ± 0.01 nM.[39] Indolylphenyl and indolinylphenyl mannosides have also been added to the collection of potent FimH antagonists.[38] Testing of the most potent indolinylphenyl mannoside **6** revealed that the administration of a low dosage such as 1 mg kg^{-1} (corresponding to approximately 25 μg per mouse) was sufficient to prevent urinary tract infection in mice for more than 8 h.[38] Bacterial colonization of the bladder could be reduced by almost four orders of magnitude, comparable to a standard antibiotic treatment. Hence the vision of an antiadhesion therapy to complement antibiotic treatment, which currently struggles with antibiotic resistance, becomes realistic.

1.5 Multivalent Inhibitors of Bacterial Adhesion

Without doubt, multivalency of molecular interactions is an important factor in carbohydrate recognition. Numerous multivalency effects have been observed in the glycosciences and interpreted according to various models.[44–48] Lee and Lee were the first to observe a multivalency effect in lectin binding employing relatively small di- and trivalent cluster glycosides and coined the term 'cluster effect' for the observations made.[49,50] At the time, they concluded that multivalent carbohydrate binding sites occur in lectins, but when the multivalency of lectin CRDs became more and more obvious based on many crystal structures, researchers became especially interested in the design of multivalent glycomimetics to produce highly potent lectin ligands or inhibitors of lectin-mediated cellular adhesion. Among others, multivalent glycomimetics based on non-carbohydrate polymer scaffolds were launched and named 'glycopolymers' to distinguish them from natural polysaccharides.[51] Roy and Kiessling and their colleagues were early pioneers

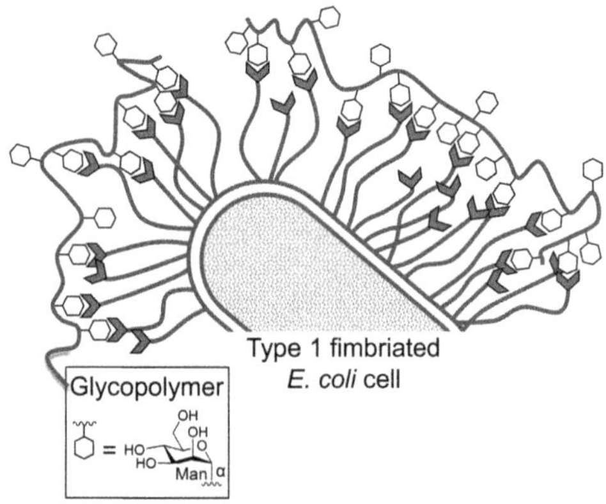

Figure 1.6 Glycopolymers equipped with specific carbohydrate residues (*e.g.* α-D-mannosyl units) could interact with, for instance, type 1-fimbriated bacteria in a multivalent fashion to effect a general deprivation of adhesive ability.

in the field,[52,53] but more recently modern polymer chemistry has further stimulated the synthesis and application of glycopolymers.[54]

Thus, glycopolymers may also be designed to inhibit fimbriae-mediated bacterial adhesion. In spite of the fact that the type 1 fimbrial lectin FimH is an explicitly monovalent lectin, multivalent mannosides have shown favourable effects as inhibitors of type 1 fimbriae-mediated bacterial adhesion.[35] Such findings might be due to statistical effects on the one hand, taking advantage of the dense packing of mannoside ligands in close proximity to one FimH carbohydrate binding site. On the other hand, appropriately dimensioned multivalent mannosides could also lead to simultaneous binding of multiple FimH units on multiple (~100–400 per bacterial cell) copies of type 1 (Figure 1.6).

Certainly, multivalent binding of glycopolymers to fimbriated bacteria or other microbes, that adhere carbohydrate-specifically, is not *per se* effective as the entropic penalty that can occur upon fixation of the formerly flexible polymer on the bacterial surface might prevent high-affinity binding. However, based on modern polymer chemistry, this problem could be circumvented. Thus, appropriately 'encoded' glycopolymer inhibitors of microbial adhesion to surfaces could indeed be provided as powerful tools for various different applications.

1.6 Carbohydrate 'Letters' for Antiadhesive Glycopolymers

For the systematic fabrication of antiadhesive glycopolymers the principal design outlined in Figure 1.7 can be envisaged. According to this approach, a polymer backbone can be designed with desired hydrophobicity, steric and

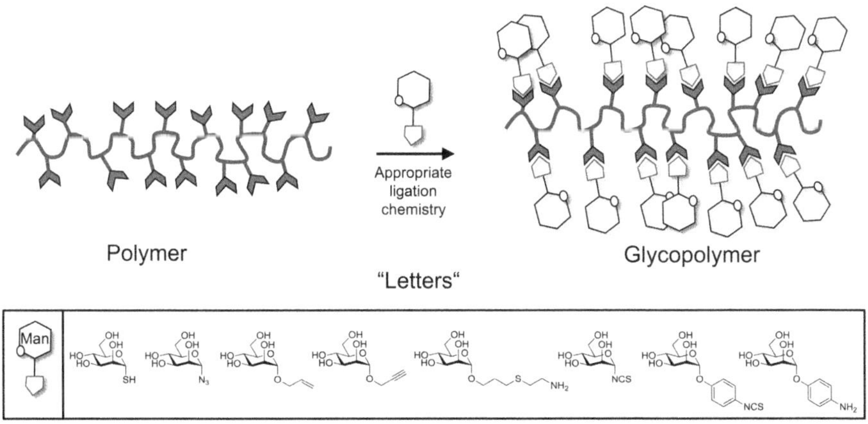

Figure 1.7 Functionalized polymers can be ligated to functional glycosides, *e.g.*, mannosides, to yield antiadhesive glycopolymers. A series of known α-D-mannosides are depicted, permitting easy ligation with appropriately functionalized polymers, such as alkyne–azide cycloaddition, thiol–ene reaction, peptide coupling or thiourea bridging. The ligated glycosides might be regarded as 'letters' and, to take the metaphor further, a resulting glycopolymer might be interpreted as a 'sentence.'

conformational properties and carrying appropriate functional groups to allow selective attachment of carbohydrate moieties. Ideally, the attachment chemistry should work according to the concept of 'click chemistry,[55] but more classical ligation methods could also be employed. For the fabrication of antiadhesive glycoarrays, we and others have often used a collection of functionalized glycosides that can be easily ligated to prefunctionalized surfaces.[56–59] Similar mannosides (*cf.* Figure 1.7) could be employed to produce glycopolymers that inhibit type 1 fimbriae-mediated adhesion, for example. Alkyne–azide cycloaddition,[60] triol–ene reactions,[61] peptide coupling[62] and thiourea bridging[63] are approved methods to ligate carbohydrates to molecular scaffolds, and even bioorthogonal chemistry[64] is certainly amenable to the synthesis of glycopolymers.

A number of structural parameters are known to influence and direct multivalent carbohydrate–lectin interactions and cell adhesion. For example, the density of carbohydrate ligands can be influential.[65,66] Moreover, the complexity of carbohydrate decoration and in particular clustering of glycosides of different nature have been found to improve ligand binding in many cases. The latter observation has been described as the 'heteroglycocluster effect'.[67] All of these aspects of carbohydrate recognition could be studied and adjusted with the aid of glycopolymers, as outlined in Figure 1.8.

With regard to inhibition of type 1 fimbriae-mediated bacterial adhesion, the identified high-performance FimH antagonists **1–6** (Figure 1.5) can be easily ligated to functionalized polymers either directly or after the introduction of a suitable linker. Some practical ideas are depicted in Figure 1.9. The mercapto-functionalized azobenzene mannoside **9**, for example, has recently

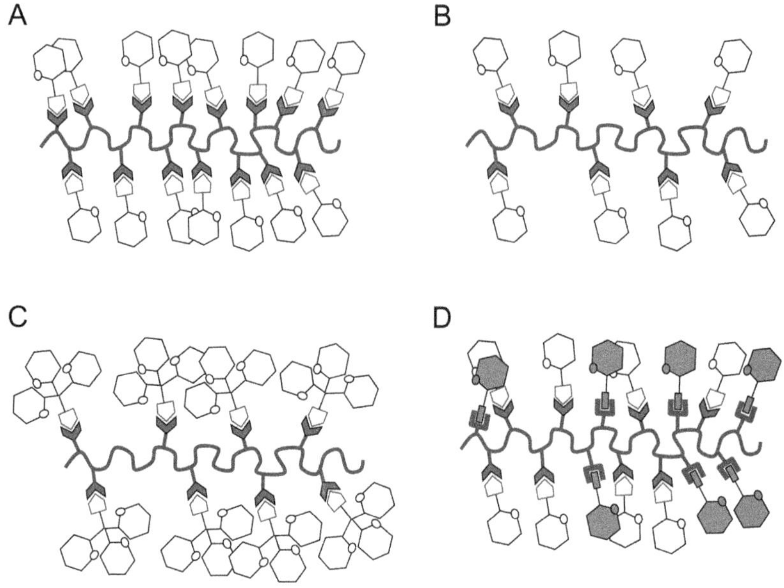

Figure 1.8 Structural parameters of glycopolymers (A) can be varied. Carbohydrate density can be influenced and varied according to cartoon B or by conjugation of cluster glycosides instead of simple monosaccharides (C). Heteroglycopolymers (D) can be achieved through an orthogonal ligation approach.

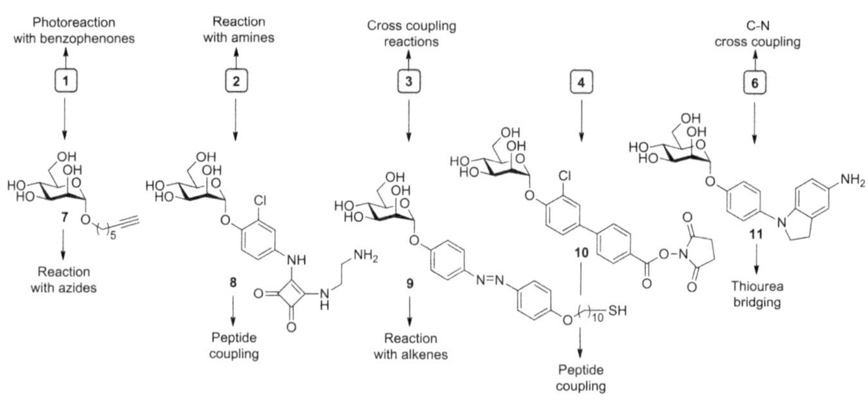

Figure 1.9 The potent FimH antagonists **1–6** (*cf.* Figure 1.5) can be used as antiadhesive carbohydrate 'letters' for the fabrication of glycopolymers either directly or after facile conversion into derivatives such as **7–11**.

been used for the fabrication of photoswitchable carbohydrate-decorated self-assembled monolayers on gold, so-called glyco-SAMs.[68] The same molecule can be utilized to make photosensitive glycopolymers, in which reversible $E \rightarrow Z \rightarrow E$ isomerization of the azobenzene N=N double bond might be used to modulate the antiadhesive power of the respective polymer.[69]

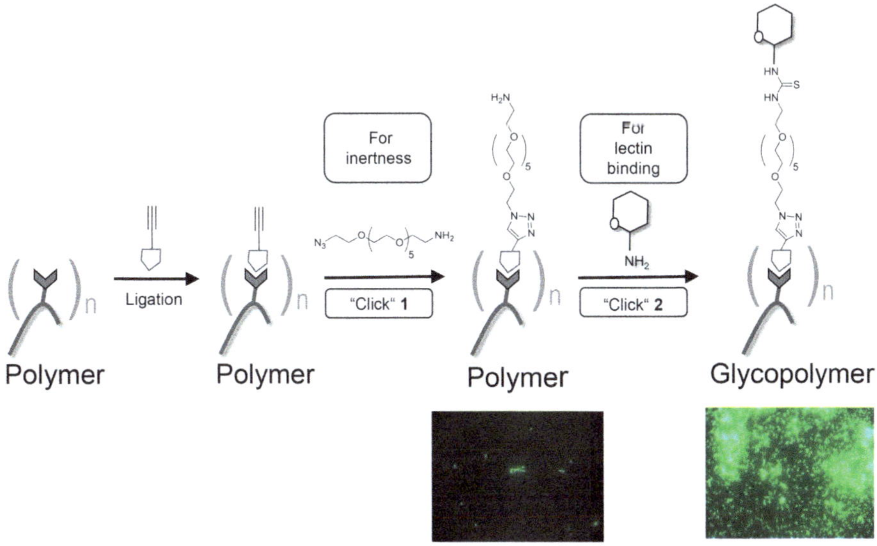

Figure 1.10 The 'dual click' approach described for the construction of glyco-SAMs,[70] could be applied in glycopolymer synthesis. An oligoethylene glycol moiety, introduced in an alkyne–azide cycloaddition ('click' 1) adds 'biorepulsive' properties (inertness) to the molecule, then thiourea bridging employing NCS-functionalized sugars ('click' 2) adds specificity for lectin binding. By adhesion of fluorescent *E. coli*, the success of this approach can be proven.

Equally, the recently introduced[70] 'dual click' approach for glyco-SAM production can enrich the field of glycopolymer synthesis (Figure 1.10). For biological adhesion studies, it is essential to use biorepulsive moieties to suppress the non-specific adsorption of proteins on a surface. Such protein-repelling properties are mediated by oligoethylene glycol (OEG) linkers, which can be introduced to a functionalized scaffold or surface by a first 'click reaction.' A second 'click reaction' allows the attachment of a carbohydrate head group, such as an α-D-mannoside, at the terminal end of the molecular construct. This 'dual click' concept has been demonstrated to be viable by systematic step-by-step assembly of glyco-SAMs and their testing as (anti)adhesive surfaces using fluorescent *E. coli* bacteria and fluorescence read-out.

1.7 No Risk, No Fun

Certainly, the use of glycopolymers in a therapeutic context, such as in antiadhesion therapy, is associated not only with potential advantages but also with risks that have to be taken seriously. As suggested in Figure 1.6, a cell could be completely knocked out by a large glycopolymer covering large parts of its surface. Physiological carbohydrate–protein interactions would then likewise be prevented. Although prevention of adhesion of pathogenic microbes is desired, inhibition of vital carbohydrate–protein and also carbohydrate–carbohydrate interactions would be highly toxic. In addition, gelation effects or haemagglutination could occur when glycopolymers are applied, which are certainly also undesired effects, especially in *in vivo* situations.

However, glycopolymers could find other fields of application. Instead of an intake (*e.g.* orally) of glycopolymers to treat patients, surface treatment of, for example, the skin can also be considered. Other areas of application concern environmentally benign prevention of biofouling or impregnation of medical surfaces that are prone to bacterial colonization.

Finally, glycopolymers could be a means to tackle the problem of redundancy in adhesion. In fact, there are many more factors than type 1 fimbriae that facilitate adhesion of microbial cells. Additional fimbriae with different carbohydrate specificities are expressed on bacteria supporting bacterial adhesion, such as the galactose-specific P-fimbriae. Possession of multiple adhesins and mechanisms of attachment is common for bacteria, permitting binding to multiple targets on a single cell. The variety of factors can cooperate to increase cellular entry, as exemplified by *Neisseria meningitidis*.[71] Here, complex, multifunctional glycopolymers could provide powerful tools to cope with redundancy in adhesion (Figure 1.11).

Furthermore, glycopolymers could even present dynamic advantages connected with bacterial adhesion by general steric hindrance of the adhesion process. In addition to the translational flexibility of bacterial cells, kinetic motion of fimbriae has to be considered, and also their conformational

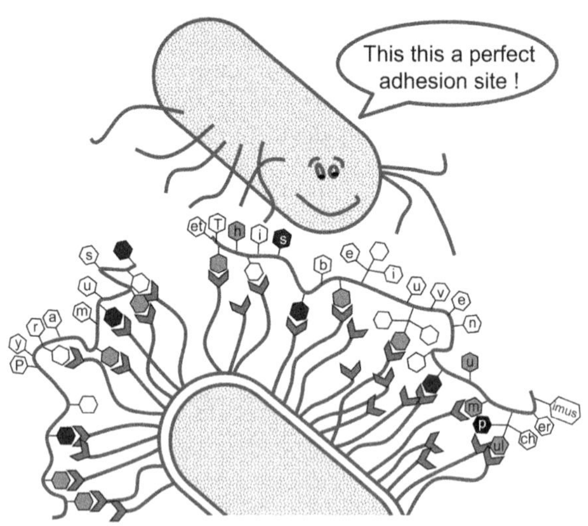

Figure 1.11 Tailor-made multifunctional glycopolymers of sophisticated complexity could provide powerful tools to cope with redundancy in microbial adhesion. The cartoon suggests some structural variations to be implemented into glycopolymers that could cooperate in a specific antiadhesion approach (*cf.* Figures 1.6 and 1.8).

flexibility, and moreover, functional flexibility of the adhesive properties due to mutations and conformational flexibility of the actual lectin are important factors in bacterial adhesion. In the case of type 1 fimbriae, a so-called 'catch bond' mechanism[72] has been discovered (see earlier), that enables bacteria to withstand mechanical force. When shear stress is applied to bacteria, an allosteric rearrangement is initiated that is mediated between the FimH domains $FimH_P$ and $FimH_L$ (*cf.* Figure 1.4, right). Thus, tightening of the mannose binding pocket at the far end of the lectin domain $FimH_L$ is effected. This leads to a high-affinity state of the lectin and longer lived interactions with mannoside ligands.[29,30] The observation that the lifetimes of some biological adhesive bonds are enhanced by tensile mechanical force certainly has important implications for inhibition of bacterial adhesion and engineering of antiadhesives.

1.8 Conclusions

Carbohydrate-specific adhesion of microbes to their host cells is often a prelude to cellular invasion, tissue penetration and disease. In the last two decades, considerable progress has been made in the development of inhibitors of adhesion, such as in case of small molecule FimH antagonists. Hence today, it appears realistic that urinary tract infections that are caused by adhesive *E. coli* bacteria (UPEC) will soon be treated by an antiadhesion therapy, applying carbohydrate inhibitors of the lectin-mediated adhesion process. However, microbial adhesion is a complex process, involving multiple adhesins and adhesive mechanisms. Therefore, there is still a need for molecular architectures that bear the conceptual flexibility to cope with the complex code that is hidden in carbohydrate–lectin and host–pathogen interactions. In this chapter, some ideas have been highlighted for the use of modern concepts of glycopolymer synthesis in the context of microbial adhesion. In this approach, small molecule lectin ligands could be applied as molecular 'letters' of an antiadhesive glycopolymer code.

References

1. I. Cho and M. J. Blaser, *Nat. Rev. Genet.*, 2012, **13**, 260.
2. A. Varki, *J. Mol. Med.*, 2012, **90**, 481.
3. H.-J. Gabius, S. André, J. Jiménez-Barbero, A. Romero and D. Solís, *Trends Biochem. Sci.*, 2011, **36**, 298.
4. T. Feizi and W. Chai, *Nat. Rev. Mol. Cell Biol.*, 2004, **5**, 582.
5. G. O'Toole, H. B. Kaplan and R. Kolter, *Annu. Rev. Microbiol.*, 2000, **54**, 49.
6. S. D. Knight and J. Bouckaert, *Top. Curr. Chem.*, 2009, **288**, 67.
7. D. Osrin, S. Vergnano and A. Costello, *Curr. Opin. Infect. Dis.*, 2004, **17**, 217.
8. N. Sharon and H. Lis, *Glycobiology*, 2004, **14**, 53R.
9. W. C. Boyd and E. Shapleigh, *Science*, 1954, **119**, 419.

10. H. Lis and N. Sharon, *Chem. Rev.*, 1998, **98**, 637.
11. M. Ambrosi, N. R. Cameron and B. G. Davis, *Org. Biomol. Chem.*, 2005, **3**, 1593.
12. K. Drickamer, *Curr. Opin. Struct. Biol.*, 1993, **3**, 393.
13. R. B. Dodd and K. Drickamer, *Glycobiology*, 2001, **11**, 71R.
14. K. Ohlsen, T. Oelschlaeger, J. Hacker and A. S. Khan, *Top. Curr. Chem.*, 2009, **288**, 109.
15. N. Sharon, *FEBS Lett.*, 1987, **217**, 145.
16. J. J. Martinez, M. A. Mulvey, J. D. Schilling, J. S. Pinkner and S. J. Hultgren, *EMBO J.*, 2000, **19**, 2803.
17. A. Ronald, *Am. J. Med.*, 2002, **113**, 14S.
18. R. J. Pieters, *Med. Res. Rev.*, 2007, **27**, 796.
19. I. Ofek, D. L. Hasty and N. Sharon, *FEMS Immunol. Med. Microbiol.*, 2003, **38**, 181.
20. D. Choudhury, A. Thompson, V. Stojanoff, S. Langerman, J. Pinkner, S. J. Hultgren and S. Knight, *Science*, 1999, **285**, 1061.
21. C. S. Hung, J. Bouckaert, D. Hung, J. Pinkner, C. Widberg, A. Defusco, C. G. Auguste, R. Strouse, S. Langermann, G. Waksman and S. J. Hultgren, *Mol. Microbiol.*, 2002, **44**, 903.
22. J. Bouckaert, J. Mackenzie, J. L. de Paz, B. Chipwaza, D. Choudhury, A. Zavialov, K. Mannerstedt, J. Anderson, D. Piérard, L. Wyns, P. H. Seeberger, S. Oscarson, H. De Greve and S. D. Knight, *Mol. Microbiol.*, 2006, **61**, 1556.
23. A. Wellens, C. Garofalo, H. Nguyen, N. Van Gerven, R. Slättegård, J. P. Hernalsteens, L. Wyns, S. Oscarson, H. De Greve, S. Hultgren and J. Bouckaert, *PLoS One*, 2008, **3**, e2040.
24. T. K. Lindhorst, in: *Synthesis and Biological Applications of Glycoconjugates*, ed. O. Renaudet, N. Spinelli, Bentham eBooks, Sharjah, 2011, 12–35.
25. D. J. Schwartz, V. Kalas, J. S. Pinkner, S. L. Chen, C. N. Spaulding, K. W. Dodson and S. J. Hultgren, *Proc. Natl. Acad. Sci. U. S. A.*, 2013, **110**, 15530.
26. G. Phan, H. Remaut, T. Wang, W. J. Allen, K. F. Pirker, A. Lebedev, N. S. Henderson, S. Geibel, E. Volkan, J. Yun, M. B. A. Kunze, J. S. Pinkner, B. Ford, C. W. M. Kay, H. Li, S. J. Hultgren, D. G. Thanassi and G. Waksman, *Nature*, 2011, **474**, 49.
27. G. Waksman and S. J. Hultgren, *Nat. Rev. Microbiol.*, 2009, **7**, 765.
28. M. D. Crespo, C. Puorger, M. A. Schärer, O. Eidam, M. G. Grütter, G. Capitani and R. Glockshuber, *Nat. Chem. Biol.*, 2012, **8**, 707.
29. I. Le Trong, P. Aprikian, B. A. Kidd, M. Forero-Shelton, V. Tchesnokova, P. Rajagopal, V. Rodgriguez, G. Interlandi, R. Klevit, V. Vogel, R. E. Stenkamp, E. V. Sokurenko and W. E. Thomas, *Cell*, 2010, **141**, 645.
30. P. Aprikian, G. Interlandi, B. Kidd, I. Le Trong, V. Tchesnokova, O. Yakovenko, M. Whitfield, E. Bullitt, R. Stenkamp, W. E. Thomas and E. V. Sokurenko, *PLoS Biol.*, 2011, **9**, e1000617.
31. J. A. Silverman, H. L. Schreiber 4th, T. M. Hooton and S. J. Hultgren, *Curr. Urol. Rep.*, 2013, **14**, 448.
32. T. K. Lindhorst, C. Kieburg and U. Krallmann-Wenzel, *Glycoconjugate J.*, 1998, **15**, 605.

33. O. Sperling, A. Fuchs and T. K. Lindhorst, *Org. Biomol. Chem.*, 2006, **4**, 3913.
34. R. U. Lemieux, *Acc. Chem. Res.*, 1996, **29**, 373.
35. M. Hartmann and T. K. Lindhorst, *Eur. J. Org. Chem.*, 2011, 3583.
36. B. Ernst and J. L. Magnani, *Nat. Rev. Drug Discovery*, 2009, **8**, 661.
37. L. Pang, S. Kleeb, K. Lemme, S. Rabbani, M. Scharenberg, A. Zalewski, F. Schädler, O. Schwardt and B. Ernst, *ChemMedChem*, 2012, **7**, 1404.
38. J. Xiaohua, D. Abgottspon, S. Kleeb, S. Rabbani, M. Scharenberg, M. Wittwer, M. Haug, O. Schwardt and B. Ernst, *J. Med. Chem.*, 2012, **55**, 4700.
39. M. Scharenberg, J. Xiaohua, L. Pang, G. Navarra, S. Rabbani, F. Binder, O. Schwardt and B. Ernst, *ChemMedChem*, 2014, **9**, 78.
40. J. Bouckaert, J. Berglund, M. Schembri, E. D. Genst, L. Cools, M. Wuhrer, C.-S. Hung, J. Pinkner, R. Slättegård, A. Zavialov, D. Choudhury, S. Langermann, S. J. Hultgren, L. Wyns, P. Klemm, S. Oscarson, S. D. Knight and H. D. Greve, *Mol. Microbiol.*, 2005, **55**, 441.
41. V. Chandrasekaran, K. Kolbe, F. Beiroth and T. K. Lindhorst, *Beilstein J. Org. Chem.*, 2013, **9**, 223.
42. M. Hartmann, H. Papavlassopoulos, V. Chandrasekaran, C. Grabosch, F. Beiroth, T. K. Lindhorst and C. Röhl, *FEBS Lett.*, 2012, **586**, 1459.
43. Z. Han, J. S. Pinkner, B. Ford, E. Chorell, J. M. Crowley, C. K. Cusumano, S. Campbell, J. P. Henderson, S. J. Hultgren and J. W. Janetka, *J. Med. Chem.*, 2012, **55**, 3945.
44. T. Christensen, D. M. Gooden, J. E. Kung and E. J. Toone, *J. Am. Chem. Soc.*, 2003, **125**, 7357.
45. M. Lahmann, *Top. Curr. Chem.*, 2009, **288**, 17.
46. L. L. Kiessling, J. E. Gestwicki and L. E. Strong, *Angew. Chem. Int. Ed.*, 2006, **45**, 2348; *Angew. Chem.*, 2006, **118**, 2408.
47. M. Mammen, S.-K. Choi and G. M. Whitesides, *Angew. Chem. Int. Ed.*, 1998, **37**, 2754; *Angew. Chem.*, 1998, **110**, 2908.
48. Y. M. Chabre and R. Roy, *Adv. Carbohydr. Chem. Biochem.*, 2010, **63**, 165.
49. R. T. Lee and Y. C. Lee, *Carbohydr. Res.*, 1974, **37**, 193.
50. Y. C. Lee and R. T. Lee, *Acc. Chem. Res.*, 1995, **28**, 321.
51. B. Voit and D. Appelhans, *Macromol. Chem. Phys.*, 2010, **211**, 727.
52. M.-G. Baek and R. Roy, *Biomacromolecules*, 2000, **1**, 768.
53. L. L. Kiessling and J. C. Grim, *Chem. Soc. Rev.*, 2013, **42**, 4476.
54. D. Ponader, P. Maffre, J. Aretz, D. Pussak, N. M. Ninnemann, S. Schmidt, P. H. Seeberger, C. Rademacher, G. U. Nienhaus and L. Hartmann, *J. Am. Chem. Soc.*, 2014, **136**, 2008.
55. H. C. Kolb, M. G. Finn and K. B. Sharpless, *Angew. Chem. Int. Ed.*, 2001, **40**, 2004; *Angew. Chem.*, 2001, **113**, 2056.
56. M. Hartmann, P. Betz, Y. Sun, S. N. Gorb, T. K. Lindhorst and A. Krueger, *Chem. Eur. J.*, 2012, **13**, 6485.
57. M. J. Weissenborn, R. Castangia, J. W. Wehner, T. K. Lindhorst and S. L. Flitsch, *Chem. Commun.*, 2012, **48**, 4444.
58. J. W. Wehner, M. J. Weissenborn, M. Hartmann, C. J. Gray, R. Šardzík, C. E. Eyers, S. L. Flitsch and T. K. Lindhorst, *Org. Biomol. Chem.*, 2012, **10**, 8919.

59. M. Hartmann, A. K. Horst, P. Klemm and T. K. Lindhorst, *Chem. Commun.*, 2010, **46**, 330.
60. C. W. Tornøe, C. Christensen and M. Meldal, *J. Org. Chem.*, 2002, **67**, 3057.
61. A. Dondoni and A. Marra, *Chem. Soc. Rev.*, 2012, **41**, 573.
62. J. Liu, W. D. Gray, M. E. Davis and Y. Luo, *Interface Focus*, 2012, **2**, 307.
63. Á. Martínez, C. Ortiz Mellet and J. M. García Fernández, *Chem. Soc. Rev.*, 2013, **42**, 4746.
64. E. M. Sletten and C. Bertozzi, *Acc. Chem. Res.*, 2011, **44**, 666.
65. J. W. Wehner, M. Hartmann and T. K. Lindhorst, *Carbohydr. Res.*, 2013, **371**, 22.
66. J. A. Deeg, I. Louban, D. Aydin, C. Selhuber-Unkel, H. Kessler and J. P. Spatz, *Nano Lett.*, 2011, **11**, 1469.
67. J. L. Jiménez Blanco, C. Ortiz Mellet and J. M. García Fernández, *Chem. Soc. Rev.*, 2013, **42**, 4518.
68. V. Chandrasekaran, H. Jacob, F. Petersen, K. Kathirvel, F. Tuczek and T. K. Lindhorst, *Chem. Eur. J.*, 2014, **20**, 8744.
69. T. Weber, V. Chandrasekaran, I. Stamer, M. B. Thygesen, A. Terfort and T. K. Lindhorst, *Angew. Chem. Int. Ed.,* 2014, **53**, 14583; *Angew. Chem.,* 2014, **126**, 14812.
70. C. Grabosch, M. Kind, Y. Gies, F. Schweighöfer, A. Terfort and T. K. Lindhorst, *Org. Biomol. Chem.*, 2013, **11**, 4006.
71. M. Virji, *Top. Curr. Chem.*, 2009, **288**, 1.
72. W. Thomas, *Annu. Rev. Biomed. Eng.*, 2008, **10**, 39.

CHAPTER 2

Direct Synthesis of Glycopolymers Using Glycomonomers

S. R. SIMON TING*[a] AND MARTINA H. STENZEL*[b]

[a]Centre for Health Technologies, University of Technology Sydney (UTS), Ultimo, NSW 2007, Australia; [b]Centre for Advanced Macromolecular Design, University of New South Wales (UNSW), Sydney, NSW 2052, Australia
*E-mail: simon.ting@uts.edu.au, m.stenzel@unsw.edu.au

2.1 History of Glycopolymers

Glycopolymers are defined as synthetic polymers with pendant carbohydrates. They have attracted extensive attention in the fields of polymer chemistry, material science and biomedicine owing to their properties, which include biocompatibility and bioactivity.

Probably the first report on glycopolymers appeared in 1944, when Nichols and Yanovsky described the formation of a viscous polymer solution after heating tetraallyl α-methylglucosides.[1] Very soon after, other reports on the polymerization of allyl ethers of carbohydrates emerged.[2–4] The first methacrylate was described by the same group in 1945,[5] but it was not until 1960 that these methacrylates were in fact polymerized by Bird *et al.*, resulting in the first methacrylate polymer.[6] Meanwhile, other polymers based on carbohydrates emerged. A substituted sugar was employed in a polycondensation

RSC Polymer Chemistry Series No. 15
Glycopolymer Code: Synthesis of Glycopolymers and Their Applications
Edited by C. Remzi Becer and Laura Hartmann

Published by the Royal Society of Chemistry, www.rsc.org

process to generate polyamides. Although this is strictly not a glycopolymer as such, these structures should be included in this discussion.[7] The work of Kimura and Imoto,[8] published in 1961, kick-started an era of synthesis of glycopolymers. A hot topic was still the polymerization of allyl sugars[9,10] and polycondensation using multifunctional sugars,[11–13] but the synthesis of glycopolymers using radical polymerization emerged slowly as the favourite technique.[14,15] Glycomonomers based on *N*-carboxyanhydrides were first reported by Rüde *et al.*,[16] which is interesting considering that this technique had been dormant for 30 years, but had recently become a topic of interest again. Radical polymerization of glycomonomers became the technique of choice in the 1970s.[17–20] The idea of using glycopolymers for their bioactivity was first conveyed in a publication by Carpino *et al.* in 1976.[17] They only reported the synthesis of a polymer based on glucose, but they discussed the possibility of designing a polymer that can be degraded by enzymes. It was not until the mid-1980s that these polymers were used because of their bioactivity. Kobayashi *et al.* coated Petri dishes with lactose containing glycopolymer and demonstrated the adhesion of hepatocytes.[21,22] The 1970s also saw the introduction of NMR techniques for the analysis of glycopolymers, which represented a quantum leap in the detailed analysis of these polymers.[19] Initially only a rather crude tool, NMR became very quickly the main approach to investigate the fine structure and even to draw conclusions about the aggregation of glycopolymers, as was done in a paper by Elias and co-workers in 1977.[18]

The following decade was dominated by the development of new glycomonomers. Radical polymerization became the technique of choice. In addition, glycopolymer synthesis was carried out in emulsions with the aim of generating glycopolymers of very high molecular weight.[23] Notable is the first ring-opening polymerization of carbohydrate anhydrides by Good and Schuerch[24] and Uryu *et al.*[25] in 1985. It took until the mid-1990s, however, for the first report on living radical polymerizations to appear. Minoda *et al.* generated glycopolymers with molecular weight increasing with conversion by cationic polymerization.[26] Another technique that became very successful in the synthesis of glycopolymers appeared in 1994, namely ring-opening metathesis polymerization (ROMP).[27] Living anionic polymerization was not reported until the late 1990s,[28] and this technique still remains a niche area.

The 1990s was also the decade for increased activity in the use of glycopolymers to perform a biological role.[27,29–31] In particular, the relationship between the structure of glycopolymers and binding to lectin was of interest, an area that was pioneered by Kiessling and co-workers.[27,30,32,33] Glycopolymers were now investigated for their ability to perform functions such as enzyme stabilization. A glycopolymer with a high density of aldehyde functionalities was conjugated to a protein, resulting most likely in the first reported glycoprotein–polymer conjugate in 1991.[34]

A substantial impact in glycopolymer synthesis was made by the advent of living radical polymerization techniques (or now termed reversible-deactivation radical polymerization). Interestingly, the first report on glycopolymer synthesis using atom-transfer radical polymerization (ATRP) was not in solution, but from a surface. Ohno *et al.* grafted 3-*O*-methacryloyl-1,2 : 5,6-di-*O*-isopropylidine-D-glucofuranose and created a system that obeys first-order kinetics with respect to monomer concentration.[35] The first well-defined glycopolymers obtained using ATRP were reported by Liang *et al.* a year later in 1999.[36] ATRP became one of the most commonly used techniques to generate well-defined glycopolymers in addition to reversible addition–fragmentation chain-transfer (RAFT) polymerization. RAFT polymerization was first employed in 2003[37] and dominated the discussion for many years owing to the robustness of the process in the presence of unprotected hydroxyl groups. Other reversible-deactivation radical polymerization techniques were investigated, such as cyanoxyl-mediated polymerization, which was first described for the synthesis of glycopolymers by Grande *et al.* in 2001,[38] and nitroxide-mediated polymerization (NMP), which was first used in the synthesis of glycopolymers by Ohno *et al.* in 1998,[39] but these techniques never reached the same importance as RAFT polymerization and ATRP. However, the introduction of these techniques saw a surge in glycopolymer synthesis.

Post-polymerization functionalization of functional polymers to generate glycopolymers has become a hot topic in the last 5 years. Although this approach was first successfully demonstrated by Kobayashi and co-workers in 1997,[31] it remained an area of low activity for the next few years, but became the centre of attention with the rise of 'click' chemistries. Haddleton and co-workers presented a series of reports on the construction of glycopolymers from alkyne backbone-functional polymers *via* Cu-catalysed azide–alkyne click (CuAAC) chemistry,[40,41] but other efficient reactions were also introduced, such as various thiol-based chemistries.[42] Post-functionalizations offer the opportunity to produce libraries of different glycopolymers on the same polymer scaffold. This is particularly advantageous when the biological function of different sugars or various sugar densities is to be tested without having to consider the effect of polymer molecular weight or architecture. The increasing interest in the bioactivity of glycopolymers has led again to an increased use of free-radical polymerizations in the last few years. Today, it is not the polymer chemist who synthesizes the glycopolymer, but researchers from other areas, especially in the biological field, who are drawn to this topic by the opportunities to design materials with specific biological functions.

From modest beginnings where glycopolymers were synthesized from vinyl-functionalized sugars with little control over the resulting polymer properties, glycopolymer synthesis has now developed into a mature area where the control over molecular weight and polymer architecture is routinely pursued and indeed accomplished. Glycopolymer synthesis has now infiltrated most known techniques of polymer synthesis and has advanced to a tool to create materials with tuneable bioactivity.

The aim of this chapter is to provide an update on the synthesis of glycopolymers using glycomonomers. The focus is on advances in the last 5 years.

For a comprehensive overview of glycopolymers prior to that time, the reader is referred to an earlier book chapter[43] and various review articles.[44–47]

2.2 Synthesis of Glycomonomers

In the following, the synthetic avenues to generate glycomonomers are summarized. The synthesis often requires several steps and good synthetic skills. However, the synthesis of glycomonomers is now a well-established field and the researcher can choose from a range of available protocols. Modification of carbohydrates became a topic of interest in the 1950s.[48–51] Most procedures rely on the modification of the anomeric carbon atom, which allows regiospecific modification without further intermediate steps. Notable also is the divide between the modifications of carbohydrates that carry protective groups and direct modification. The choice between these different pathways is dependent on the ease of stereospecific functionalization of the sugar, the solubility of the monomer and polymer, the potential incompleteness of the removal of the protective group, and the ease of purification.

2.2.1 Synthesis of Glycomonomers for Radical Polymerizations

2.2.1.1 *Monomers from Protected Carbohydrates*

The commercial availability of a range of carbohydrates provides access to a wide array of different glycomonomers, and significant efforts have been dedicated to the synthesis of polymerizable vinyl sugars. Wulff *et al.*[52] reviewed this field in 1996. The most common synthetic approaches are outlined below, but they are discussed in more detail elsewhere.[52] In general, these monomers follow a specific template, based on four parts that can affect the bioactivity of the polymer (Scheme 2.1).

The type of vinyl functionality affects parameters of the polymerization such as the rate of propagation, but also influences the decision on the choice of reversible-deactivation radical polymerization techniques. It can furthermore affect the physical properties of the polymer such as the glass transition temperature. The bonding unit connects the polymer backbone with the carbohydrate and can affect the stability. Whereas amide bonds are stable under most circumstances, other bonds such as ester[8] or glycosidic bonds[53] are susceptible to hydrolysis. The spacer can have an important role in enhancing the flexibly of the carbohydrate to ensure good binding

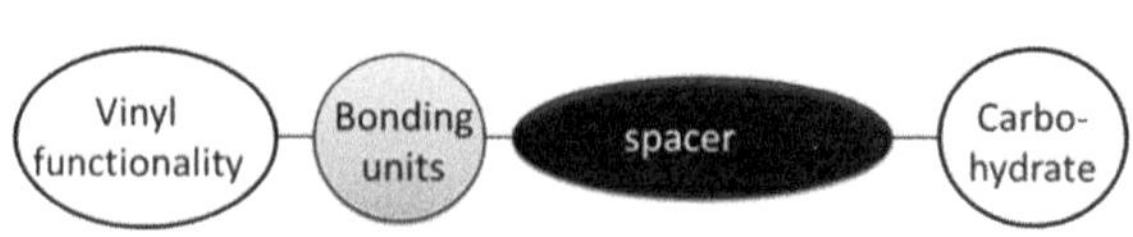

Scheme 2.1 Schematic representation of glycomonomers.

with lectins,[54] but can also influence the overall polarity of the polymer. The choice of carbohydrate is usually determined by the end application.

2.2.1.1.1 Reactions Using Isopropylidene-Protected Sugars. Many sugars can easily be protected using acetone to form isopropylidene derivatives (Figure 2.1, Table 2.1). The process is cheap and typically has yields greater than 80%. The two adjacent hydroxyl groups are protected, leaving the last hydroxyl group for further functionalization. This approach is suitable for a range of sugars, but the type of sugar determines the position of the free hydroxyl groups, which could be detrimental to the bioactivity of polymer.

2.2.1.1.2 Chemical Glycosylation

Glycosides from Halo Sugars. Halo sugars are another source of reactive sugars that can be converted into monomers by reaction with hydroxyl functionalities (Figure 2.2, Table 2.2). The technique is often employed when longer spacers are required. In addition, the α : β ratio on the anomeric carbon atom

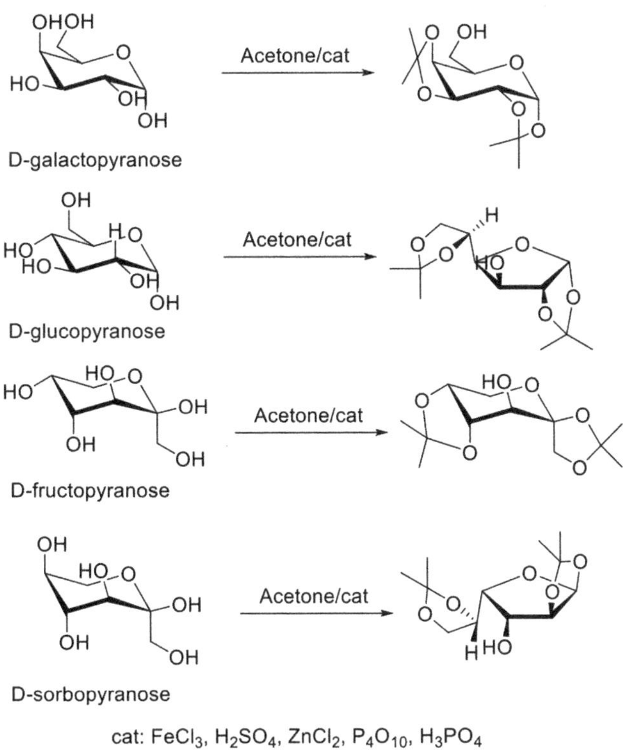

Figure 2.1 Monosaccharides protected with isopropylidene.

Table 2.1 Monomers Prepared from Isopropylidene-Protected Sugars.

Carbohydrate	Vinyl Functionality	References
Glucose	Methacryl	6,8,23,35,55–65
	Acryl	66–71
	4-Vinylbenzyl	72,73
	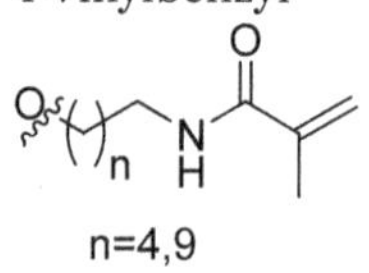	74
Fructose	Methacryl	65,67
	Acryl	67
Galactose	Methacrylate	64,75–81
	Acrylate	75,82
	$-CH_2C(COOCH_3)=CH_2$	83
	Vinyl	83
	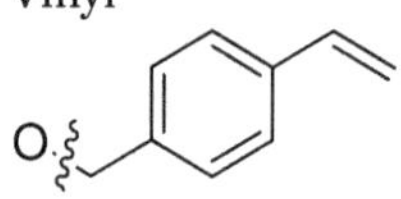	80,84,85
	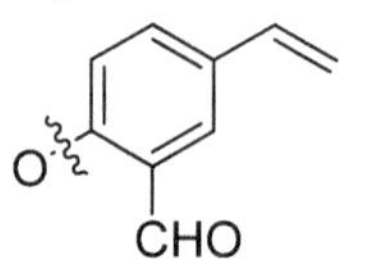	86
Sorbose	Methacrylate	55,67
	Acrylate	67

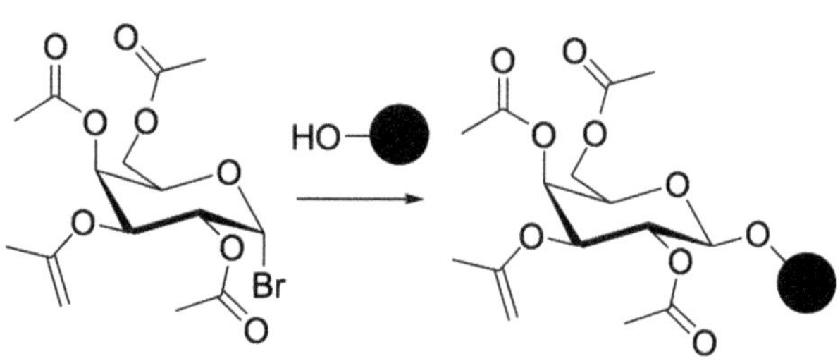

Figure 2.2 Synthesis of glycomonomers by reaction of vinyl compounds with halo sugars.

can in general be well controlled. The advantage is that the cleavage of the acetyl protecting groups in alkaline media would not destroy the glycoside bond; however, it needs to be considered that the ester functionality of the polymerizable group, such as in methacrylates, might be affected.[17,18,50,87]

Glycosides from Trichloroacetimidate Sugars. Related to the approach with halo sugars is the use of acetylated trichloroacetimidate sugars, which are obtained from the reaction between carbohydrates and trichloroacetonitrile, catalysed using NaH (Figure 2.3). Interestingly, the ratio between the

Table 2.2 Monomers Prepared from Halo Sugars.

Carbohydrate	Vinyl Functionality	References
D-Glucose	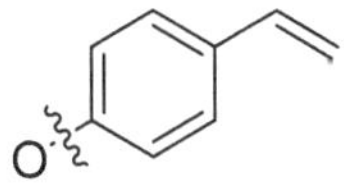	88
	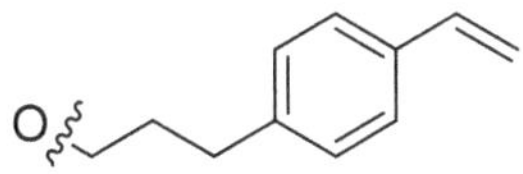	18
	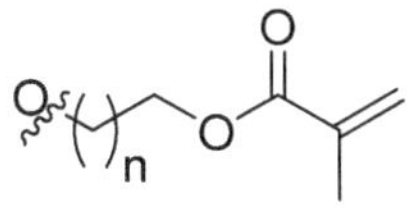	19,87,89
	Methacrylate	6,23,55,56
	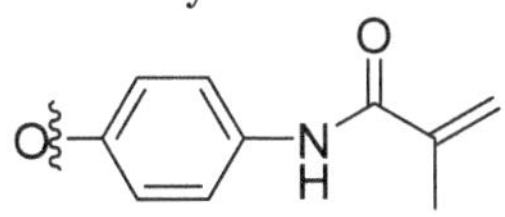	17
	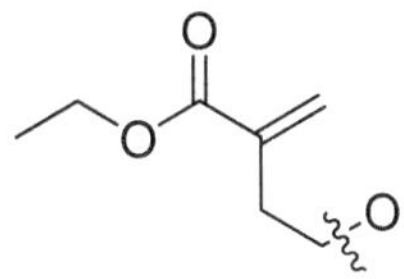	90
	Acrylate	91,92
	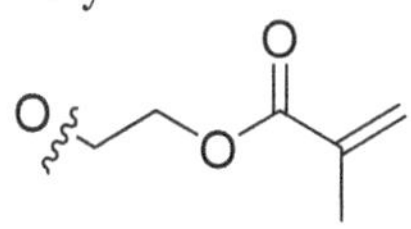	89
Galactose	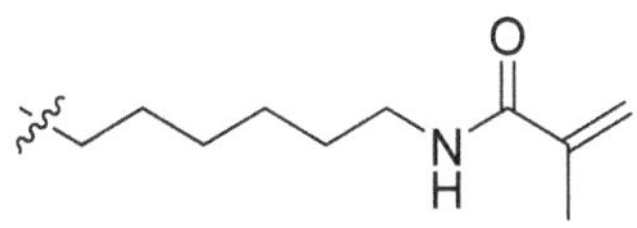	93
	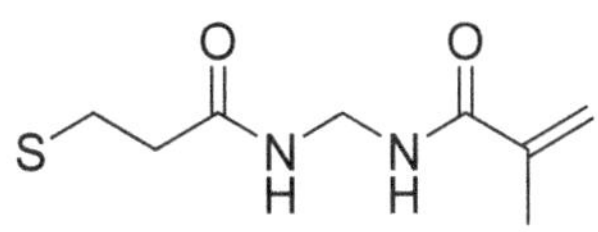	94
	Thioglycoside	
	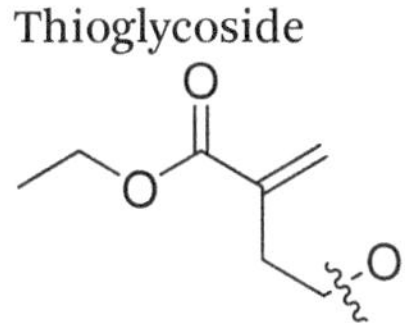	90
	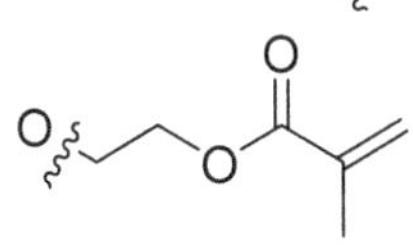	89,95
Cellobiose	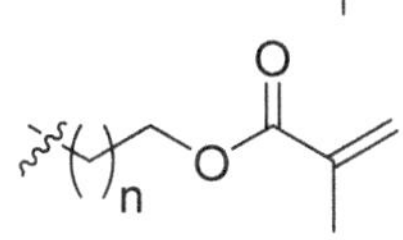	96

(continued)

Table 2.2 *(continued)*

Carbohydrate	Vinyl Functionality	References
Lactose	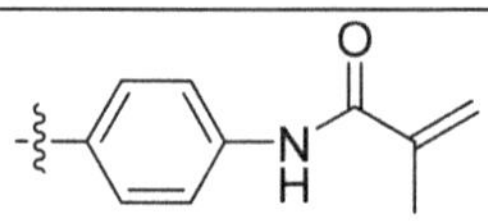	97
Mannose	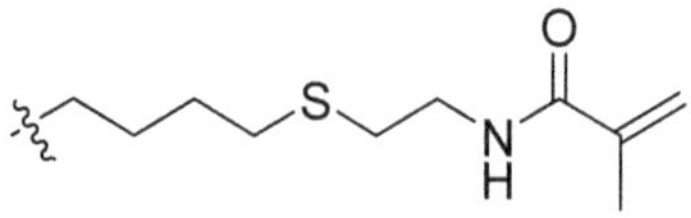	98
Mannose	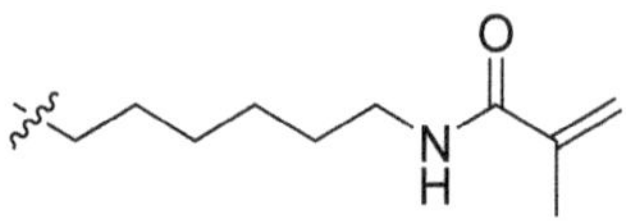	98

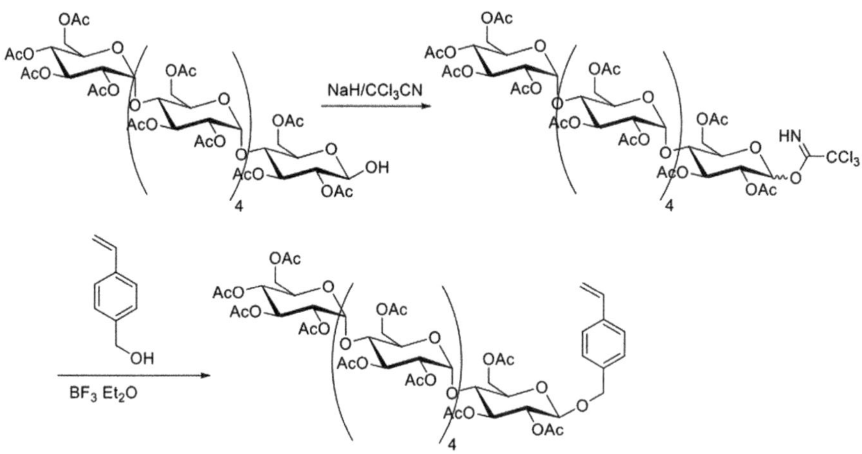

Figure 2.3 Synthesis of glycomonomers using trichloroacetimidate sugars.

α- and β-anomers can be adjusted by the reaction time. Whereas the β-anomer is formed initially, the α-anomer is thermodynamically the more stable product.[99] Subsequent reaction with 4-vinylbenzyl alcohol in the presence of $BF_3{\cdot}OEt_2$ led to the desired monomer in high yields.[100,101]

Glycosides from Carbohydrate Esters. An increasingly popular route is the use of acetyl esters of carbohydrates as substrates (Figure 2.4). These esters are usually accessible in large quantities and are often commercially available. Alcohols such as 2-hydroxyethyl methacrylate act as nucleophilic donors and react with the anomeric position, often catalysed with $BF_3{\cdot}OEt_2$ or similar. The reaction can lead to a mixture of α- and β-anomers, but the influence of the neighbouring group is detrimental and can result in a preference for just one stereo centre. This simple two-step approach (esterification of

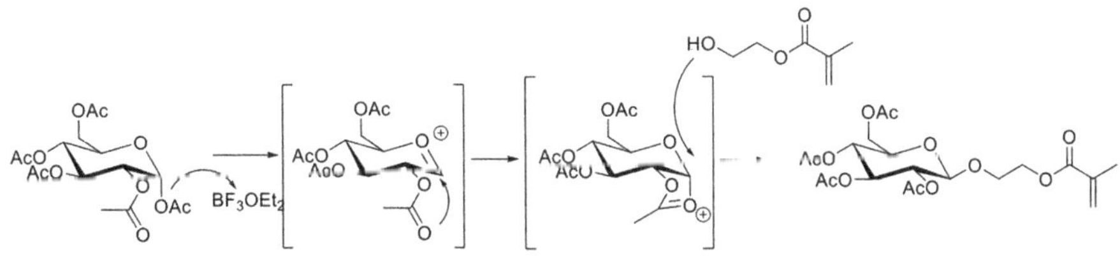

Figure 2.4 Synthesis of glycomonomers using fully acetylated carbohydrates.

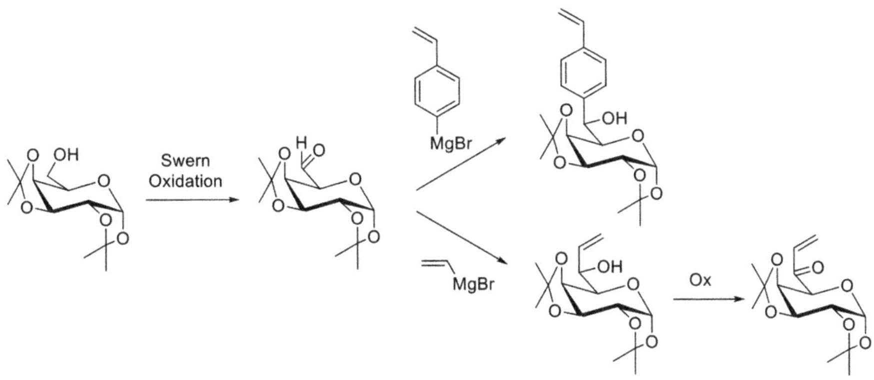

Figure 2.5 Synthesis of glycopolymers *via* Grignard reaction using oxidized sugars.

carbohydrate and glycoside synthesis) allows the synthesis of several grams of monomers with overall yields of around 60% depending on the carbohydrate. Monomers based on galactose[102,103] or lactose[104] have been prepared using this route.

2.2.1.1.3 Grignard Reactions. A rather unusual pathway is *via* the use of Grignard agents (Figure 2.5). This requires the oxidation of the hydroxyl group of the carbohydrate while the protecting groups need to be stable under the conditions employed with Grignard reagents. Wulff and co-workers prepared a library of such glycomonomers based on fructose, galactose and arabinose and provided evidence for their stereochemistry.[105,106]

2.2.1.1.4 Synthesis Using Oxazoline Derivatives. Protected amino sugars can be a further platform for targeted functionalization at the anomeric position. Glycosylation of an oxazoline derivative, which was derived from an *N*-acetylated amino sugar and *N*-protected amino alcohol, was carried out by hydrogenation and subsequent *N*-acryloylation (Figure 2.6).[107,108] Glucosamine, chitobiose and lactosamine were utilized,[29,109] but also more complex carbohydrates such as the lactosamine-based Lewis x (Lex) carbohydrate.[29]

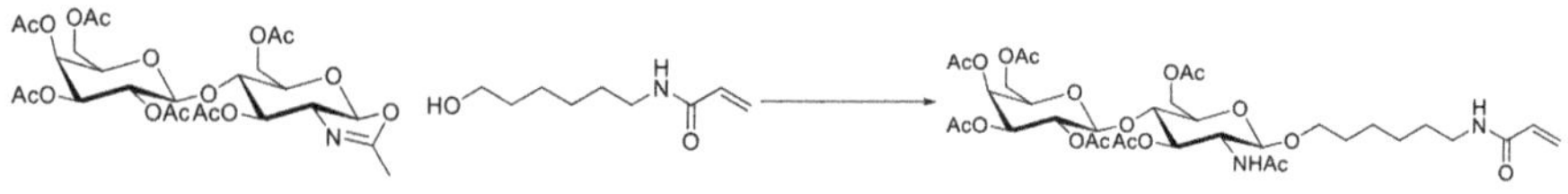

Figure 2.6 Synthesis of a glycomonomer using an oxazoline derivative.

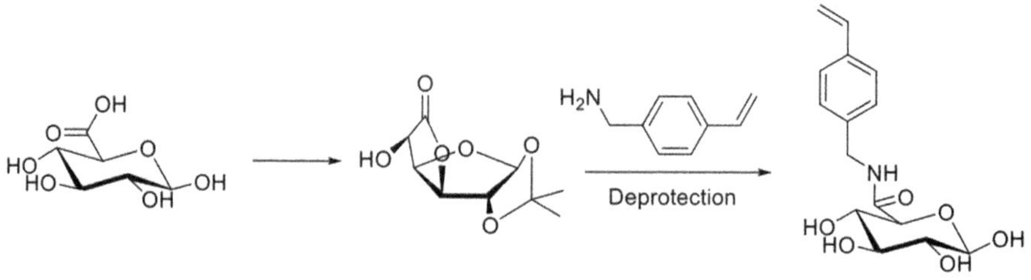

Figure 2.7 Synthesis of a glycomonomer using an oxidized sugar.

2.2.1.1.5 Synthesis Using Oxidized Carbohydrates. Although the oxidation of unprotected reducing sugars usually leads to the ring-opening reaction at the anomeric position, prior protection can direct oxidation in order to maintain the ring structure. The isopropylidene-protected glucuronolactone reacts in a short period of time with amines. Upon deprotection, the pyranose form is obtained (Figure 2.7).[110]

2.2.1.2 *Monomers from Unprotected Sugars*

Direct synthesis of glycomonomers is desirable to eliminate the need for protection and deprotection steps. However, direct reaction with (meth)acryloyl chloride would lead to a mixture of products. In the following, several robust pathways that allow glycomonomers to be synthesized using unprotected carbohydrate chemistry are discussed.

2.2.1.2.1 Enzymatic Transesterification. An elegant way to obtain glycomonomers is with the help of enzymes (Figure 2.8). Although lipases are responsible for ester hydrolysis, they catalyse the formation of esters in dry organic solvents. In a typical reaction, an ester is employed instead of a carboxylic group as substrate. The resulting transesterification is accelerated by the formation of a by-product such as vinyl alcohol, which is immediately converted into acetaldehyde, shifting the equilibrium to high conversions. This highly efficient and regioselective reaction for obtaining glycomonomers is catalysed by lipases such as *Candida antarctica*, usually yielding derivatives functionalized in the 6-position,[111] although efficient functionalization in the 1-position has also been reported.[112] This reaction has been tested with glucose[113–119] and mannose,[115,120,121] mainly to introduce methacrylate or acrylate functionalities, but 6-*O*-vinyladipoyl groups have also been

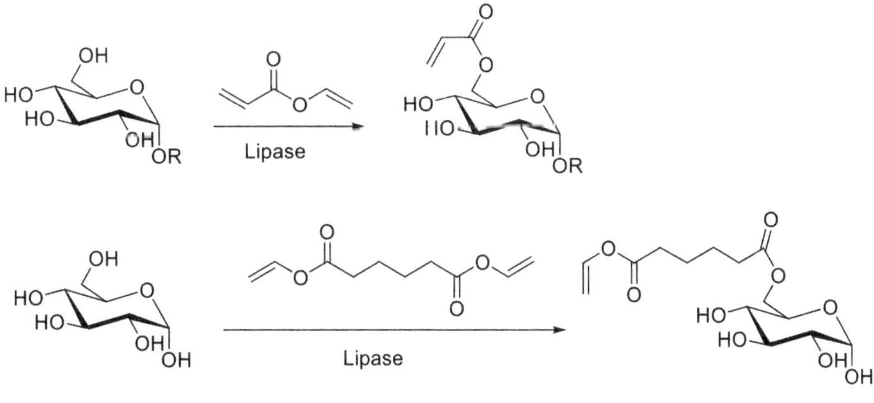

Figure 2.8 Regioselective modification of carbohydrates using enzymes.

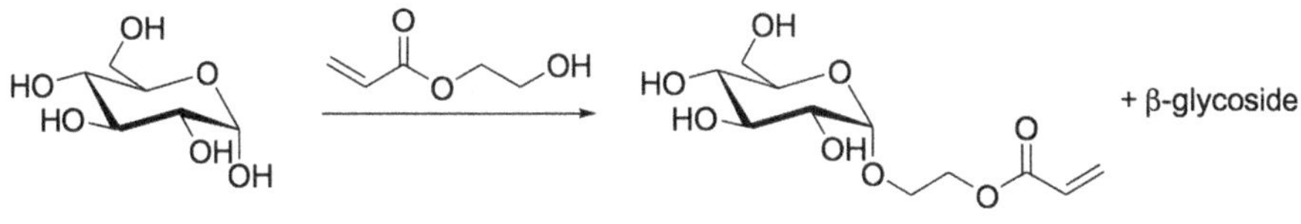

Figure 2.9 Direct monomer synthesis using unprotected sugars and hydroxyl groups, catalysed by acids.

attached to the 6-position either on glucose[113] and maltitol or lactitol, two reduced disaccharides.[122] A prerequisite is the solubility of the carbohydrate in specific organic solvents such as acetone since the type of solvent can influence the outcome significantly. In most cases, the disadvantage of the reaction is that it takes place at the 6-position, which can lead to the loss of bioactivity.[120]

2.2.1.2.2 Glycoside Synthesis Using Unprotected Sugars. Direct mono-substitution at the anomeric (C-1) position without recourse to protective chemistry can be achieved by the reaction of a large excess of hydroxyl groups, such as in hydroxyethyl acrylate, with the sugar in the presence of phosphomolybdic acid as catalyst[123] or 10-camphorsulfonic acid (Figure 2.9).[124–126] The types of carbohydrate studied included glucose, galactose, mannose and xylose, using either the sugar or methyl glycoside as substrate.[123] Related to the oxazoline derivatives pathway described above, *N*-acetylglucosamine can react with various alcohols in the presence of 10-camphorsulfonic acid.[38,124,127,128]

2.2.1.2.3 Synthesis *via* Barbituric Acid. Barbituric acid reacts readily with the C-1 position of an unprotected sugar to generate a reactive salt (Figure 2.10). Subsequent reaction with bromides such as 4-vinylbenzyl bromide leads to polymerizable monomers. Conversion of the barbituric acid ring to a diamide further improves water solubility.[129]

2.2.1.2.4 Conversion of Amino Sugars. A popular route to glycomonomers is the fast reaction between amino sugars and acyl halides or anhydrides. The high reactivity of the amine group ensures its preferential reaction even in the presence of unprotected hydroxyl groups. Reactions of acryloyl chloride and methacryloyl chloride[14,15,130–132] but also of isocyanates[133,134] and epoxides with various amino sugars have been explored to obtain glycomonomers in high yields (Figure 2.11). In particular, the use of glucosamine is advantageous thanks to the commercial availability of this substrate at very low cost.[132,135–137]

Whereas some amino sugars such as glucosamine and galactosamine are readily available, many amino sugars can only be obtained *via* multi-step synthesis, frequently requiring the use of protective group chemistry. The introduction of a vinyl functionality was then often carried out while the remaining functional groups were still protected, although this was theoretically not necessary. A range of galacto-trehalose monomers were successfully prepared using this route (Figure 2.12).[138]

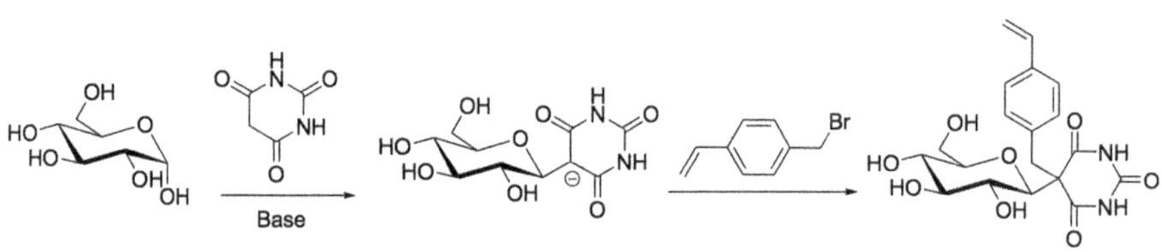

Figure 2.10 Glycomonomer synthesis using barbituric acid.

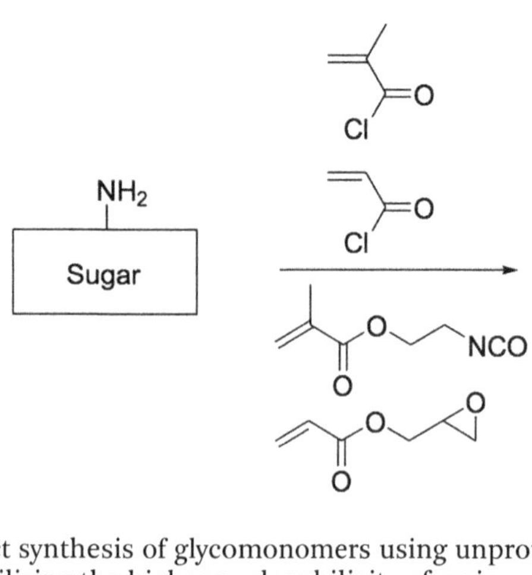

Figure 2.11 Direct synthesis of glycomonomers using unprotected carbohydrates by utilizing the higher nucleophilicity of amino groups

Additional amino functionalities can be introduced to the reducing end of carbohydrates. An example is the reaction between lactose or *N*,*N*′-diacetylchitobiose with ammonium hydrogencarbonate. The amine was then reacted with *p*-vinylbenzoyl chloride.[31]

2.2.1.2.5 Monomer Synthesis by Ring-Opening Reactions. A range of amide-linked glycomonomers are accessible from sugars that have been oxidized to their corresponding lactones and can therefore be reacted with vinyl-functionalized amines.[21,22,73,139] The intermediate oxidation step is not always necessary and oxyamines with the help of an ion-exchange resin can react directly with the carbohydrate.[140] This process is also suitable for the simultaneous ring-opening reaction, while introducing an amine using hydrazine. The amine can be used for efficient functionalization steps such as the attachment of polymerizable functionalities with isocyanates.[141] The ring-opening approach is in particular popular for obtaining monomers base on glucose[142–151] and lactose,[140,141,143,149,151–158] but maltose and maltotriose have also been explored (Figure 2.13).[152,153]

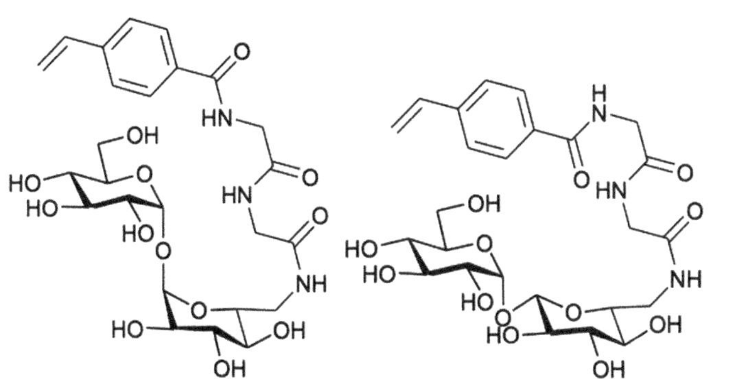

Figure 2.12 Examples of glycomonomers prepared from amino sugars.[138]

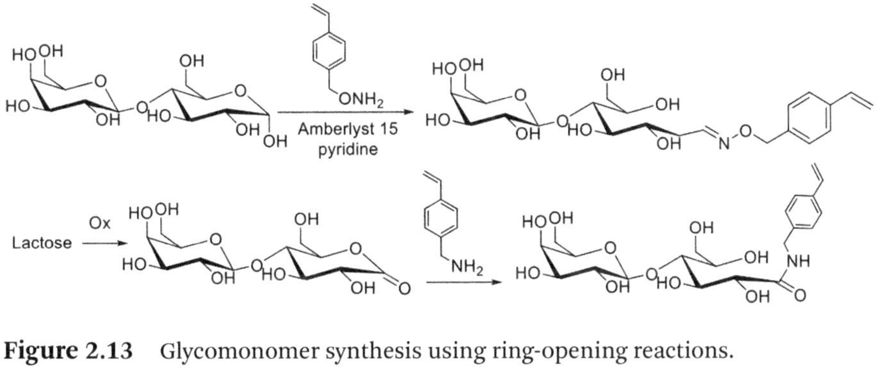

Figure 2.13 Glycomonomer synthesis using ring-opening reactions.

2.2.1.2.6 Other Approaches. While the above are the most common strategies used for glycomonomer synthesis, other pathways are available. Most of the monomers described above are based on styrene, methacrylates or acrylates, which is in contrast to the rather unusual glycomonomer based on 4-vinyl-1,2,3-triazole, which is related in reactivity to acrylates (Figure 2.14). The advantage of this monomer is the absence of hydrolysable functionalities.[159]

An alternative route involves the functionalization of carbohydrates with *p*-nitrophenyl groups, which are occasionally commercially available. Subsequent reduction to amines and reaction with acryloyl chloride leads to the monomer depicted in Figure 2.15.[160]

In summary, the large number of available structures reflects the interest in the use of glycomonomers in radical polymerization processes. Many of the monomers depicted above were employed in conventional free-radical polymerization, but they were also polymerized in a controlled manner using ATRP, NMP, RAFT and cyanoxy-mediated polymerization. There is a lack of more complex carbohydrates, *e.g.* di- and trisaccharides. Whereas many monosaccharides have been used in this process, other saccharides – apart from lactose – are notably absent.

2.2.2 Synthesis of Glycomonomers for Anionic and Cationic Polymerization

Most monomers utilized to prepare glycopolymers *via* the ionic route have already been mentioned above. The choice is limited, however, to protected monomers, in particular isopropylidene-protected monomers for anionic polymerization (Figure 2.16).[28]

Monomers for cationic polymerization are based on vinyloxy functional groups. Although the monomer still needs to be protected to achieve living characteristics, the choice of protecting group can now be extended to ester functionalities and others. Some of the monomers have been copolymerized

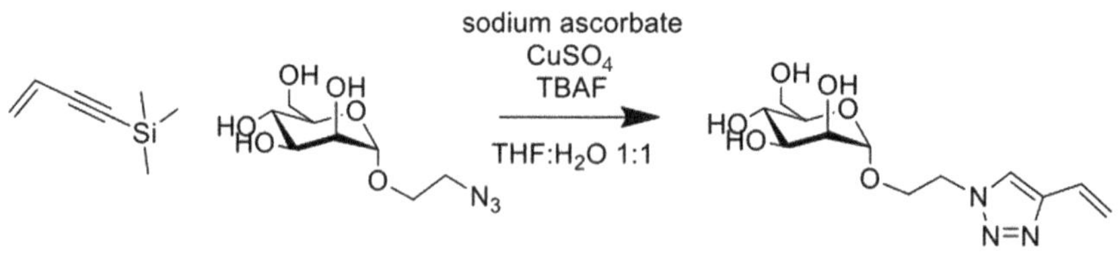

Figure 2.14 Glycomonomer bearing a 4-vinyl-1,2,3-triazole group.

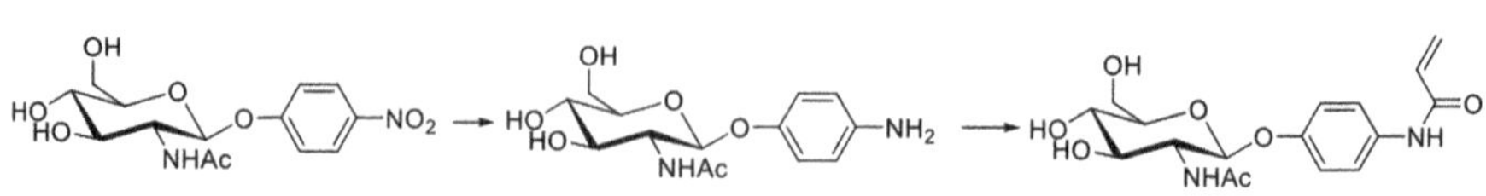

Figure 2.15 Glycomonomer using a *p*-nitrophenyl derivative of a carbohydrate.

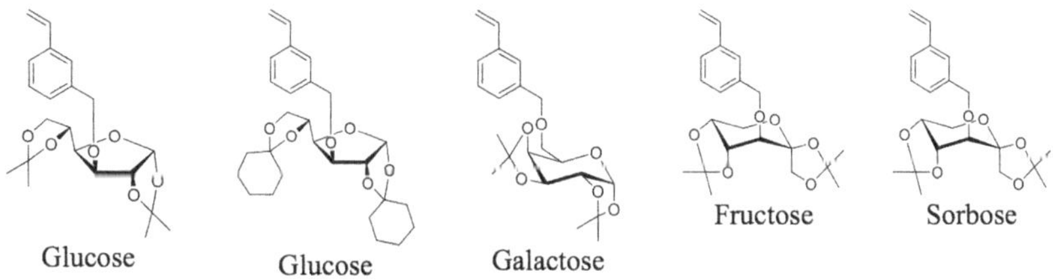

Figure 2.16 Monomers for anionic polymerization.[28]

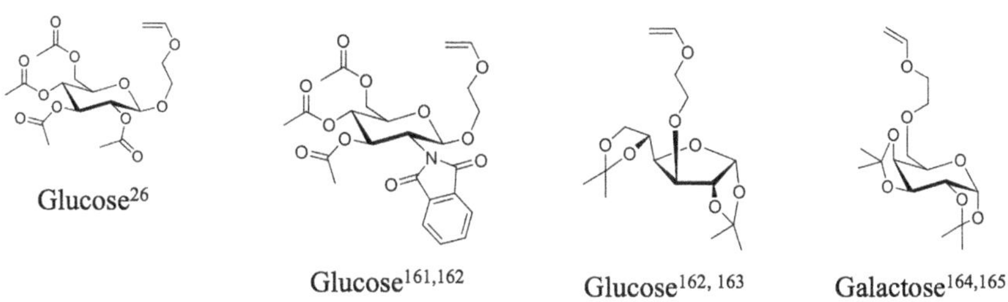

Figure 2.17 Monomers for cationic polymerization.

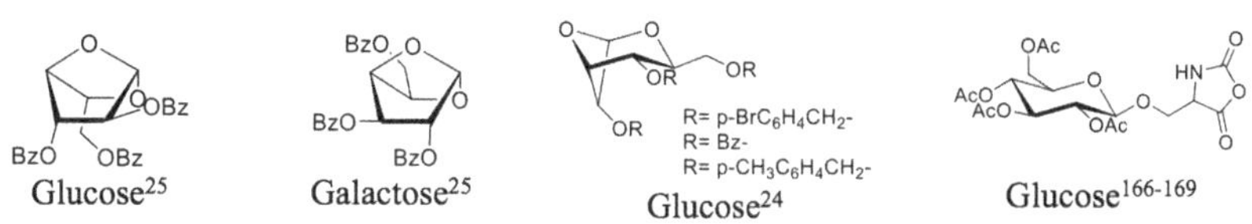

Figure 2.18 Monomers for ring-opening polymerization.

in a radical process, but only cationic polymerization allows homopolymerization (Figure 2.17).

2.2.3 Glycomonomers for Ring-Opening Polymerization

Ring-opening polymerization is unique among all the methods employed to design glycopolymers since the resulting glycopolymers are potentially fully degradable. Glycomonomers for ring-opening polymerization can be divided into two groups: one group carries the reactive cyclic group as the side group (*e.g.* carboxyanhydride), whereas the other group is derived from the anhydride of the corresponding carbohydrates (Figure 2.18).

2.2.4 Glycomonomers for Ring-Opening Metathesis Polymerization (ROMP)

Monomers for ROMP that have been employed so far are based on either 7-oxobornene or norbornene (Figure 2.19). Monomers can be obtained from the corresponding norbornenecarbonyl chlorides (racemic in the case of

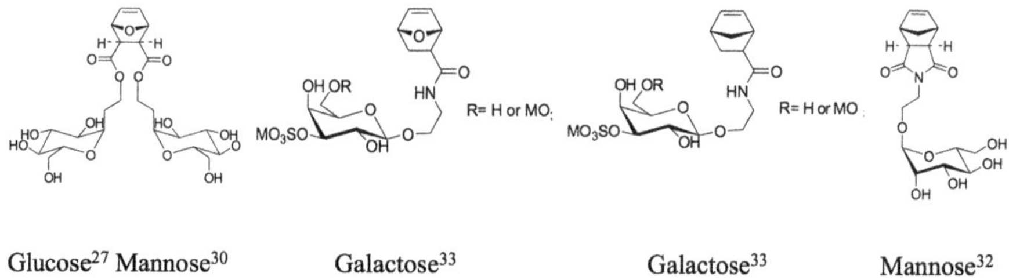

Figure 2.19 Monomers for ROMP.

the 5-norbornene-*trans*-2,3-dicarbonyl chloride) and the protected sugar. Isopropylidene-protected sugars have been identified to accommodate the conditions during ROMP polymerization. 3,6-Oxy-l,2,3,6-tetrahydrophthalic anhydride was reacted with a range of triethylsilyl-protected carbohydrates, which were deprotected prior to ROMP polymerization.

2.3 Conventional Free-Radical Polymerization

Despite the success of reversible-deactivation radical polymerization techniques, free-radical polymerization remains the favourite technique for generating glycopolymers. Although interest declined with the advent of controlled techniques, it is notable that free-radical polymerization has started to become the technique of choice again. The reason may be that direct synthesis has often been replaced by post-modification, which is discussed in a later chapter. More often it is not the length or the molecular weight distribution that is of interest to the researcher, but the type of carbohydrate. It is certainly notable that biological investigations are becoming more and more sophisticated and the application takes centre stage and not the control over the polymer architecture. Therefore, most publications in the last few years were dedicated not to the synthesis, but to the activity of the glycopolymer and its ability to cure diseases.

In the following, only the latest publications have been considered. The reader is referred to a recent book chapter that presents a more comprehensive summary.[43]

2.3.1 Acrylamide Monomers

Acrylamide monomers are often the result of the direct synthesis of amino sugars with (meth)acrylate in water. Their polymerization is often carried out in water as a non-toxic and environmentally benign solvent. These types of monomers have been copolymerized in the past with other water-soluble monomers such as acrylamide,[97] and homopolymers were also generated in water.[108,170] Aggregate formation was often achieved by employing monomers with long hydrophobic spacers between the vinyl functionality and the carbohydrate.[98] Occasionally the polymerization was carried out using protected glycomonomers, but this required deprotection after synthesis.[171]

Acrylamide monomers have been one of the favourites in the last few years as a means to generate glycopolymers with radical polymerization, usually with a view to generating new materials. Applications include protein conjugation, toxin neutralization, drug delivery and hydrogels.

Protein conjugation has been a hot topic for many years. Although most glycopolymers for protein conjugation were typically generated using controlled techniques, radical polymerization can still be an easy pathway to generate end functional polymers, albeit with a broader molecular weight distribution. An unprotected acrylamide monomer based on a β-mannan trisaccharide was copolymerized with acrylamide in the presence of a chain-transfer agent, cysteamine. The purpose of the chain-transfer agent was not only to control the molecular weight, but also to generate end functional polymers that were conjugated to albumin. The resulting glycopolymer vaccine was shown to induce a significant immune response with higher trisaccharide-specific antibody titres.[172]

The ability of glycopolymers to bind to lectins makes these polymers attractive for toxin neutralization. For example, *p*-acrylamidophenyl β-lactoside and *p*-acrylamidophenyl β-D-glucopyranoside (Figure 2.15) were polymerized with acrylamides to generate various glycopolymers that were active against ricin, a toxin that binds efficiently to galactose units.[173] The ability of these polymers to bind to ricin led to further developments aimed at the removal of ricin from solutions. For this purpose, a monomer with a pendant biotin group was added during the polymerization of a lactose-based monomer. After binding with ricin, the complex could be removed from the solution with magnetic streptavidin-coated nanoparticles.[174]

A set of sulfated monomers based on the monomer depicted in Figure 2.15 was prepared to study the inhibitory effects on the Alzheimer's β-secretase (BACE-1). The various monomers showed different sulfation patterns, which could significantly influence binding to the protein.[175]

A methacrylamide similar to that depicted in Figure 2.15, *p*-acrylamidophenyl α-D-mannopyranoside (Man), was copolymerized with poly-*N*-isopropylacrylamide (PNIPAm) *via* emulsion polymerization in the presence of a crosslinker. The resulting nanoparticles could recognize concanavalin A (ConA) through multipoint interactions. Depending of the swelling of the nanoparticle, the kinetic of protein binding could be fine-tuned.[176] A monomer from the same family was also copolymerized with 3-(trimethoxysilyl) propyl methacrylate. This time, the acetyl-protected monomer was employed, most likely to accommodate the solubility of the hydrophobic 3-(trimethoxysilyl)propyl methacrylate. After immobilization of the polymer on a silica surface, a sensor that could detect ConA was obtained.[177]

Radical polymerization has frequently been applied to generate glycopolymers for hydrogels since molecular weight control is often not the main concern. An example is the copolymerization of 2-hydroxyethyl methacrylate (HEMA) with (1 → 4)-β-D-mannuronan- or (1 → 4)-α-L-guluronan-functionalized (meth)acrylamides. The consumption of HEMA with different glycomonomers was monitored, showing the preferred incorporation of HEMA into the polymer. owing to the resemblance of the glycomonomer to alginate, the hydrogels could be crosslinked using calcium ions.[178]

Notable is the interest in more complex carbohydrate structures. Whereas reports on the polymerization of trisaccharides were almost absent until 5 years ago, there has been a clear surge to create glycomonomers based on more complex structures. Fucosyl chitobiose [GlcNAcβ1 → 4(Fucα1 → 3)GlcNAc], an allergy-associated carbohydrate epitope, has been derivatized into a glycomonomer (Figure 2.20). The resulting polymer showed good binding to the lectin WGA.[179]

2.3.2 (Meth)acrylate Monomers

The biggest group of monomers is based on methacrylates and acrylates, which is also evident considering the list of glycomonomers above. Interestingly, there has been a strong tendency in the past to carry out the polymerization in organic solvents employing protected monomers.[78,89,95,180] Dimethylformamide (DMF),[78,171] benzene[181] and chloroform[89,95] are popular solvents and azobisisobutyronitrile (AIBN) is the most common form of initiator used. Some monomers were deprotected prior to polymerization, which allows synthesis in an aqueous solution, initiated using $K_2S_2O_8$.[89,95] The use of acrylates was not very widespread in the past, mainly because of the low stability of the monomer and maybe the tendency to polymerize during storage. Although maybe less popular, the use of acrylates is still common.

Methacrylate monomers were frequently used in recent years to create biomaterials for various applications such as hydrogels: 6-methacryloyl-α-D-galactopyranose, for example, was copolymerized with the crosslinker poly(ethylene glycol) diacrylate in the presence of collagen to generate an interpenetrating hydrogel network for corneal applications. Since the polymer could not be processed after crosslinking, the mixture was directly polymerized with the help of UV irradiation in a contact lens mould.[182]

Drug delivery carriers are also increasingly based on glycopolymers. A range of statistical copolymers, poly[2-(α-D-mannopyranosyloxy)ethyl-*co*-2-dimethylaminoethyl methacrylate]s, were prepared by radical polymerization (monomer reactivity ratios r_{ManEMA} and r_{DMAEMA} were 0.98 and 1.22, respectively). The polymers were capable of binding with plasmid DNA while displaying a cluster glycoside effect.[183] To understand better the cellular uptake

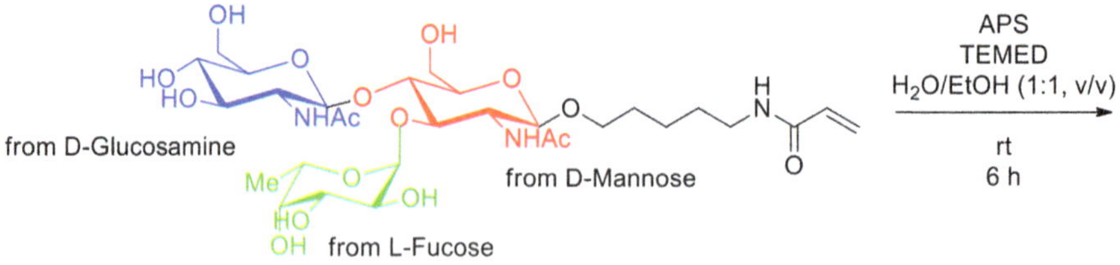

Figure 2.20 Fucosylated trisaccharide monomer.[179]

of these polymers, the same monomer, 2-(α-D-mannopyranosyloxy)ethyl methacrylate, was now polymerized with HEMA. Prior to biological evaluation, reactivity ratios were determined and from those the sequence distribution. The polymers were then quickly taken up by HeLa cells, a cell line that has a high affinity to mannose.[184]

Glucose-responsive nanoparticles were prepared from polymers obtained by copolymerization of 3-acrylamidophenylboronic acid, 2-lactobionamidoethyl methacrylate (Figure 2.13) and a poly(ethylene glycol) (PEG)-based monomer. Self-assembly of these polymers led to nanoparticles that can encapsulate insulin for on-demand release, which was triggered by the interaction between glucose and boronic acid.[185] The polymer nanoparticles were mainly developed for oral administration,[185] although it was reported later that it can also be used to deliver insulin *via* the nasal route.[186]

2-[(D-Glucosamin-2-*N*-yl)carbonyl]oxyethyl acrylate, a monomer prepared from an unprotected amino-functionalized carbohydrate and 2-{[(4-nitrophenoxy)carbonyl]oxy}ethyl methacrylate was employed to coat magnetic nanoparticles with glycopolymers.[187] The monomer was initially immobilized on the surface of Fe_3O_4 nanoparticles, followed by dispersion of the particles with further addition of glycomonomers in DMF (Figure 2.21).

The following polymerization, which was initiated with AIBN, led to magnetic nanoparticles with good bioactivity. The same monomer was then also used to prepare porous films. To achieve this, the monomer was copolymerized in DMF with a sufficient amount of styrene to yield polymers that can be dissolved in an organic solvent for further processing.[188]

2.3.3 Styrene-Based Monomers

Styrene-based monomers were particularly attractive in the 1990s and most reports using this class of monomers stem from that era.[105,106] These monomers were occasionally polymerized in water,[140,189] but dimethyl sulfoxide (DMSO) or a mixture of DMSO[138,153] was more common, probably owing to the hydrophobic nature of the benzene ring. Thermal properties of the polymers were studied by differential scanning calorimetry (DSC). In the same report, the monomer was copolymerized with styrene in DMF using AIBN as initiator. The attraction of styrene monomers lies in the high glass transition

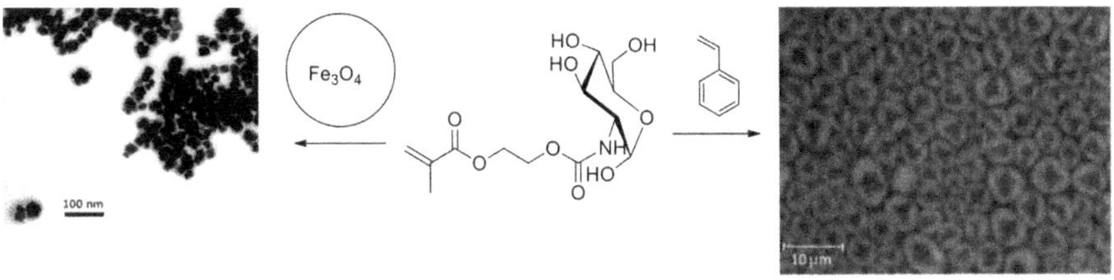

Figure 2.21 Glycopolymers used to stabilize Fe_3O_4 nanoparticles[187] or to prepare porous films.[188]

temperature, which was occasionally increased by the presence of interacting groups that form strong H-bonds such as urea.[141]

Although styrenic monomers have been less popular in the last few years, they are still widely used, such as in applications where stable glycopolymer surfaces are targeted. Poly(*N*-*p*-vinylbenzyl-4-*O*-β-D-galactopyranosyl-D-gluconamide) (related to the monomer depicted in Figure 2.13) was coated on a polystyrene surface to act as a substrate for isolation and enrichment of embryonic stem (ES) cell-derived hepatocytes with adequate homogeneity and functionality.[190]

The same polymer was also employed to coat polythiophene to create water-soluble nanowires. The glycopolymer not only ensured stability in aqueous solutions, but also introduced bioactivity, as demonstrate by the ability of these nanowires to bind to a lectin.[191]

2.3.3.1 *Vinyl Esters*

Polyvinyl esters of carbohydrates are accessible thanks to the chemoenzymatic synthesis of the monomer (Figure 2.8).[122,192] The resulting polymers showed good bioactivity when disaccharides such as maltose or lactose were employed.[122]

In recent years, more reports emerged of studies of the material properties of these polymers. For example, a statistical copolymer of NIPAAm and 6-*O*-vinyladipoyl-D-glucose was observed to form loose nano-sized aggregated or collapsed spheres depending on the temperature. Detail light scattering investigations revealed changes in the R_g/R_h values dependent not only on the temperature, but also the concentration.[193]

Poly(6-*O*-vinylsebacoyl-D-glucose) could be used as a substrate for enzyme immobilization. Embedded in fibres made from polyacrylonitrile, the blended glycopolymer was used to tailor the hydrophilicity of fibres, but also to act as a reactive substrate for the binding of catalase *via* epichlorohydrin. Furthermore, the glycopolymer enhanced the stability of the enzyme.[194]

2.3.4 Other Vinyl-Containing Glycomonomers

A range of other vinyl-containing glycomonomers have been polymerized, but often only as comonomers because of the inability of the monomer to undergo radical homopolymerization. Glycomonomers with 1-alkenyl

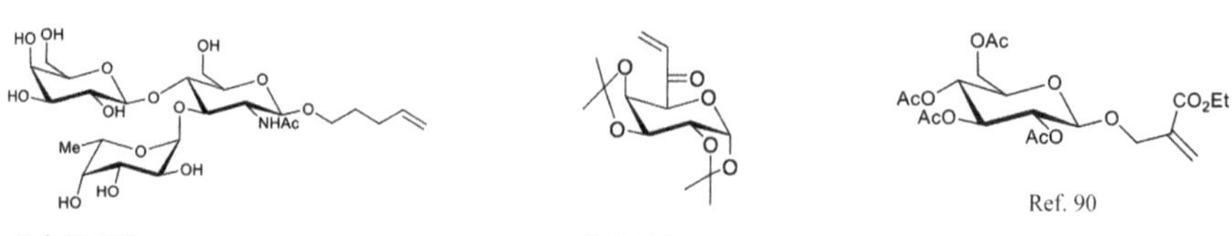

Figure 2.22 Other monomers employed in radical polymerization.

functionalities as polymerizing groups were copolymerized with acrylamides to yield water-soluble polymers with high activity.[29,107] A vinyl ketone monomer that is incapable of homopolymerization was copolymerized with MMA, styrene and acrylonitrile.[195] Also unique is the sterically hindered monomer depicted in Figure 2.22.[90] These types of monomers are usually more sluggish during polymerization, but they have the added benefit of creating a backbone with a stiffer structure, which could potentially influence binding with lectins.

2.4 Reversible-Deactivation Radical Polymerization (RDRP)

Since 2010, it has become more obvious that the term controlled/living radical polymerization (CLRP) is now better replaced with reversible-deactivation radical polymerization (RDRP).[196] In most cases, the term 'controlled/living' radical polymerization refers to polymerization processes in which irreversible chain-transfer and termination reactions will not occur. However, this condition is not the case for many radical polymerization techniques employed using mediating/control agents, as termination events are usually inevitable. Hence the term 'controlled/living' radical polymerization is now more correctly denoted 'reversible-deactivation radical polymerization'. RDRP comprises nitroxide-mediated polymerization (NMP) and cyanoxyl-mediated polymerization, both of which fall into the category of stable free-radical polymerization (SFRP), atom-transfer radical polymerization (ATRP), single-electron transfer–living radical polymerization (SET-LRP) and reversible addition–fragmentation transfer (RAFT) polymerization. Figure 2.23 shows a simplified mechanism of RDRP, where $P^{\cdot}$ = propagating radical, M = monomer, X = mediating agent and k = rate constants.

There are two main methods of using glycomonomers in the RDRP process. One method involves the polymerization of unprotected glycomonomers in an aqueous environment and the other involves the polymerization of protected glycomonomers in organic solvents. The terms 'random' and 'statistical' copolymers are used interchangeably throughout this section.

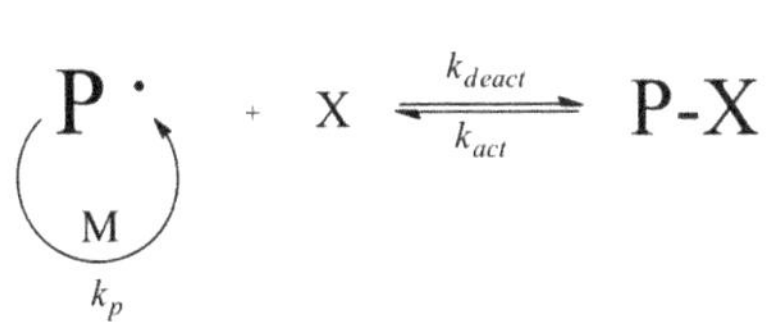

Figure 2.23 Mechanistic pathway of RDRP.[197]

2.4.1 Stable Free-Radical Polymerization (SFRP)

2.4.1.1 Nitroxide-Mediated Polymerization (NMP)

Being the earliest technique in reversible-deactivation radical polymerization, there have been numerous studies of the synthesis of glycopolymers using NMP. It requires the use of a nitroxide mediating agent, which consists of a bulky R group attached to a nitrogen atom linking the alkoxyamine. Figure 2.24 shows an example of a typical nitroxide SG1 and an SG1-based alkoxyamine (BlocBuilder). During polymerization, homolytic cleavage of alkoxyamine at C–O at the neighbouring nitrogen atom generates initiating radicals for polymerization. Based on the persistent radical effect, reversible capping of growing radical chains is consistently occurring during the course of polymerization.[198]

Recently, combined work by the Centre for Advanced Macromolecular Design (CAMD) and Equipe de Physique et Chimie des Polymères (EPCP) synthesized galactose-based glycopolymers for micelles and honeycomb structure formation. Studies showed that the copolymerization of acetylated galactose methacrylate monomer (90 mol%) and styrene (10 mol%) in dioxane at 85 °C resulted in well-defined glycopolymers with a polydispersity of 1.26. Based on the penultimate effect, a low percentage of styrene allowed SG-1-mediated polymerization to take place without compromising the polydispersity of the final glycopolymers. However, chain extension of the galactose/styrene first polymer block did not result in complete conversion to the targeted diblock copolymer. Phosphorus NMR analysis revealed that only 81% of the copolymer first block retained SG-1 end functionality. Figure 2.25 shows the structure of the nitroxide-mediated diblock glycopolymer. In further studies, the polymerization sequence was reversed. Chain-extension polymerization of galactose monomer (90 mol%) with styrene (10 mol%) with a 0.5 h initiation time with polystyrene macroinitiator at 120 °C and subsequent polymerization at 85 °C led to well-defined amphiphilic diblock copolymers. Acetylated galactose moieties on glycopolymers were deacetylated using sodium methoxide. Micelles and honeycomb structures formed by the amphiphilic glycopolymers were found to interact with peanut agglutinin (PNA).[102]

Babiuch *et al.* synthesized a linear glycopolymer from pentafluorostyrene glucose-based monomers (Figure 2.26).[199] The alkoxyamine initiator,

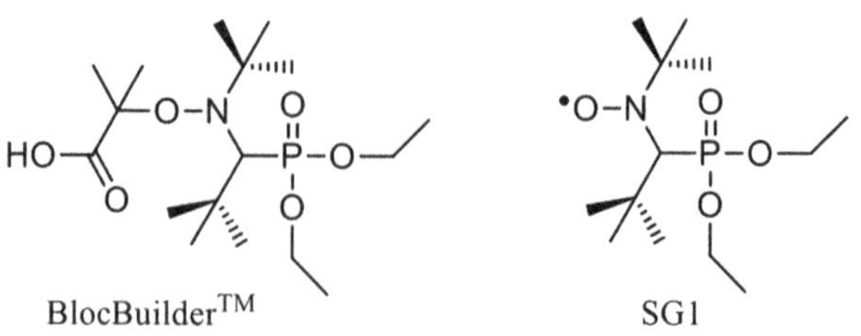

Figure 2.24 Structures of nitroxide SG1 and SG1-based alkoxyamine (BlocBuilder).[102]

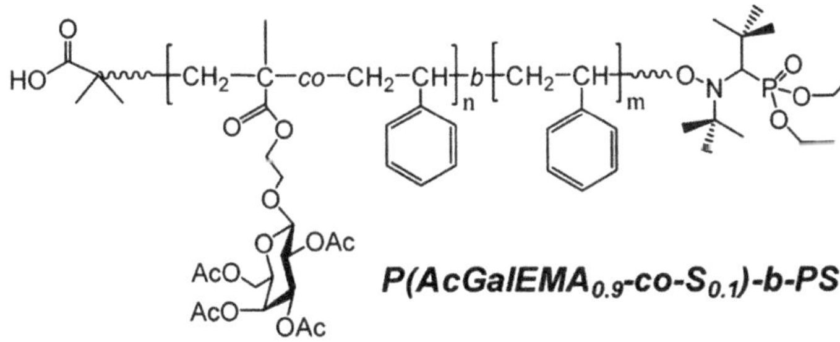

Figure 2.25 Schematic representation of an acetylated glycopolymer with poly-galactose-*co*-polystyrene first block and polystyrene second block.[102] Reproduced with permission of the American Chemical Society, Copyright 2009.

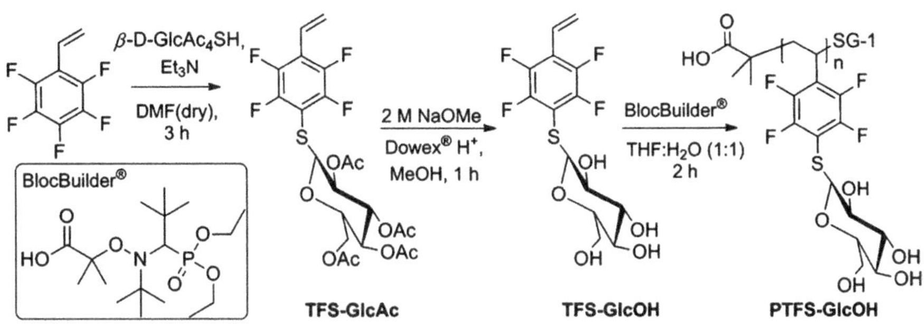

Figure 2.26 Chemical reaction pathway of glycomonomer synthesis and its nitroxide-mediated polymerization.[199] Reproduced with permission of the American Chemical Society, Copyright 2011.

BlocBuilder, was used as mediating agent to yield well-defined glucose-based glycopolymers. Owing to the water-soluble deprotected glucose pentafluorostyrene monomer used, polymerizations were carried out in tetrahydrofuran (THF)–water (1 : 1) mixture to solubilize the reactants. The reaction vessel was pressurized, purged with argon and polymerized at 110 °C. The monomer:initiator ratio was 50. Fluorine NMR spectroscopy was used to characterize the *ortho*- and *meta*-fluorines of the tetrafluorostyryl units and gel permeation chromatography (GPC) showed that glycopolymers of molar mass 24 000 g mol^{-1} and polydispersity 1.16 were obtained. The glycopolymers were used as coating materials to stabilize magnetic iron oxide nanoparticles and cell–carbohydrate interactions.[199]

2.4.1.2 *Cyanoxyl-Mediated Polymerization*

In the case of cyanoxyl-mediated polymerization, cyanoxyl persistent radicals (•OC≡N) are generated during polymerization and the reversible capping of growing radical chains dramatically minimizes termination events. One

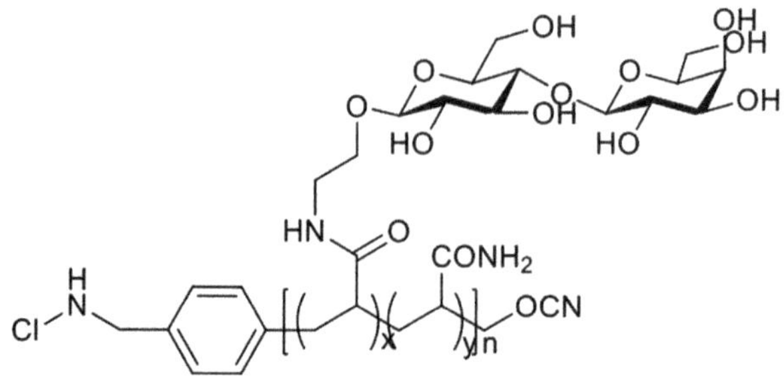

Figure 2.27 Structure of OCN chain-end-functionalized lactose-based glycopolymer.[201]

advantage of cyanoxyl-mediated polymerization over NMP is the wide range of the low reaction temperature (0–80 °C).[200]

Following Chaikof and co-workers' studies,[126] chain-end functionalized glycopolymer was synthesized *via* cyanoxyl-mediated free-radical polymerization by Narla and Sun.[201] A disaccharide glycomonomer comprised of lactose residues was employed to copolymerize acrylamide using 4-chloroaniline as initiator. The polymerization was carried out in THF–water (1:1) mixture. Unlike nitroxide-mediated polymerization, reaction in cyanoxyl-mediated polymerization was performed at a lower temperature of 65 °C for 16 h. OCN chain-end functional groups on the glycopolymers (Figure 2.27) were used as anchoring points for immobilization on amine-modified silica gel *via* the formation of isourea bonds. Glycopolymers on the silica gel and glass slides were found to bind specifically to peanut agglutinin, also known as *Arachis hypogaea* lectin.[201] In another study by Narla and Sun, *O*-cyanoxyl-mediated glycopolymers were further reacted with α-2,3-sialyltransferase to yield sialyllactose-containing glycopolymers.[202]

2.4.2 Atom-Transfer Radical Polymerization (ATRP)

2.4.2.1 *Homopolymers and Statistical Copolymers*

Multivalent and highly specific BODIPY-conjugated glycopolymers were synthesized using ATRP. A monomer:initiator:CuBr:amine ligand concentration ratio of 100:1:1:1 was employed for the polymerization. The monomer concentration ratio of 2-*O*-methacryloyloxyethyl (2,3,4,6-tetra-*O*-acetyl-β-D-galactopyranoside) (AcGEMA) to 4,4-difluoro-4-bora-3α,4α-diaza-*s*-indacene (BODIPY) was 100:1. After deacetylation using hydrazine in DMSO, the resulting random copolymers were found to be highly water soluble and they self-assembled into micelles.[203] Qu *et al.* used an alkyl halide–copper complex system to polymerize 3-*O*-methacryloyldiacetone-D-glucose. In this study, parameters such as different types of initiators, catalysts and ligands and conducting polymerizations at different temperatures were investigated. It was found that a complementary match between the different reactions conditions was needed to achieve a well-controlled polymerization with homoglycopolymers of low polydispersity.[204]

2.4.2.2 *Diblock Copolymers*

Linear diblock copolymers were synthesized by Wan and co-workers using atom transfer radical polymerization (Figure 2.28).[205] Polystyrene was first synthesized as a macroinitiator serving as the hydrophobic segment of the diblock copolymer during honeycomb formation. Amphiphilic diblock glycopolymers were eventually synthesized by chain extending polystyrene first homopolymer with 2-(2,3,4,6-tetra-*O*-acetyl-β-D-glucosyloxy)ethyl methacrylate (AcGEMA) glycomonomer. Polymerization was conducted using a copper bromide–pentamethyldiethylenetriamine catalyst system with chlorobenzene as the solvent. The polymerization reaction was conducted at 80 °C and yields between 52 and 62% were obtained. Polydispersities of the polymers ranged from 1.08 to 1.32 and the average molecular weights measured by GPC were between 20 300 and 27 600 g mol^{-1}.

Leon *et al.* synthesized both amphiphilic diblock and triblock glycopolymers from hydrophobic poly(butyl acrylate) (PBA) first block.[206] Butyl acrylate monomer conversion was kept below 50% in order to retain efficient bromide end functionalities for further chain-extension polymerizations. ATRP macroinitiator, poly(butyl acrylate), with monofunctional, PBA–Br, and difunctional, Br–PBA–Br, bromide groups was subsequently chain extended by polymerization with unprotected glycomonomer, 2-{[(D-glucosamin-2-*N*-yl) carbonyl]oxy}ethyl methacrylate (HEMAGl). ATRP was carried out in a polar single solvent, DMF, system with copper chloride (CuCl)–pentamethyldiethylenetriamine (PMDETA) as catalyst. However, the solubility of HEMAGI in poly(butyl acrylate) containing DMF was poor owing to the highly hydrophilic nature of HEMAGI. A homogeneous reaction mixture was eventually achieved at 90 °C, the temperature at which the polymerization was performed.

Diblock copolymers bearing a hydrophobic colourless spiropyran (SP) block and a glucose block were synthesized by Das and co-workers (Figure 2.29).[207] 3-*O*-4-Vinylbenzoyl-D-glucopyranose (BipG) was first polymerized at 60 °C to generate a PBipG ATRP macroinitiator. PBip was later chain extended with spiropyran methacrylate (SPM) to yield PSP-*b*-PBipG diblock glycopolymer. In another study by the same group, photoresponsive fluorescent diblock glycopolymers

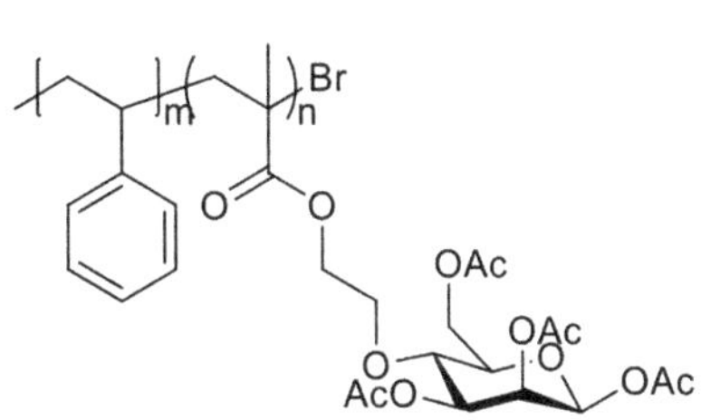

Figure 2.28 Glycopolymer synthesis by atom-transfer radical polymerization.[205]

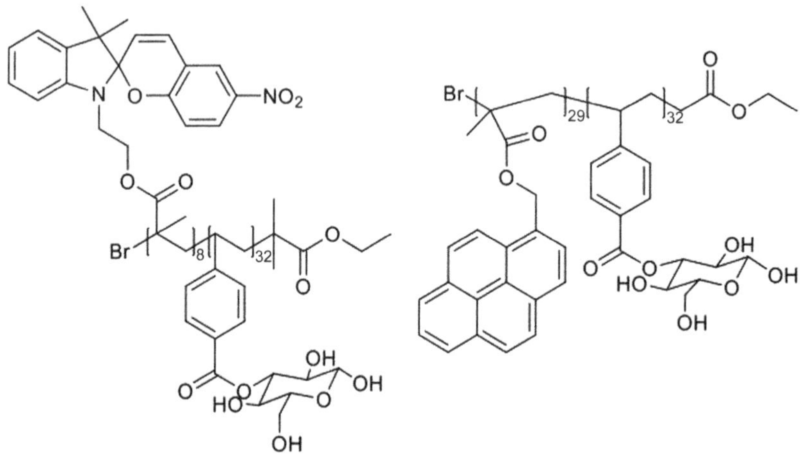

Figure 2.29 Photoresponsive glycopolymers. By alternating UV and visible irradiation, poly(spiropyran methacrylate)-based glycopolymer (left) can undergo isomerization.[207] Photo-solvolysis will occur when UV light is irradiated on poly(pyrenylmethyl methacrylate)-based glycopolymer (right).[208]

were synthesized (Figure 2.29).[208] First, BipG, an isopropylidene-protected glucose monomer, was homopolymerized under ATRP conditions using copper bromide (CuBr)–PMDETA with ethyl 2-bromoisobutyrate at 60 °C in THF for 24 h. The resulting PBipG glycopolymer gave an experimental number-average molecular weight (M_n) of 12 297 g mol^{-1} and a polydispersity index (PDI) of 1.19. Chain-extension copolymerization of pyrenylmethyl methacrylate (PyM) with PBipG macroinitiator gave an M_n of 21 194 and a PDI of 1.46.

Munisso *et al.* synthesized diblock copolymers starting with a galactose or glucose initial block followed by the copolymerization of 2-(dimethylamino) ethyl methacrylate (DMAEMA) second block.[209] Pendant carbohydrate groups from the galactose combined with cationic groups from DMAEMA reduced the toxicity of the overall polymers, and pendant galactose groups on the polymers were able to target hepatocytes in a ligand-specific manner. Li and co-workers prepared diblock glycopolymers comprised of poly(ethylene glycol)-*b*-poly(2-acryloxyethylgalactose) (PEG-*b*-PAEG).[210] Diblock glycopolymers were crosslinked using a poly(3-methacrylamidophenylboronic acid) (PMAPBA) homopolymer prepared by RAFT polymerization. Owing to the shielding of negative charges by phenylboronic acid moieties, the aggregation of boronate-crosslinked nanoparticles was found to increase with increasing salt concentration.

2.4.2.3 *Multi-Block Copolymers*

ABA-type triblock glycopolymers were synthesized by three-step procedures. Melt condensation of L-lactic acid was performed using stannous chloride ($SnCl_2{\cdot}2H_2O$) as a catalyst. This reaction led to a hydroxyl-terminated prepolymer by alkyd condensation with 1,4-butanediol. In the second step, esterification of polylactide (PLA) hydroxyl end-groups with 2-bromoisobutyryl bromide was carried out. A PLA macroinitiator with dibromide groups at both ends was obtained. The final step was the copolymerization of 6-*O*-methacryloyl-1,2; 3,4-di-*O*-isopropylidene-D-galactopyranose (MAIPGal) using PLA macroinitiator obtained from the previous step. It was proposed that the hydrolysis of PMAIPGal blocks can result in triblock copolymers with two hydrophilic galactose blocks at both ends of the polymer. The proposed final amphiphilic polymer self-assembled into micelles.[211]

Biswas and co-workers synthesized an ABA-type triblock glycopolymer *via* a combination of step-growth condensation and copper-catalysed atom-transfer radical polymerization dual-step reactions.[212] Dihydroxyl polyurethane (HO–PU–OH) was initially synthesized by condensation polymerization of poly(tetramethylene glycol) (PTMG, M_n = 1400 g mol^{-1}) and 1,6-hexamethylene diisocyanate (HDMI) in a 2.1 : 1 molar ratio in the molten state at 70 °C. HO-PU-OH was converted into 2-bromoisobutyryl-terminated PU (Br–PU–Br) by reaction with 2-bromoisobutyryl bromide in the presence of triethylamine. The end-functionalized Br–PU–Br can function as an ATRP macroinitiator for copolymerizaton of D-gluconamidoethyl methacrylate (GAMA).

To generate an ABC-type triblock glycopolymer (Figure 2.30), Wang *et al.* first esterified poly(ethylene glycol) monomethyl ether (CH_3O–PEG_{45}–OH) (M_w = 2000 and M_w/M_n = 1.05) with 2-bromo-2-methylpropionyl bromide

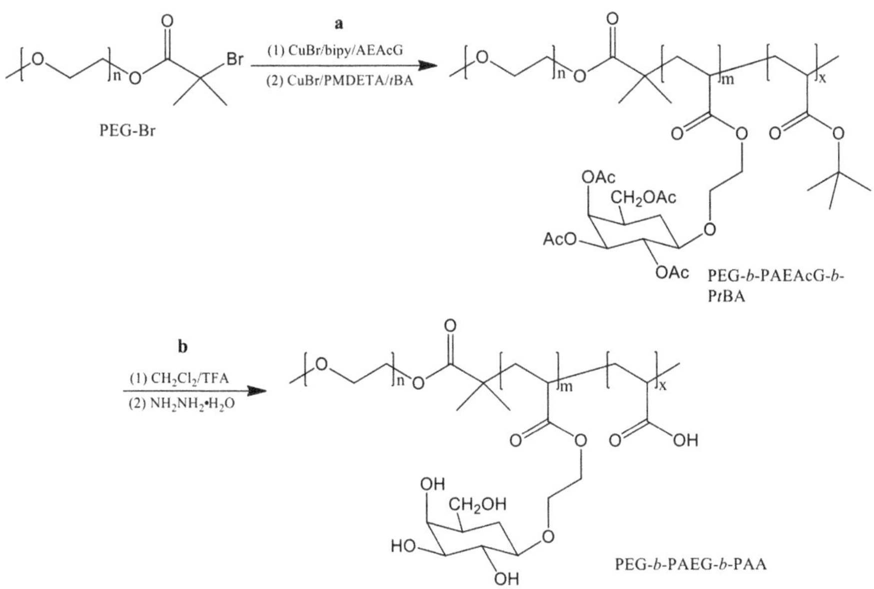

Figure 2.30 Reaction pathway of PEG-*b*-PAEG-*b*-PAA: (a) triblock glycopolymer synthesis and (b) deprotection of galactose moieties and *tert*-butyl acrylate units.[213]

(BMPB) catalyst with triethylamine in dichloromethane.[213] The reaction resulted in a PEG-Br macroinitiator as the starting homopolymer. Diblock copolymers were further synthesized by copolymerizing PEG-Br with protected 2-*O*-acryloyloxyethyl(2,3,4,6-tetra-*O*-acetyl-β-D-galactopyranoside) (AEAcG) glycomonomer in chlorobenzene at 80 °C for 24 h. ATRP of AEAcG with PEG-Br macroinitiator yielded PEG-*b*-PAEAcG with an experimental molecular weight of 8800 g ml^{-1} and a narrow molecular weight distribution of 1.20. A solvent mixture of 2-butanone and 2-propanol (7 : 3, v/v) was eventually used for the triblock copolymer synthesis. *tert*-Butyl acrylate was employed as the third polymer block after the second galactose block. The reaction mixture was degassed by three freeze–pump–thaw cycles, sealed under vacuum and placed in an oil bath at 90 °C. Finally, the reaction mixture was passed through an aluminium oxide column to remove the copper catalyst. The *tert*-butyl acrylate block was hydrolysed and the protected galactose block was deacetylated. The resulting ABC-type triblock copolymers consisted of (a) a PEG block that prevents the interactions of polymeric micelles with serum protein and cells, (b) a galactose targeting block that recognizes liver cell asialoglycoprotein receptors (ASGP-R) and (c) a polyacrylic acid block that facilitates the synthesize of degradable crosslinked micelles.

By combining the knowledge of the pH and glucose responsiveness of phenylboronic acids used in diblock copolymers,[210] Li and co-workers synthesized ABA-type triblock copolymers with a polygalactose block in between two poly(acrylic acid-*co*-acrylamidophenylboronic acid) blocks.[214] Poly(2-acryloxyethylgalactose) (PAEG) homopolymer was first synthesized by using a protected 2-*O*-acryloyloxyethyl(2,3,4,6-tetra-*O*-acetyl-β-D-galactopyranoside) (AEAcG) glycomonomer with dibromo-*p*-xylene (DBX) as initiator. This was followed by copolymerization of *tert*-butyl acrylate to achieve P*t*BA-*b*-PAEAcG-*b*-P*t*BA triblock copolymer. After the hydrolysis of *tert*-butyl acrylate groups to acrylic acid groups on the polymer, the latter was further reacted with amino groups of 3-aminophenylboronic acid (APBA) monohydrate to yield the final glycopolymer poly(acrylic acid-*co*-acrylamidophenylboronic acid)-*b*-poly(2-acryloyloxyethyl pentaacetylgalactoside)-*b*-poly(acrylic acid-*co*-acrylamidophenylboronic acid) {[(PAA-*co*-PAAPBA)-*b*-$]_2$ PAEAcG}.

2.4.2.4 Complex Polymer Structures and Glycomaterials

Disulfide-linked linear homoglycopolymers comprised of glucose moieties were prepared by Kitano *et al.*[215] A special ATRP initiator, 2-(2-bromoisobutyroyloxy)ethyl disulfide (DT-Br), with disulfide groups was employed during polymerization in methanol–water (5 : 1) and *N*-methyl-2-pyrrolidone (NMP)–water (7 : 3) mixtures at room temperature for 24 h. Poly(methyl α-D-glucopyranoside) glycopolymer bearing disulfide allowed thiol exchange reactions to take place on the surface of gold colloids. Sulfated linear glycopolymers were deposited and bound on the surfaces of gold colloids, giving a polymer brush. The sulfated polymer brushes were found to interact with β-amyloid protein (Aβ) whereas non-sulfated brushes did not interact.

Novel glycopolymer brushes were synthesized *via* a combination of RAFT polymerization and ATRP. Glycopolymers have become increasingly popular in the field of polymer chemistry as they are able to mimic important biological molecules and they have proved to be very useful biomaterials. Glycopolymers are highly hydrophilic and interact well with biological environments. They can therefore be used for artificial materials such as those for biological, pharmaceutical and biomedical use. To develop the glycopolymer brushes as biomimics, Klumperman and co-workers first used RAFT polymerization to generate four well-defined ATRP macroinitiators as the backbone of glycopolymer brushes.[216] The macroinitiators were subsequently employed to grow glycopolymer chains *via* ATRP from the RAFT-generated polymer backbone. During the 'grafting from' approach, it is crucial that a low concentration of active species is present in the reaction, as termination events could occur by intramolecular and intermolecular coupling, which will lead to crosslinked polymers and multimodal distributions of polymers. Low temperature and low active species (copper iodide) were used to lower the concentration of radicals during polymerization.

Carbohydrates were used as initiators for the ring-opening polymerization (ROP) of L-lactide and ε-caprolactone to yield brush-like copolymers.[217] Three methacrylate monomers, methyl methacrylate (MMA), α-methoxy, ω-methacrylate poly(ethylene oxide) (MAPEO) and 1,2:3,4-di-*O*-isopropylidene-6-*O*-methacryloyl-D-galactopyranose (MAIGP), were first polymerized *via* the ATRP process in toluene at 80 °C. CuBr ligand with 1,1,4,7,10,10 hexamethyltriethylenetetramine (HMTETA) was used as a catalytic complex. The protected galactose moieties on the glycopolymers were selectively deprotected by reacting with 80% formic acid for 48 h to yield hydroxyl groups for initiating ROP of L-lactide and ε-caprolactone. Owing to the degradation of galactose moieties when tin octoate at a high temperature of 80 °C was used, dibutyltin dimethoxide [$Bu_2Sn(OMe)_2$] and triethylaluminium catalysts were eventually used to replace tin octoate. The replacement catalysts allowed ROP of L-lactide and ε-caprolactone to take place at room temperature and prevented degradation of galactose units on the polymers.[217]

Comb-like glycopolymers were also synthesized by Wan and co-workers.[205] With the knowledge of converting hydroxyl groups into ATRP initiators, polystyrene-*b*-poly(hydroxyethyl methacrylate) (PS-*b*-PHEMA) was initially synthesized by ATRP. The polymers were subsequently converted into pendant ATRP initiators at the PHEMA block to form polystyrene-*b*-poly[2-(2-bromoisobutyryloxy)ethyl methacrylate] (PS-*b*-PHEMA-Br). Comb-like glycopolymers were finally synthesized by using the PS-*b*-PHEMA-Br macroinitiator to initiate the polymerization of AcGEMA by a routine ATRP procedure.

The diblock glycopolymer poly(2-{[(D-glucosamin-2-*N*-yl)carbonyl]oxy}ethyl methacrylate)-*b*-poly(butyl methacrylate) (PHEMAGl-*b*-PBMA) was synthesized as a polymeric surfactant in the emulsion polymerization of butyl methacrylate. Glycosylated core–shell polymer particles were developed using the technique. The soft core of the nanoparticle consisted of poly(butyl methacrylate) and the shell of the nanoparticle consisted of glucose moieties from the polymeric surfactant. The latex glycoparticles were used to prepare polymer

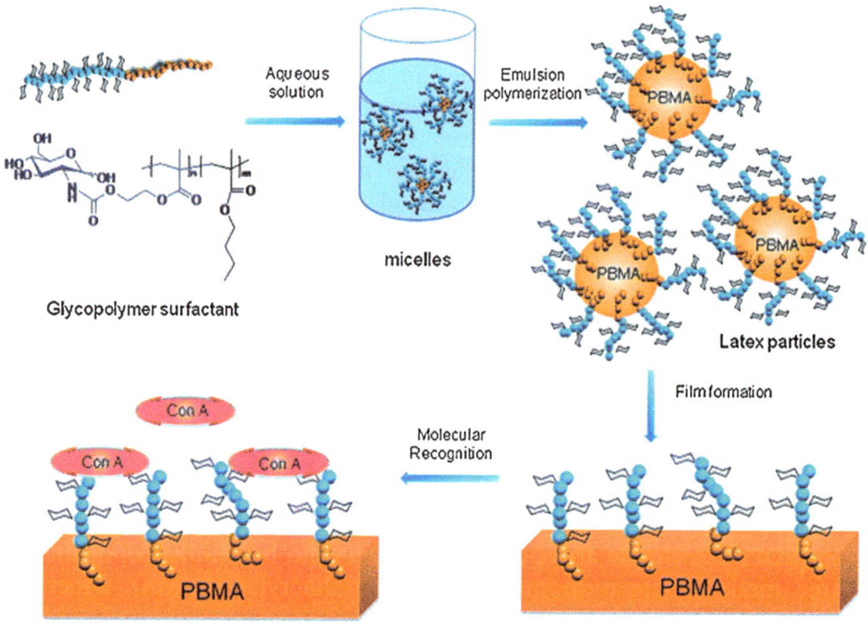

Figure 2.31 Synthetic route for preparing glycosylated polymeric particles and films and their interactions with ConA.[218] Reproduced with permission from The Royal Society of Chemistry, Copyright 2011.

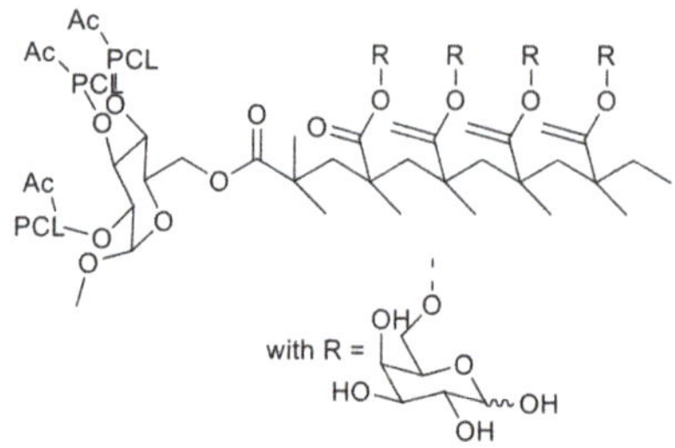

Figure 2.32 Structure of amphiphilic mikto-arm copolymer. R represents 1,2;3,4-di-*O*-isopropylidene-D-galactopyranose.[219]

films with bioactive surfaces that bound with ConA from *Canavalia ensiformis*, a lectin that recognizes and binds glucose and mannose (Figure 2.31).[218]

Suriano *et al.* developed novel amphiphilic A_3B mikto-arm copolymers from glucose molecules.[219] *tert*-Butyldiphenylsilyl-based methylglucoside derivative was used as initiator for the polymerization of ε-caprolactone, resulting in three-arm star-shaped structures. *tert*-Butyldiphenylsilyl (TBDPS) ether was employed to protect selectively primary alcohols of the pyranosidic sugars. In this case, the 6-position of the glucose molecule was protected before ROP of ε-caprolactone took place. Star-shaped poly(ε-caprolactone) macroinitiator was obtained after several post-polymerization steps. Atom transfer radical polymerization of diisopropylidene galactose methacrylate yielded A_2B mikto-arm copolymers with different compositions and molecular weights. The ATRP process was carried out in THF at 60 °C using CuBr ligated with 1,1,4,7,10,10-hexamethyltriethylenetetramine (HMTETA) as catalytic complex (Figure 2.32).

2.4.3 Single-Electron Transfer–Living Radical Polymerization (SET-LRP)

Diblock and triblock glycopolymers were synthesized by single-electron transfer–living radical polymerization (SET-LRP). Homopolymerization of unprotected acrylate-based glycomonomer was first investigated. 2-{[(D-Glucosamin-2-*N*-yl)carbonyl]oxy}ethyl acrylate (HEAGI) was polymerized at 40 °C employing CuBr–PMDETA as catalyst system at a monomer:initiator:CuBr:ligand ratio of 100 : 1:1 : 1 in DMF–water mixture (80 : 20, w/w). Diblock copolymers poly(butyl acrylate)-*b*-poly(2-{[(D-glucosamin-2-*N*-yl)carbonyl] oxy}ethyl acrylate) (PBA-*b*-PHEAGI) and poly(methacrylic acid)-*b*-poly(2-{[(D-glucosamin-2-*N*-yl)carbonyl]oxy}ethyl acrylate) (PMMA-*b*-PHEAGI) were synthesized, and also triblock copolymers PHEAGl-*b*-PBA-*b*-PHEAGl and PHEAGl-*b*-PMMA-*b*-PHEAGl.[220]

2.4.4 Reversible-Addition Fragmentation Transfer (RAFT)

2.4.4.1 Linear Polymers

RAFT linear glycopolymers are initial building blocks of more complex macromolecules. There have been many studies of the synthesis of linear RAFT glycopolymers, including homopolymers, diblock copolymers, statistical copolymers and gradient copolymers. Table 2.3 shows a series of linear glycopolymers synthesized to date. Reaction conditions reported are based on the polymerization of glycomonomer block or statistical glycopolymers.

Novel acrylamide-based unprotected glycomonomers were used to generate well-defined clickable glycopolymers in water–methanol mixtures by employing 3-benzylsulfanylthiocarbonylsulfanylpropionic acid trithiocarbonate RAFT agent at 70 and 30 °C. Three glycomonomers, namely *N*-{2-[(α-D-glucopyranosyloxy)acetamido]ethyl}acrylamide (without azide group), (3,4,6-tri-*O*-acetyl-α-D-glucopyranosyloxy)-*N*-(*tert*-butylcarbonyloxy) acetamide (with azide group at the 2-position) and *N*-{2-[(6-deoxy-6-*N*-azido-α-D-glucopyranosyloxy)acetamido]ethyl}acrylamide (with azide group at the 6-position) were employed. RAFT glycopolymers synthesized with glycomonomer without azide group resulted in characteristics of well-defined homopolymers and block copolymers. However, RAFT glycopolymers synthesized with the other two glycomonomers only led to well-defined polymers

Table 2.3 Linear Glycopolymers Prepared from Reversible Addition–Fragmentation Transfer (RAFT) Polymerization.

Carbohydrate	Polymer	Solvent	Temperature/°C	Initiator	References
Glucose	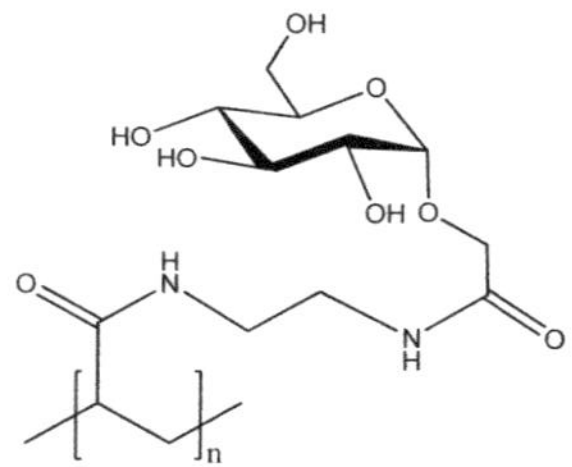	H_2O–MeOH (5 : 1)	70/30	ACPA–V-70	221
Glucose (furanose)	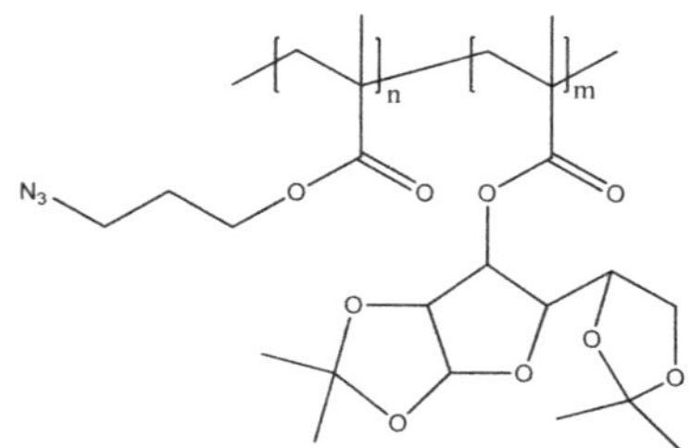	Anhydrous anisole	70	AIBN	222
Glucose (furanose)	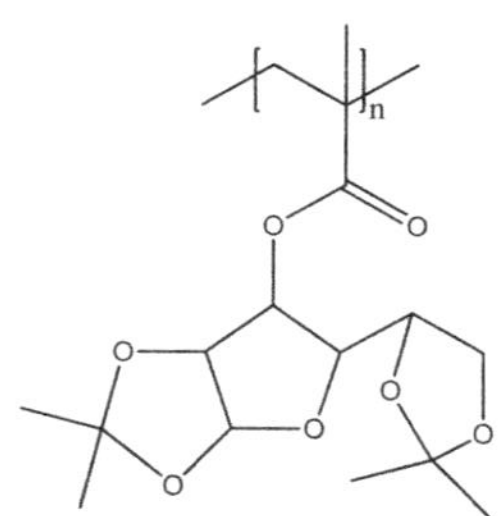	1,4-Dioxane	70	ACPA	223

Galactose	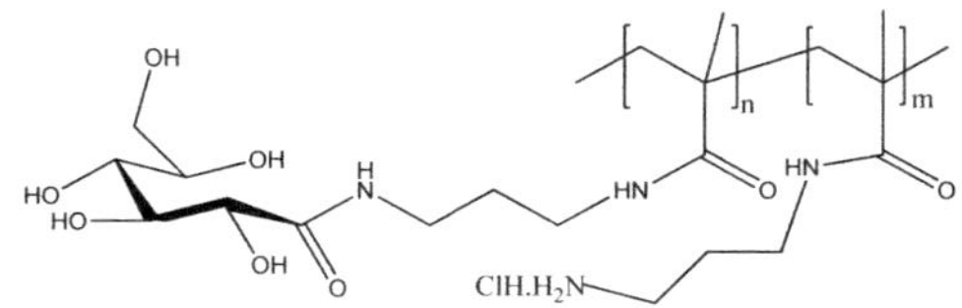	H_2O–DMF (5:1)	70	ACPA	224
Galactose	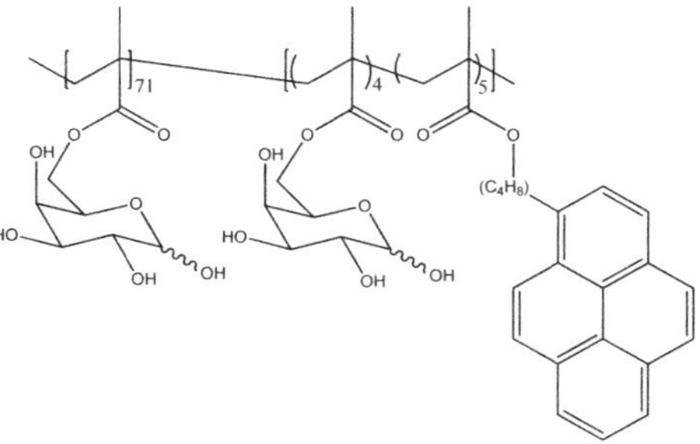	DMF	70	AIBN	225
Galactose	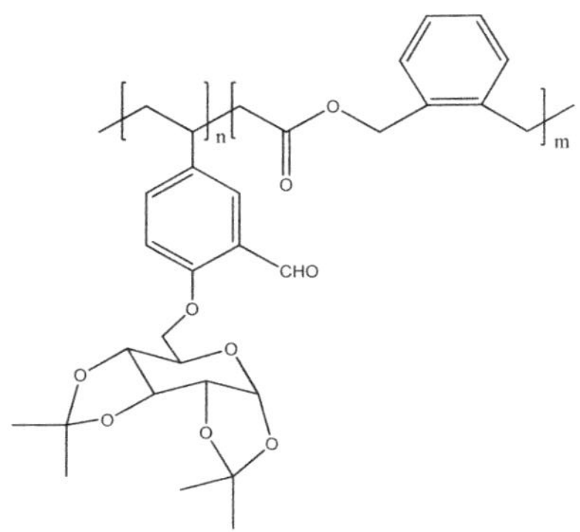	Anisole	130	DCP	226

(continued)

Table 2.3 (*continued*)

Carbohydrate	Polymer	Solvent	Temperature/°C	Initiator	References
Glucose	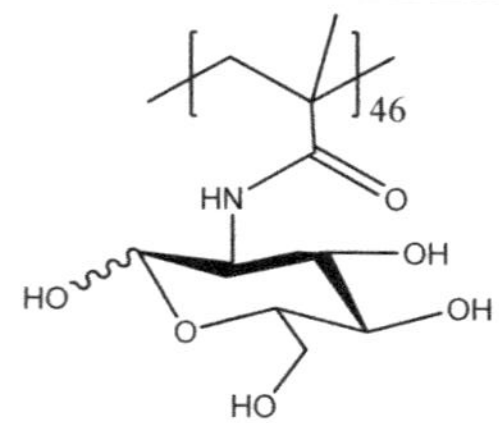	Acetate buffer–EtOH (4:1)	70	ACPA	227
Lactose	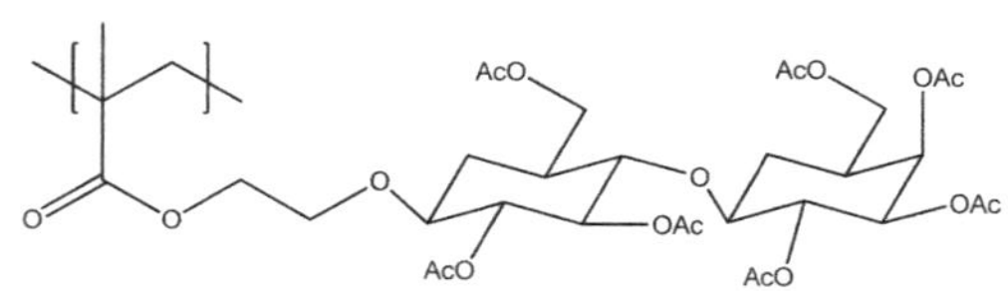	Chloromethane	70	AIBN	228
Glucose	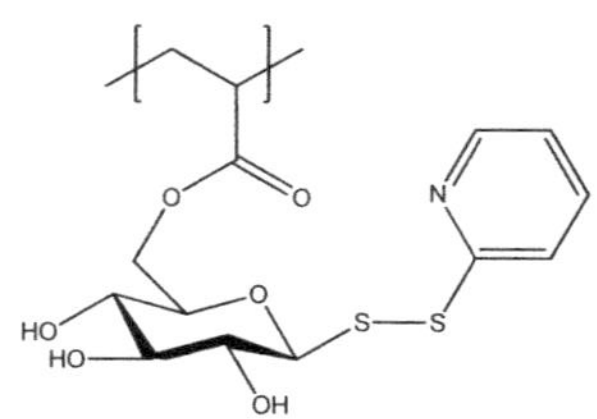	Chlorobenzene	60	AIBN	229

Trehalose	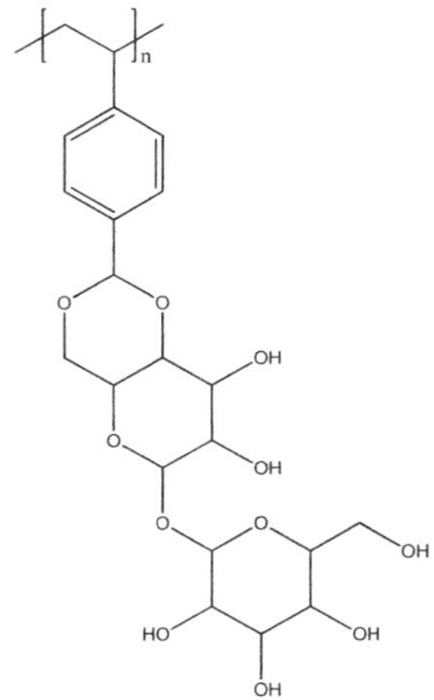	DMF	80	AIBN	230
Galactose	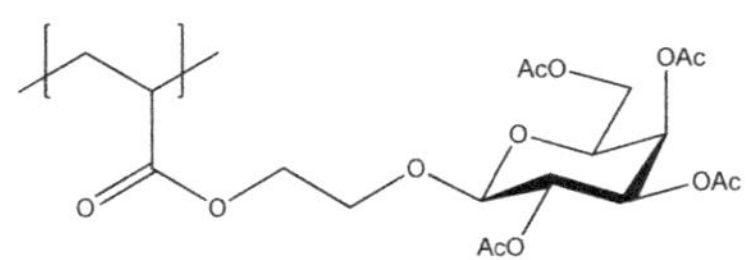	DMAc	60	AIBN	231

(continued)

Table 2.3 (*continued*)

Carbohydrate	Polymer	Solvent	Temperature/°C	Initiator	References
Mannose	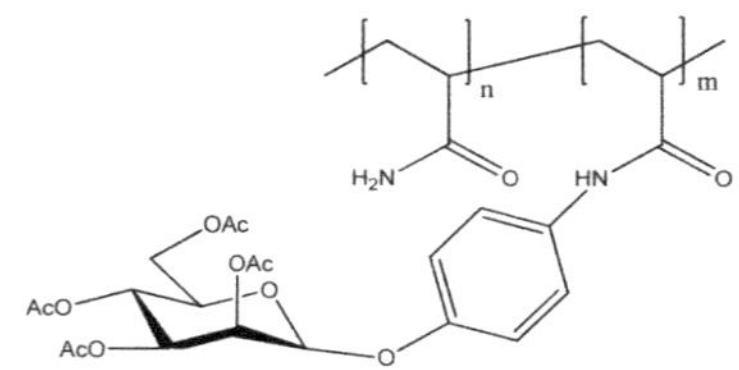	DMAc	70	AIBN	232
Glucose	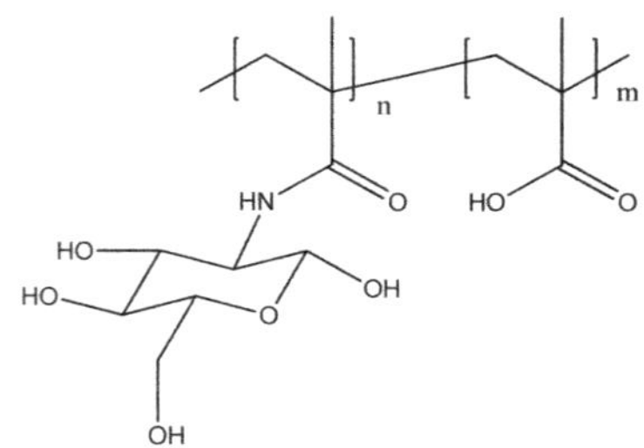	DMF	70	AIBN	233

when glycomonomer (3,4,6-tri-*O*-acetyl-α-D-glucopyranosyloxy)-*N*-(*tert*-butylcarbonyloxy)acetamide (with azide group at the 2-position) was employed under low-temperature polymerization conditions (30 °C) in the presence of 2,2′-azobis(4-methoxy-2,4-dimethylvaleronitrile) (V-70) as initiator. Unsuccessful polymer synthesis was assigned to the 1,3-cycloaddition side reactions between the vinyl and the azide groups within the glycomonomers.[221]

Linear statistical glycopolymers of poly(styrene-*co*-3-azidopropyl methacrylate) [poly(styrene-*co*-AzPMA)] and poly(3-*O*-methacryloyl-1,2 : 5,6-di-O-isopropylidene-D-glucofuranose-*co*-AzPMA) [poly(MAIpGlc-*co*-AzPMA)] were synthesized.[222] RAFT glycopolymers were synthesized with high molecular weight and narrow polydispersities using two trithiocarbonate RAFT agents, namely *S*,*S*′-bis(α,α′-dimethyl-α′-acetic acid)trithiocarbonate and *S*-1-dodecyl-*S*′-(α,α′-dimethyl-α′-acetic acid)trithiocarbonate. During RAFT polymerization at 70 °C, anhydrous anisole was used as solvent and AIBN was used a radical initiator. Glycopolymers were eventually conjugated with urea peptoids using copper-catalysed azide–alkyne cycloaddition (CuAAC) coupling reactions. Liu *et al.* synthesized pH-sensitive diblock glycopolymers comprising poly[2-(diethylamino)ethyl methacrylate] (PDEAEMA) and poly(3-*O*-methacryloyl-α,β-D-glucopyranose) (PMAGlc) blocks.[223] 4-Cyanopentanoic acid dithiobenzoate was used as a RAFT chain-transfer agent. Polymerizations were performed using 4,4′-azobis(4-cyanopentanoic acid) (ACPA) as initiator; the solvent employed was 1,4-dioxane and the reaction was carried out at 70 °C. DEAEMA and MAGlc were homopolymerized successfully. PDEAEMA macroRAFT was employed to synthesize PDEAEMA-*b*-PMAGlc diblock copolymers.

Ahmed and Narain synthesized a variety of cationic glycopolymers of predetermined molar masses and narrow molecular weight distributions ranging from 3000 to 30 000 g mol^{-1} in the absence of protecting group chemistry.[224] Polymerizations were carried out at 70 °C with 4-cyanopentanoic acid dithiobenzoate (CTP) as the RAFT agent and 4,4′-azobis(4-cyanopentanoic acid) (ACPA) as the initiator. Anti-cancer agents were derived from glycopolymers in Narain's laboratory.[234] Gold(I) phosphine conjugated with linear glycopolymers bearing dithiocarbamates (DTC) compounds yielded polymeric gold glycoconjugates. In another study by Ahmed and Narain, 3-gluconamidopropyl methacrylamide (GAPMA) was copolymerized with methacrylic acid (MA) to yield statistical glycopolymers.[235] More galactose-based diblock glycopolymers with fluorescent 4-(pyrenyl)butyl methacrylate (PyMA) were synthesized by Muller and co-workers.[225] They employed sequential RAFT polymerization with 2-cyanoprop-2-yl dithiobenzoate (CPDB) by initially polymerizing 6-*O*-methacryloyl-1,2 : 3,4-di-*O*-isopropylidene galactopyranose (MAIGal) in DMF for 22 h at 70 °C. This was followed by the introduction of MAIGal–PyMA (6 : 1) into the reaction and polymerization lasted for a further 6 h to yield the fluorescent glycopolymer PMAIGal-*b*-(PMAIGal-*co*-PPyMA).

Protection chemistry was employed in the synthesis of well-defined aldehyde-functionalized glycopolymers. Statistical glycopolymers were synthesized comprising poly[1,2 : 3,4-di-*O*-isopropylidene-6-*O*-(2′-formyl-4′-vinylphenyl)-D-galactopyranose] (PIVDG) and poly(5,6-benzo-2-methylene-1,3-dioxepane) (PBMDO) blocks.[226] Polymerizations were carried out in anisole at 130 °C using dicumyl peroxide (DCP) as initiator and 1-phenylethyl phenyldithioacetate as RAFT agent. Reineke's group at Virginia Tech designed glycopolymers containing poly(2-deoxy-2-methacrylamidoglucopyranose) (PMAG) and primary amine-containing poly[*N*-(2-aminoethyl) methacrylamide] (PAEMA) polymer blocks using unprotected glucose-based glycomonomer.[227] Polymers were synthesized *via* aqueous RAFT polymerization using 4-cyano-4-(propylsulfanylthiocarbonyl)sulfanylpentanoic acid as the RAFT agent. The glycopolymers were used as siRNA delivery systems. In an another study by the same group, an adamantane RAFT agent was used for the polymerization of 2-methacrylamido-2-deoxyglucopyranose (MAG) to achieve an adamantane-conjugated glycopolymer, poly(2-methacrylamido-2-deoxyglucopyranose) (Ad-pMAG).[236] Polymerizations were conducted in a mixture of water, acetate buffer and DMSO using ACPA initiator at 70 °C. The polymers were assembled through non-covalent interactions, forming an inclusion complex between hydrophobic adamantane and the β-cyclodextrin (βCD) core.

Glycopolymers have been synthesized to couple with poly(L-lysine) (PLL) *via* the formation of amide linkages. Disaccharide-based glycopolymers were reported to reduce the cytotoxicity of cationic PLL. Polymers were synthesized with 4-cyanopentanoic acid dithiobenzoate (CPADB) as the RAFT agent in chloromethane (CH_3Cl) and initiated using AIBN at 70 °C. Polymerizations were performed in a water bath. The resulting glycopolymers had narrow molecular weight distributions and could be deprotected using sodium methoxide and methanol.[228] Stenzel and co-workers developed polymeric versions of deacetylated auranofin for the treatment of ovarian cancer.[229] Glycomonomers bearing pyridyl disulfide bonds were synthesized for RAFT polymerization by employing 3-benzylsulfanylthiocarbonylsulfanylpropionic acid as the RAFT agent. A halogenated solvent was another popular choice in the RAFT polymerization. Chlorobenzene was the solvent, AIBN was the initiator and reactions were performed at 60 °C for 7 h.

Glycopolymers were used for the stabilization of protein conjugates to environmental stressors. Mancini *et al.* synthesized glycopolymers based on trehalose carbohydrate moieties. Polymers were synthesized using thiol-reactive chain-transfer agents in DMF and initiated by AIBN at 80 °C.[230] Mannose-, *N*-acetylglucosamine- and galactose-based glycopolymers were synthesized by RAFT polymerization using 4-cyano-4-(ethylsulfanylthiocarbonyl)sulfanylpentanoic acid (ECT) as the RAFT agent and 4,4′-azobis(4-cyanovaleric acid) (V501, known as ACPA) as the radical initiator. Polymerizations were conducted in a heterogeneous solvent system of water–ethanol (3 : 1, v/v) at 70 °C. Mannose- and *N*-acetylglucosamine-containing glycopolymers were found to interact well with mouse bone marrow-derived macrophages (BMDMs), while the galactose-containing glycopolymers showed much lower uptake by the macrophages.[237]

There have been only a few articles reporting the synthesis of gradient glycopolymer systems. The combined work led by Billon, Stenzel and co-workers successfully synthesized block to gradient copolymers based on acryloylgalactose monomer using RAFT polymerization. Polymerizations were performed in a polar aprotic solvent, dimethylacetamide (DMAc), with AIBN as the initiator at 60 °C. Gradient copolymers were synthesized using acryloylgalactose monomer with styrene as the second monomer. One-pot polymerization was carried out at 90 °C by first using styrene as the initial monomer with *S*-methoxycarbonylphenylmethyl dodecyltrithiocarbonate (MCPDT) as the RAFT agent, followed by the semi-batch dropwise addition of 2-(2′,3′,4′,6′-tetra-*O*-acetyl-β-D-galactosyloxy)ethyl acrylate (AcGalEA), dissolved in DMAc, to the reaction mixture for the polymerization of styrene.[231]

Polyacrylamide-derived glycopolymers were synthesized using *S*-2-cyano-2-propyl-*S*-butyl trithiocarbonate (CPBTC) as the RAFT agent and reactions were performed in DMAc at 70 °C for 48 h.[232] The final glycopolymers showed relatively narrow molecular weight distributions of ≤1.5 and the polymers were deacetylated by stirring with sodium methoxide in methanol at pH 10 for 30 min. RAFT end-groups were reduced by the addition of sodium borohydride ($NaBH_4$). Thiolated glycopolymers were eventually immobilized on gold nanoparticles.

Mawhinney and co-workers synthesized three-component statistical fluorescent glycopolymers by one-step polymerization in a lectin-mediated bacterial binding study. Galactose moieties from the three-component glycopolymers have long spacer units from the polymer backbone.[238] Chen and co-workers recently synthesized statistical copolymers comprising 2-(methacrylamido)glucopyranose (MAG) and methacrylic acid (MAA).[233] Polymerizations were performed in DMF at 70 °C using 2-cyanoprop-2-yl α-dithionaphthalate (CPDN) as the RAFT mediating agent and a conventional AIBN radical initiator was employed. By employing microwave irradiation, glycopolymers were used as templates to synthesize glycopolymer-functionalized silver nanoclusters (Gly-Ag NCs).

2.4.4.2 Multi-Block Copolymers

ABA triblock glycopolymers were synthesized by aqueous RAFT polymerization. Initially, a symmetric trithiocarbonate, *S*,*S*′-bis(2-propanoic acid) trithiocarbonate, RAFT agent was employed to polymerize *N*-acryloylmorpholine (NAM) to achieve a poly(NAM)trithiocarbonate macro-CTA. In the second step, chain extension of *N*-acryloyl-β-D-glucopyranosylamine (NAGlc) from both side (*R* and *Z* positions) of the macroRAFT agent resulted in an ABA-type triblock glycopolymer, poly(NAM)-*b*-poly(NAGlc)-*b*-poly(NAM) (Figure 2.33). In order to monitor conversions closely with NMR spectroscopy, polymerizations were carried out *in situ* in deuterated water–deuterated DMSO (1 : 1) mixture in an NMR tube fitted with a Young's valve at 60 °C.[239]

Xing *et al.* synthesized triblock glycopolymers from a trithiocarbonate RAFT agent terminated by poly(ε-caprolactone) (PCL). This was followed by two-step polymerizations to obtain triblock glycopolymers.[240] RAFT-terminated PCL was first polymerized with an acrylate-based galactose glycomonomer and further chain extended with poly(ethylene glycol) methyl ether

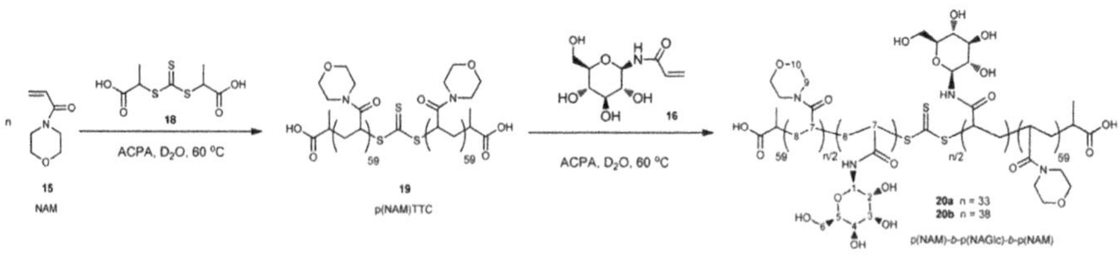

Figure 2.33 Polymerization pathways of ABA-type triblock glycopolymer synthesized using a symmetric trithiocarbonate RAFT agent.[239] Reproduced with permission from Wiley, Copyright 2012.

acrylate (OGEA). Glycomonomer was polymerized in toluene using AIBN as initiator at 70 °C for 12 h.

Thermoresponsive micellar aggregates were self-assembled from 'coil–comb–coil' triblock glycopolymers.[241] Triblock glycopolymers were synthesized by employing RAFT and ROP. The whole process of fabrication required five steps. Poly(3-*O*-methacryloyl-1,2 : 5,6-di-*O*-isopropylidiene-D-glucofuranose) (PMAIpGlc) macroRAFT agent was first synthesized, followed by two consecutive chain-extension polymerizations of 2-hydroxyethyl methacrylate (HEMA) and *N*-isopropylacrylamide. The triblock glycopolymers were grafted with ε-caprolactone at the middle block (HEMA section) to yield poly(3-*O*-methacryloyl-1,2 : 5,6-di-*O*-isopropylidiene-D-glucofuranose)-*b*-poly[2-hydroxyethyl methacrylate-*g*-poly(3-caprolactone)]-*b*-poly(*N*-isopropylacrylamide) [PMAIpGlc-*b*-P(HEMA-*g*-PCL)-*b*-PNIPAM].

2.4.4.3 *Complex Polymer Structures and Glycomaterials*

Hyperbranched polymers have also attracted significant interest in glycopolymer synthesis. This is due to their highly branched structure, which allows greater interactions with desired molecules or receptors on cell surfaces. Random cationic glycopolymers were synthesized using the RAFT process without protecting group chemistry in one reaction step. Linear random glycopolymers are generated when the crosslinker *N*,*N*′-methylenebisacrylamide was not employed. The linear polymers showed characteristics of polymers synthesized by the RAFT process. Cationic galactose-based hyperbranched polymers were synthesized with 1.7 : 1 and 3 : 1 amine to galactose ratios by varying the concentration of *N*,*N*′-methylenebisacrylamide (Figure 2.34).[242]

Hughes and co-workers at the Commonwealth Scientific and Industrial Research Organization (CSIRO) in Australia synthesized galactose-based hyperbranched glycopolymers in an organic solvent *via* one-pot polymerization from glycan monomers and two novel RAFT inimers, 2-(methacryloyloxy)ethyl 4-cyano-4-(phenylcarbonothioylthio)pentanoate (MAE-CPP) and 2-(3-(benzylthiocarbonothioylthio)propanoyloxy)ethyl acrylate (BCP-EA) without the use of any crosslinkers.[243] Polymerizations were performed employing AIBN as the initiator to RAFT inimer with a ratio of 1:3. The

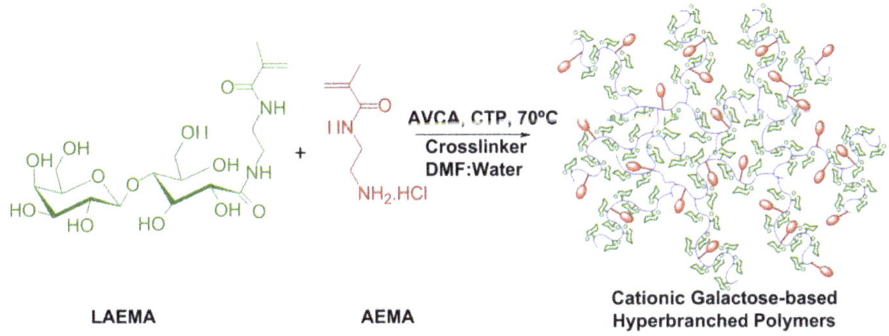

Figure 2.34 Polymerization of 2-lactobionamidoethylmethacrylamide (LAEMA) and 2-aminoethylmethacrylamide (AEMA) using 4-cyanopentanoic acid dithiobenzoate (CTP) as the RAFT agent and 4,4′-azobis(4-cyanopentanoic acid) (ACPA) as the initiator.[242] Reproduced with permission from Elsevier, Copyright 2012.

polymers were subsequently deprotected in trifluoroacetic acid to afford water-soluble hyperbranched glycopolymers.

Lu *et al.* synthesized a glycopolymer–porphyrin conjugate as photosensitizer for imaging in cancer and photodynamic treatment.[244] Linear RAFT glycopolymers of poly[2-(methacrylamido)glucopyranose] (PMAG) were synthesized using 2-cyanoprop-2-yl-α-dithionaphthalate (CPDN) as RAFT agent and AIBN as initiator. Glycopolymers were subsequently functionalized onto protoporphyrinogen by first reducing the RAFT glycopolymers into reactive thiol end-functionalized glycopolymers. In the second step, protoporphyrin was converted to protoporphyrinogen. Finally, thiolene chemistry was employed to couple PMAG onto converted protoporphyrinogen.

The use of linear glycopolymers to attach to the surface of gold nanoparticles and other substrates has also been an attractive avenue in the design of complex-structured materials for advanced applications. Glycopolymers were synthesized from acrylamide-based mannose monomer and acrylamide. The polymers were functionalized onto the surface of gold nanoparticles *via* thiol-exchanged chemistry.[245] Barner-Kowollik and co-workers developed glycomicrospheres by employing mild pericyclic reactions.[246] Poly(3-*O*-acryloyl-1,2 : 5,6-di-*O*-isopropylidene-α-D-glucofuranoside) (pAIpGlc) glcyopolymers were first synthesized using RAFT polymerization. Initially, microspheres were functionalized with a very reactive diene. Owing to the presence of RAFT end-groups on protected glycopolymers, hetero-Diels–Alder (HDA) reactions could be performed between glycopolymers and surface reactive dienes on microspheres. Parry *et al.* developed ‘multicopy–multivalent’ nanoscale glycoconjugates as potential synthetic cancer vaccines.[247] Linear statistical glycopolymers were first synthesized by polymerizing α-*N*-acetyl-D-galactosamine-based monomer and poly(ethylene glycol) methacrylate using ACPA as initiator at 70 °C for 48 h. The polymers were subsequently reduced and functionalized onto the surface of gold nanoparticles. The glycopolymer-decorated nanoparticles showed significant immune response *in vivo*, moreover, synthetic antigen glycans synthesized displayed great potential in recognizing natural Tn-antigen glycans and mammalian-mucin glycoproteins. Glycopolymers were grafted onto gold nanoparticles using the reduced RAFT end-group of the glycopolymer. By adjusting the glycopolymer–polyacrylamide mixture, Miura and co-workers were able to control the stability of polymer-coated gold nanoparticles.[232] An immunochromatographic assay was carried out using the glycoparticles.

Min *et al.* grafted statistical glycopolymers composed of *N*-isopropylacrylamide (NIPAAm) and *N*-acryloylglucosamine (AGA) repeating units onto honeycomb-structured porous films.[248] Solution RAFT polymerizations were carried out with free RAFT agent in solution and RAFT agent immobilized on the surface of honeycomb films. After aminolysis, the molecular weights of the surface-grafted glycopolymer chains were much higher than those polymerized in solution. Polymerization of monomers occurred mostly on the films rather than in solution. The responsiveness of the polymer chains was examined by incubating functionalized honeycomb films with ConA. At temperatures below the lower critical solution temperature (LCST), thermally responsive glycopolymers precipitated and switched off, preventing interactions with ConA, and at temperatures above the LCST, the polymers switched on and interacted with ConA (Figure 2.35).

Sugar-based nanogels were synthesized as a delivery system to deliver DNA and enzyme into cells. MacroRAFT agents of poly(3-gluconamidopropylmethacrylamide) (PGAPMA) and poly(2-lactobionamidoethylmethacrylamide) (PLAEMA) were initially synthesized using a dithioester RAFT agent. The homoglycopolymers were chain extended with PEGMA and an

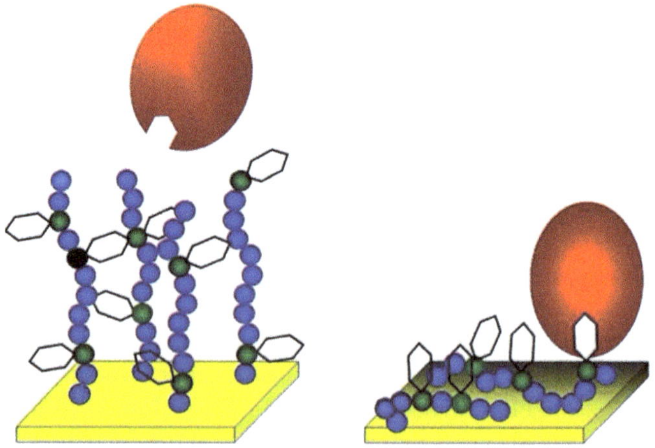

Figure 2.35 Poly(*N*-isopropylacrylamide) (PNIPAAm) with LCST composed in statistical glycopolymers were employed as an on (above LCST)/off (below LCST) switch during ConA interactions.[248] Reproduced with permission from Wiley, Copyright 2010.

acid-degradable crosslinker was also added during the aqueous polymerization. The polymerization resulted in glycopolymer nanogels.[249] Glycoparticles were also synthesized in an aqueous system by Ting *et al.*[250] The particles were capable of degrading into linear polymer chains in a reducing environment. MacroRAFT agent of PMAG was first synthesized, followed by the chain-extension polymerization of styrene and a disulfide acrylate-based crosslinker in emulsion. Glucosamine-based glycoparticles were found to interact specifically with ConA lectin.

2.5 Ionic Chain Polymerization

Neither anionic nor cationic chain polymerizations have been of much interest in the last few years, probably owing to the rise of other controlled polymerization techniques. Especially anionic polymerization requires very stringent reaction conditions compared with other polymerization techniques. In addition, the choice of monomers is limited to protected carbohydrates and the protecting groups need to meet further requirements.

Anionic polymerization is still the method of choice to generate polymers with the lowest possible molecular weight distribution. Hirao and co-workers achieved well-defined glycohomo and block copolymers using styrene-based glycomonomers. A prerequisite was substitution in the *meta*-position; polymerization of *para*-substituted glycopolymers did not proceed.[28,84]

Whereas reports on anionic polymerizations are rare, the synthesis of glycopolymers *via* living cationic polymerization was frequently described in the 1990s. The initiating system was found to be crucial as the HCl–ZnI_2 system only ensures good control up to conversions of around 80%.[26,161] Changes to the solvent or temperature did not improve the outcome, which subsequently led to a search for a more suitable initiation system. Trifluoroacetic acid with ethylaluminium dichloride ensured living behaviour up to 100% monomer conversion[161] and it was suitable for the synthesis of block copolymers such as poly(3,4,6-tri-*O*-acetyl-2-deoxy-2-phthalimide-β-D-glucopyranoside)-*b*-poly(isobutyl vinyl ether),[251] which showed superior bioactivity in interaction with wheat germ agglutinin (WGA) compared with the homopolymer.

It is important to mention that living cationic polymerization allowed for the first time the synthesis of a library of block copolymers. Block copolymers based on poly(isobutyl vinyl ether) and poly[3-*O*-(vinyloxy)ethyl-1,2 : 5,6-di-*O*-isopropylidene-D-glucofuranose], with the number of repeating units varying between 20 and 90, were found to self-assemble into various morphologies in the solid state, including spheres, cylinders and lamellae.[162] For comparison, statistical copolymers were prepared that showed solution properties significantly different to those of the block copolymer. NMR spin-lattice relaxation times confirmed the aggregate formation of the block copolymer in solution whereas the statistical copolymer was present in its unimolecular state.[252]

Glycopolymers prepared from cationic polymerization were also investigated for their ability to react with amino-terminated oligonucleotides. The aldehyde functionality of the acrylic form of the galactose group underwent Schiff base formation with the amine, which was followed by reductive amination.[165] The products could be further analysed using MALDI-TOF mass spectrometry, which confirmed the nature of the end functionalities, making this probably one of the first reports in which the MALDI-TOF technique was used to investigate chemical structures further.

2.6 Ring-Opening Metathesis Polymerization (ROMP)

ROMP, which can be carried out using molybdenum- and tungsten-based Schrock initiators and the ruthenium-based Grubbs initiator was one of the favourite methods in the 1990s to create bioactive glycopolymers, but research activity has declined in the last few years. A significant part of the knowledge- based on glycopolymer–lectin interactions that is available today was established during that period, thanks to the pioneering work of Kiessling and co-workers. In their initial work, various 7-oxobornene derivatives with carbohydrate functionalities (α-*O*-glucoside, α-*O*-mannoside, α-*C*-glucoside and α-*C*-mannoside) were polymerized in water in the presence of $RuCl_3$ at 55 °C, resulting in molecular weights of up to 10^6 g mol^{-1} (Figure 2.36).[27,30] This study showed that polymers based on mannose were highly efficient in binding to ConA, an interaction that could further be enhanced using stereochemically diverse glycopolymers by utilizing the *cis*- and *trans*-isomers in the backbone.[253] With a view to creating non-toxic biomaterials, toxicity and efficient removal pathways for the catalyst were investigated.[253]

The catalyst $RuCl_3$ resulted in high molecular weights in polymerizations, but the polymerization was not living. In order to be able to generate polymers of different molecular weights, the Grubbs ruthenium alkylidene catalyst was employed. The polymerization was carried out in an emulsion with dodecyltrimethylammonium bromide (DTAB) in a 1,2-dichloroethane–water mixture to accommodate the different solubilities of the monomer and catalyst.[33] These conditions were also suitable for polymerizing norbornene derivatives with pendant disulfated trisaccharide side groups, resulting in a polymer with high selectivity towards L-selectin.[254] Detailed studies by Grubbs and co-workers confirmed that the ruthenium carbene-catalysed polymerization of glycomonomers is indeed living, depending on the choice of catalyst.[32,255]

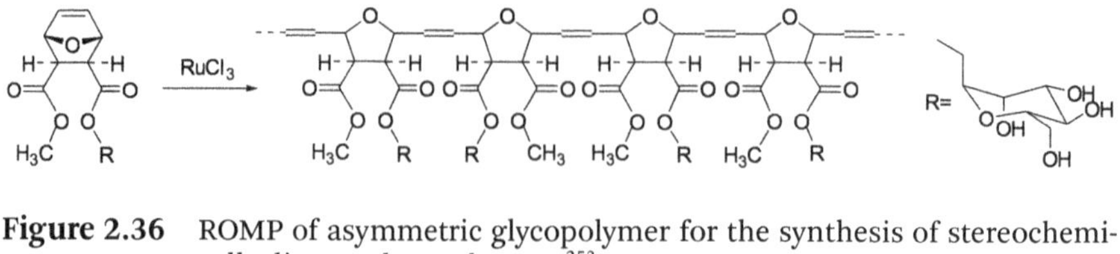

Figure 2.36 ROMP of asymmetric glycopolymer for the synthesis of stereochemically diverse glycopolymers.[253]

A molybdenum-based Schrock-type catalyst was also shown to convey living characteristics to the polymerization of a selection of norbornene-based glycomonomers.[256] The polymerization was living and block copolymers could be obtained by the addition of further monomers. Although the polymers had narrow molecular weight distributions, broadening can occur when attempting the removal of the metal from the polymer in a Wittig-type reaction.

2.7 Ring-Opening Polymerization (ROP)

The attraction of ROP lies in the formation of degradable backbones, which contrasts with all the other techniques described above. Anionic or cationic ROP of cyclic monomers such as cyclic ethers, acetals, amides (lactams), esters (lactones) and siloxanes results in the formation of similar products to step polymerization products, but usually with better control over molecular weight and smaller molecular weight distributions. The discussion on the polymerization of carbohydrates with cyclic reactive groups should make a clear distinction between polymerization that results in glycopolymers (polymers with pendant carbohydrates) and polymers that resemble polysaccharides, which are connected by glycosidic bonds.

The first report on the ROP of glycomonomers belonged to the latter group. The monomers depicted in Figure 2.19 were polymerized in dichloromethane at low temperature, initiated with Lewis acids such as PF_5, $SbCl_5$, $BF_3{\cdot}O(C_2H_5)_2$, $SnCl_4$, $(CF_3SO_2)_2O$ and $(i\text{-Bu})_3Al{-}H_2O$,[24,25] resulting in polymers with molecular weights of up to 11 kDa (Figure 2.37, reaction 1). Depending on the initiator, stereoregular glycopolymers were obtained, as evidenced by optical rotation and NMR spectroscopy.

This area of synthetic polysaccharides seemed dormant until recently, when some unusual polymer structures were reported, which were obtained by the ROP of glycomonomers. Polysaccharides with amide functionality between the carbohydrates were prepared by anionic ROP of a β-lactam sugar monomer (Figure 2.37, reaction 2). The ROP was initiated with *p*-nitrobenzoyl chloride and LiHMDS in THF at 0 °C to yield polymer with up to 100 repeating units and polydispersity indices below 1.2. The galactose-based polymer was tested for its ability to interact selectively with asialoglycoprotein, a receptor that is located on hepatocytes and responds to galactose. However, the structure of the polymer led to the loss of bioactivity and the polymers can therefore be considered as neutral in terms of bioactivity.[257] Interestingly, the equivalent polyamidosaccharide based on glucose can still bind to ConA, a lectin responsive to glucose and mannose. This highlights how subtle changes to carbohydrates can render them inactive while other binding modes are not affected.[258]

Polysaccharides with orthoester linkages were obtained from 2,3,4-tri-*O*-acetyl-α-D-glucopyranosyl bromide, resulting in polymers that can be degraded in acidic conditions. It is initially not evident that the reaction is indeed an ROP, but detailed inspection of the mechanism reveals that the ring-closing process was followed by a ring-opening process *in situ*. The reaction was promoted by tetra-*n*-butylammonium iodide in dichloromethane under reflux (Figure 2.37, reaction 3).[259]

Monomers with epoxide functionalities can in theory lead to glycopolymers or to synthetic polysaccharides depending on the amount of protective groups present on the glycopolymer. Monomers such as 5,6-anhydro-1,2-*O*-isopropylidene-α-D-glucofuranose (Figure 2.37, reaction 4) can be polymerized *via* epoxide ROP or can involve the unprotected hydroxyl group, which leads to branching. This AB_2-type monomer can be polymerized by anionic or cationic ROP using potassium *tert*-butoxide (*t*-BuOK) or boron trifluoride diethyletherate ($BF_3{\cdot}OEt_2$), respectively, as initiator. In particular, the cationic pathway was found to be very efficient for generating hyperbranched glycopolymers with high molecular weights and highly branched structures.[260]

A topic of increasing interest is the area of glycomonomers based on *N*-carboxyanhydrides, which were first reported by Rüde *et al.*[16] Okada and co-workers accomplished the living polymerization of these monomers

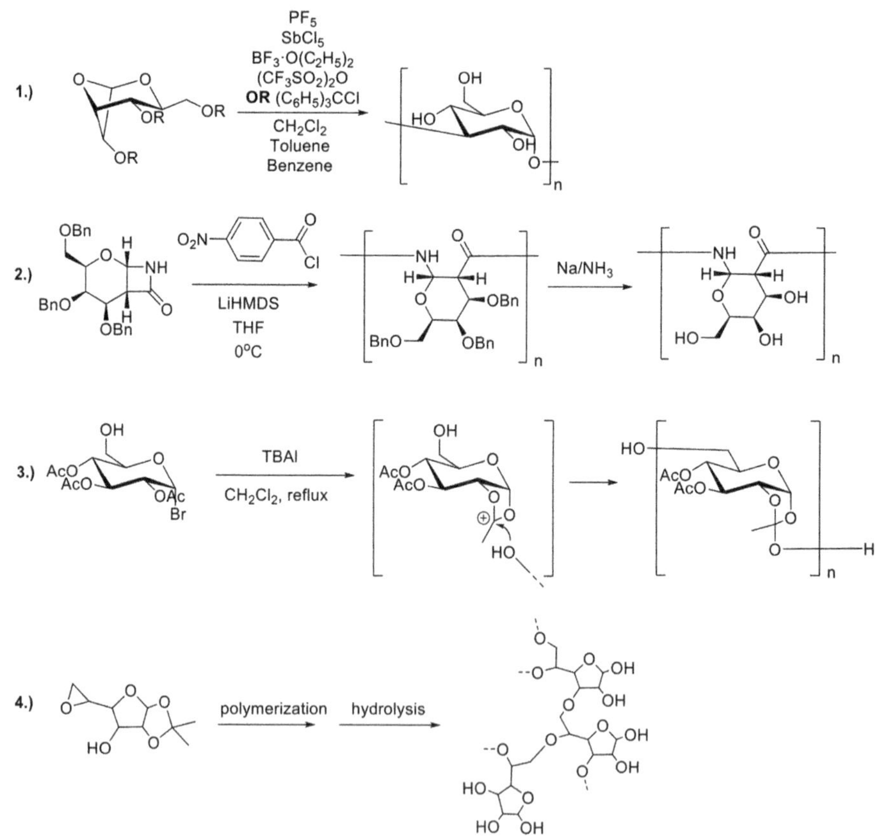

Figure 2.37 Synthesis of synthetic polysaccharides by ring-opening polymerization.

using primary amines as initiators.[167] The use of amines was a significant key to this process since it opened up the way to more complex architectures. The amino end functionality of the product could be utilized, for example, to generate block copolymers in a subsequent step,[166] but block copolymer formation using amino-terminated polyoxazolines as macroinitiators to initiate the polymerization of *O*-(tetra-*O*-acetyl-β-D-glucopyranosyl)-L-serine *N*-carboxyanhydride is also possible (Figure 2.38(a)).[169] A related polymer could also be obtained by adding polyoxazoline as termination agent to the anionic ROP of *O*-(tetra-*O*-acetyl-β-D-glucopyranosyl)-L-serine *N*-carboxyanhydride), thus yielding block copolymers with PDIs less than 1.1.[261] Evidence for the bioactivity of these types of polymers was provided with a comb polymer. The ROP of *O*-(tetra-*O*-acetyl-β-D-glucopyranosyl)-L-serine *N*-carboxyanhydride was initiated with *p*-vinylbenzylamine, resulting in the formation of macromonomers that could be copolymerized with acrylamide. The resulting polymer showed good lectin binding with wheat germ agglutinin.[168]

An alternative backbone for glycopolymers can be obtained by the polymerization of oxazolines. Glucosamine was modified to create an oxazoline functionality, which was then used as an initiator in the ROP of various 2-oxazolines (Figure 2.38(b)). The acrylate functionality was introduced during the termination of the polymerization, which resulted in macromonomers that were copolymerized with styrene.[262]

This area has recently been expanded and syntheses of new 2-oxazolines containing *S*-galactosyl substituents with different spacer lengths were recently reported (Figure 2.38(c)).[263] The initiator was based on a 1 : 1 adduct of 2-methyl-2-oxazoline and methyl triflate. The resulting poly(*S*-glycooxazoline)s

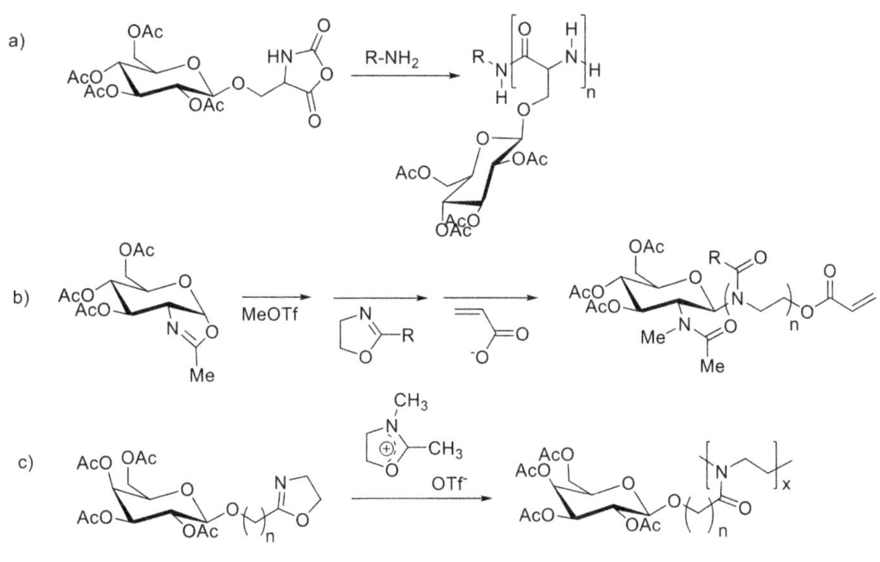

Figure 2.38 Ring-opening polymerization of glycomonomers resulting in polypetides (a) or polyoxazolines (b and c).

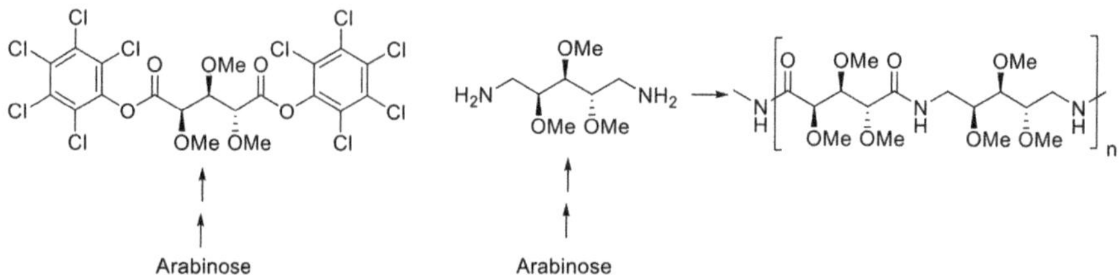

Figure 2.39 Synthesis of linear polyamides derived from L-arabinitol.[265]

showed excellent binding abilities to the lectin RCA120 and *in vivo* expression of green fluorescent protein in *Escherichia coli* host was observed.

2.8 Polycondensation

Polycondensation utilizes the abundance of hydroxyl groups on carbohydrates to create polymers derived from sugars.[264,265] These polymers are not glycopolymers in a strict sense since the process results in the permanent opening of the ring and associated loss of bioactivity. These polymers are more of interest as a pathway to materials from renewable resources and as potentially biocompatible polymers that are neutral in behaviour. These polymers are only included here for completion, but also because they are currently making some impact as alternative biomaterials.[266,267] The carbohydrate needs to be modified, usually by oxidation or amination, while the remaining hydroxyl groups remain protected to avoid crosslinking (Figure 2.39).[265]

2.9 Conclusions

The application of glycopolymers in biology, bionanotechnology and medicine has strongly attracted material/polymer scientists and chemists to designing a myriad of fundamental to complex glycopolymer architectures. Notably, some biology-based research groups are also starting to engineer glycopolymers for their own studies. In this chapter, focus has been placed on synthesizing glycopolymers directly from their monomeric building blocks. It is crucial that the design and development of novel glycomonomers should be thoroughly analysed and addressed. Particularly, modifications of various mono-, di- and trisaccharides for the attachment of vinyl polymerizable groups present risks of losing carbohydrate bioactivities.

The synthesis of glycomonomers can involve single-step to multiple-step reactions depending on the complexity of the final products. Especially in the synthesis of trisaccharide-based glycomonomers such as trichloroacetimidate and sialyl-Lewis X-based glycomonomers and position-specific modifications of carbohydrates, *e.g.* sulfation and inorganic functionalization, detailed structural analysis *via* NMR spectroscopy with complementary characterization techniques is required. Two very distinct chemistries in the synthesis of glycomonomer are protecting group organic-phase reactions and direct modifications in aqueous environments.

Conventional free-radical polymerization led the synthesis of glycopolymers. Since the advent of reversible-deactivation radical polymerization (RDRP), attention has shifted towards the design of well-defined glycopolymer architectures and diverse carbohydrate-based materials. Among the three best known RDRP techniques, the number of reports on the synthesis of glycopolymers in ATRP and RAFT has outweighed that of NMP. The high temperature required in NMP has limited this area of study. Nonetheless, novel nitroxide mediating agents for low-temperature polymerization are currently under detailed investigation. RDRP offers numerous options in designing complex glycopolymers and glycomaterials tailored for research needs. However, there are still limitations to these techniques, which create room for non-radical-based techniques, namely ionic chain polymerization, ROMP, ROP and polycondensation.

In a nutshell, the direct synthesis of glycopolymers from glycomonomers has proven to be a powerful strategy in obtaining functional glycopolymers. It remains an indispensable technique even though new avenues to obtaining glycopolymers, such as post-polymerization glycosylation, have emerged.

Acknowledgements

S. R. S. Ting acknowledges a National Health and Medical Research Council (NHMRC) Early Career Fellowship. M. H. Stenzel thanks the Australian Research Council (ARC) for funding.

References

1. P. L. Nichols and E. Yanovsky, *J. Am. Chem. Soc.*, 1944, **66**, 1625–1627.
2. P. L. Nichols and E. Yanovsky, *J. Am. Chem. Soc.*, 1945, **67**, 46–49.
3. E. A. Talley, M. D. Vale and E. Yanovsky, *J. Am. Chem. Soc.*, 1945, **67**, 2037–2039.
4. P. L. Nichols, A. N. Wrigley and E. Yanovsky, *J. Am. Chem. Soc.*, 1946, **68**, 2020–2022.
5. R. H. Treadway and E. Yanovsky, *J. Am. Chem. Soc.*, 1945, **67**, 1038–1039.
6. T. P. Bird, W. A. P. Black, E. T. Dewar and D. Rutherford, *Chem. Ind.*, 1960, 1331–1332.
7. M. L. Wolfrom, M. S. Toy and A. Chaney, *J. Am. Chem. Soc.*, 1958, **80**, 6328–6330.
8. S. Kimura and M. Imoto, *Makromol. Chem.*, 1961, **50**, 155–160.
9. W. A. P. Black, E. T. Dewar and D. Rutherford, *Chem. Ind.*, 1962, 1624–1624.
10. R. G. Schweige, *J. Polym. Sci., Part A: Gen. Pap.*, 1964, **2**, 2471–2480.
11. Y. Iwakura, K. Hayashi and K. Inagaki, *Makromol. Chem.*, 1967, **104**, 56.
12. T. P. Bird, W. A. P. Black, E. T. Dewar and J. B. Hare, *Chem. Ind.*, 1961, 1077–1077.

13. W. E. Dick and R. L. Whistler, *J. Polym. Sci., Part A: Gen. Pap.*, 1964, **2**, 1833–1838.
14. R. L. Whistler and J. L. Goatley, *J. Polym. Sci.*, 1961, **50**, 127–132.
15. R. L. Whistler, H. P. Panzer and H. J. Roberts, *J. Org. Chem.*, 1961, **26**, 1583–1588.
16. E. Rüde, O. Westphal, E. Hurwitz, S. Fuchs and M. Sela, *Immunochemistry*, 1966, **3**, 137–151.
17. L. A. Carpino, H. Ringsdorf and H. Ritter, *Makromol. Chem. Macromol. Chem. Phys.*, 1976, **177**, 1631–1635.
18. D. C. Chung, R. J. Kostelnik and H. G. Elias, *Makromol. Chem. Macromol. Chem. Phys.*, 1977, **178**, 691–700.
19. K. Nishio, T. Nakaya and M. Imoto, *Makromol. Chem. Macromol. Chem. Phys.*, 1974, **175**, 3319–3323.
20. H. Gruber, *Monatsh. Chem.*, 1981, **112**, 273–285.
21. K. Kobayashi, H. Sumitomo, A. Kobayashi, M. Ishida, S. Nishizawa and T. Akaike, *Kobunshi Ronbunshu*, 1985, **42**, 719–724.
22. A. Kobayashi, T. Akaike, K. Kobayashi and H. Sumitomo, *Makromol. Chem. Rapid Commun.*, 1986, **7**, 645–650.
23. J. Klein, D. Herzog and A. Hajibegli, *Makromol. Chem. Rapid Commun.*, 1985, **6**, 675–678.
24. F. J. Good and C. Schuerch, *Macromolecules*, 1985, **18**, 595–599.
25. T. Uryu, C. Yamaguchi, K. Morikawa, K. Terui, T. Kanai and K. Matsuzaki, *Macromolecules*, 1985, **18**, 599–605.
26. M. Minoda, K. Yamaoka, K. Yamada, A. Takaragi and T. Miyamoto, *Macromol. Symp.*, 1995, **99**, 169–177.
27. K. H. Mortell, M. Gingras and L. L. Kiessling, *J. Am. Chem. Soc.*, 1994, **116**, 12053–12054.
28. S. Loykulnant, M. Hayashi and A. Hirao, *Macromolecules*, 1998, **31**, 9121–9126.
29. S. Nishimura, K. Matsuoka, T. Furuike, N. Nishi, S. Tokura, K. Nagami, S. Murayama and K. Kurita, *Macromolecules*, 1994, **27**, 157–163.
30. K. H. Mortell, R. V. Weatherman and L. L. Kiessling, *J. Am. Chem. Soc.*, 1996, **118**, 2297–2298.
31. K. Kobayashi, A. Tsuchida, T. Usui and T. Akaike, *Macromolecules*, 1997, **30**, 2016–2020.
32. M. Kanai, K. H. Mortell and L. L. Kiessling, *J. Am. Chem. Soc.*, 1997, **119**, 9931–9932.
33. D. D. Manning, X. Hu, P. Beck and L. L. Kiessling, *J. Am. Chem. Soc.*, 1997, **119**, 3161–3162.
34. T. G. Hill, P. Wang, M. E. Huston, C. A. Wartchow, L. M. Oehler, M. B. Smith, M. D. Bednarski and M. R. Callstrom, *Tetrahedron Lett.*, 1991, **32**, 6823–6826.
35. K. Ohno, Y. Tsujii and T. Fukuda, *J. Polym. Sci., Part A: Polym. Chem.*, 1998, **36**, 2473–2481.
36. Y.-Z. Liang, Z.-C. Li, G.-Q. Chen and F.-M. Li, *Polym. Int.*, 1999, **48**, 739–742.

37. A. B. Lowe, B. S. Sumerlin and C. L. McCormick, *Polymer*, 2003, **44**, 6761–6765.
38. D. Grande, S. Baskaran and E. L. Chaikof, *Macromolecules*, 2001, **34**, 1640–1646.
39. K. Ohno, Y. Tsujii, T. Miyamoto, T. Fukuda, M. Goto, K. Kobayashi and T. Akaike, *Macromolecules*, 1998, **31**, 1064–1069.
40. V. Ladmiral, G. Mantovani, G. J. Clarkson, S. Cauet, J. L. Irwin and D. M. Haddleton, *J. Am. Chem. Soc.*, 2006, **128**, 4823–4830.
41. G. Chen, L. Tao, G. Mantovani, J. Geng, D. Nystroem and D. M. Haddleton, *Macromolecules*, 2007, **40**, 7513–7520.
42. M. H. Stenzel, *ACS Macro Lett.*, 2013, **2**, 14–18.
43. S. Pearson, G. Chen and M. H. Stenzel, in *Engineered Carbohydrate-Based Materials for Biomedical Applications*, ed. R. Narain, Wiley, Hoboken, NJ, 2011, pp. 1–118.
44. V. Ladmiral, E. Melia and D. M. Haddleton, *Eur. Polym. J.*, 2004, **40**, 431–449.
45. C. R. Becer, *Macromol. Rapid Commun.*, 2012, **33**, 742–752.
46. Y. Miura, *Polym. J.*, 2012, **44**, 679–689.
47. Q. Wang, J. S. Dordick and R. J. Linhardt, *Chem. Mater.*, 2002, **14**, 3232–3244.
48. H. B. Wood and H. G. Fletcher, *J. Am. Chem. Soc.*, 1956, **78**, 2849–2851.
49. B. Helferich and D. V. Kashelikar, *Chem. Ber. Recl.*, 1957, **90**, 2094–2096.
50. B. Helferich and G. Duve, *Chem. Ber. Recl.*, 1958, **91**, 1790–1793.
51. B. Helferich and W. Piel, *Justus Liebigs Ann. Chem.*, 1959, **623**, 124–128.
52. G. Wulff, J. Schmid and T. Venhoff, *Macromol. Chem. Phys.*, 1996, **197**, 259–274.
53. S. Mikkola and M. Oivanen, *Arkivoc*, 2009, 39–53.
54. J. Kumar, L. McDowall, G. J. Chen and M. H. Stenzel, *Polym. Chem.*, 2011, **2**, 1879–1886.
55. S. Kimura and K. Hirai, *Makromol. Chem.*, 1962, **58**, 232–236.
56. W. A. P. Black, E. T. Dewar and D. Rutherford, *J. Chem. Soc.*, 1963, 4433.
57. S. Muthukrishnan, M. Zhang, M. Burkhardt, M. Drechsler, H. Mori and A. H. E. Muller, *Macromolecules*, 2005, **38**, 7926–7934.
58. S. Muthukrishnan, H. Mori and A. H. E. Muller, *Macromolecules*, 2005, **38**, 3108–3119.
59. S. Muthukrishnan, F. Plamper, H. Mori and A. H. E. Muller, *Macromolecules*, 2005, **38**, 10631–10642.
60. J. Wang, Y. Qian, F. Zhang, B. Zhu and Y. Xu, *Chin. Sci. Bull.*, 2008, **53**, 1343–1351.
61. S. Muthukrishnan, D. P. Erhard, H. Mori and A. H. E. Muller, *Macromolecules*, 2006, **39**, 2743–2750.
62. M. Ejaz, K. Ohno, Y. Tsujii and T. Fukuda, *Macromolecules*, 2000, **33**, 2870–2874.
63. C. Gao, S. Muthukrishnan, W. Li, J. Yuan, Y. Xu and A. H. E. Muller, *Macromolecules*, 2007, **40**, 1803–1815.

64. V. Ladmiral, L. Monaghan, G. Mantovani and D. M. Haddleton, *Polymer*, 2005, **46**, 8536–8545.
65. M. Al-Bagoury, K. Buchholz and E.-J. Yaacoub, *Polym. Adv. Technol.*, 2007, **18**, 313–322.
66. T. Ouchi, S. Jokei and H. Chikashita, *J. Heterocycl. Chem.*, 1982, **19**, 935–936.
67. G. Kossmehl and J. Volkheimer, *Liebigs Ann. Chem.*, 1989, 1127–1130.
68. G. Kossmehl, J. Volkheimerl and H. Schafer, *Angew. Makromol. Chem.*, 1995, **228**, 59–72.
69. H. Götz, E. Harth, S. M. Schiller, C. W. Frank, W. Knoll and C. J. Hawker, *J Polym. Sci., Part A: Polym. Chem.*, 2002, **40**, 3379–3391.
70. K. Ohno, Y. Izu, S. Yamamoto, T. Miyamoto and T. Fukuda, *Macromol. Chem. Phys.*, 1999, **200**, 1619–1625.
71. S. Muthukrishnan, G. Jutz, X. Andre, H. Mori and A. H. E. Muller, *Macromolecules*, 2004, **38**, 9–18.
72. K. Kobayashi and H. Sumitomo, *Macromolecules*, 1980, **13**, 234–239.
73. K. Kobayashi, H. Sumitomo and Y. Ina, *Polym. J.*, 1983, **15**, 667–671.
74. Z. Özyürek, H. Komber, S. Gramm, D. Schmaljohann, A. H. E. Müller and B. Voit, *Macromol. Chem. Phys.*, 2007, **208**, 1035–1049.
75. T. P. Bird, W. A. P. Black, J. A. Colquhoun, E. T. Dewar and D. Rutherford, *Chem. Ind.*, 1965, 1073.
76. W. A. P. Black, J. A. Colquhoun and E. T. Dewar, *Makromol. Chem.*, 1969, **122**, 244–260.
77. Y. Iwakura, Y. Imai and K. Yagi, *J. Polym. Sci., Part A: Polym. Chem.*, 1968, **6**, 1625–1632.
78. M. J. Carneiro, A. Fernandes, C. M. Figueiredo, A. G. Fortes and A. M. Freitas, *Carbohydr. Polym.*, 2001, **45**, 135–138.
79. Y.-M. Chen and G. Wulff, *Macromol. Rapid Commun.*, 2002, **23**, 59–63.
80. F.-Z. Lu, J.-Q. Meng, F.-S. Du, Z.-C. Li and B.-Y. Zhang, *Macromol. Chem. Phys.*, 2005, **206**, 513–520.
81. F. Suriano, O. Coulembier, P. Degée and P. Dubois, *J. Polym. Sci. Part A: Polym. Chem.*, 2008, **46**, 3662–3672.
82. S. R. S. Ting, A. M. Gregory and M. H. Stenzel, *Biomacromolecules*, 2009, **10**, 342–352.
83. J. Klein and K. Blumenberg, *Makromol. Chem. Macromol. Chem. Phys.*, 1988, **189**, 805–813.
84. S. Loykulnant and A. Hirao, *Macromolecules*, 2000, **33**, 4757–4764.
85. J. Wang, X. Zhu, Z. Cheng, Z. Zhang and J. Zhu, *J. Polym. Sci., Part A: Polym. Chem.*, 2007, **45**, 3788–3797.
86. N.-Y. Xiao, A.-L. Li, H. Liang and J. Lu, *Macromolecules*, 2008, **41**, 2374–2380.
87. T. Nakaya, M. Memita and M. Kang, *J. Macromol. Sci. Pure Appl. Chem.*, 1993, **A30**, 349–355.
88. B. Helferich and H. J. Hofmann, *Chem. Ber. Recl.*, 1952, **85**, 175–180.
89. M. Ambrosi, A. S. Batsanov, N. R. Cameron, B. G. Davis, J. A. K. Howard and R. Hunter, *J. Chem. Soc., Perkin Trans.*, 2002, **1**, 45–52.

90. R. Cuervo-Rodríguez, V. Bordegé and M. Fernández-García, *Carbohydr. Polym.*, 2007, **68**, 89–94.
91. Z.-C. Li, Y.-Z. Liang and F.-M. Li, *Chem. Commun.*, 1999, 1557–1558.
92. Y. Z. Liang, Z. C. Li and F. M. Li, *J. Colloid Interface Sci.*, 2000, **224**, 84–90.
93. P. H. Weigel, R. L. Schnaar, S. Roseman and Y. C. Lee, *Methods Enzymol.*, 1982, **83**, 294–299.
94. R. T. Lee, S. Cascio and Y. C. Lee, *Anal. Biochem.*, 1979, **95**, 260–269.
95. M. Ambrosi, N. R. Cameron, B. G. Davis and S. Stolnik, *Org. Biomol. Chem.*, 2005, **3**, 1476–1480.
96. M. T. Charreyre, P. Boullanger, C. Pichot, T. Delair, B. Mandrand and M. F. Llauro, *Makromol. Chem. Macromol. Chem. Phys.*, 1993, **194**, 117–135.
97. R. Roy, F. D. Tropper and A. Romanowska, *Bioconjugate Chem.*, 1992, **3**, 256–261.
98. K. Tagawa, N. Sendai, K. Ohno, T. Kawaguchi, H. Kitano and T. Matsunaga, *Bioconjugate Chem.*, 1999, **10**, 354–360.
99. R. R. Schmidt, J. Michel and M. Roos, *Liebigs Ann. Chem.*, 1984, 1343–1357.
100. A. Narumi, T. Satoh, H. Kaga and T. Kakuchi, *Macromolecules*, 2001, **35**, 699–705.
101. A. Narumi, T. Matsuda, H. Kaga, T. Satoh and T. Kakuchi, *Polymer*, 2002, **43**, 4835–4840.
102. S. R. S. Ting, E. H. Min, P. Escale, M. Save, L. Billon and M. H. Stenzel, *Macromolecules*, 2009, **42**, 9422–9434.
103. S. G. Spain, L. Albertin and N. R. Cameron, *Chem. Commun.*, 2006, 4198–4200.
104. T.-Y. Guo, P. Liu, J.-W. Zhu, M.-D. Song and B.-H. Zhang, *Biomacromolecules*, 2006, **7**, 1196–1202.
105. G. Wulff, J. Schmid and T. P. Venhoff, *Macromol. Chem. Phys.*, 1996, **197**, 1285–1299.
106. G. Wulff and H. Diederichs, *Macromol. Chem. Phys.*, 1998, **199**, 141–147.
107. S. Nishimura, K. Matsuoka, T. Furuike, S. Ishii, K. Kurita and K. M. Nishimura, *Macromolecules*, 1991, **24**, 4236–4241.
108. T. Furuike, N. Nishi, S. Tokura and S.-I. Nishimura, *Macromolecules*, 1995, **28**, 7241–7247.
109. V. Vazquez-Dorbatt and H. D. Maynard, *Biomacromolecules*, 2006, **7**, 2297–2302.
110. S.-H. Kim, M. Goto, C.-S. Cho and T. Akaike, *Biotechnol. Lett.*, 2000, **22**, 1049–1057.
111. A. Degoede, M. Vanoosterom, M. P. J. Vandeurzen, R. A. Sheldon, H. Vanbekkum and F. Vanrantwijk, in *Heterogeneous Catalysis and Fine Chemicals Iii*, ed. M. Guisnet, J. Barbier, J. Barrault, C. Bouchoule, D. Duprez, G. Perot and C. Montassier, Elsevier, Amsterdam, 1993, pp. 513–520.
112. I. Gill and R. Valivety, *Angew. Chem. Int. Ed.*, 2000, **39**, 3804–3808.
113. L. Albertin, C. Kohlert, M. Stenzel, L. J. R. Foster and T. P. Davis, *Biomacromolecules*, 2004, **5**, 255–260.

114. L. Albertin, M. Stenzel, C. Barner-Kowollik, L. J. R. Foster and T. P. Davis, *Macromolecules*, 2004, **37**, 7530–7537.
115. L. Albertin, M. H. Stenzel, C. Barner-Kowollik, L. J. R. Foster and T. P. Davis, *Macromolecules*, 2005, **38**, 9075–9084.
116. N. R. Cameron, S. G. Spain, J. A. Kingham, S. Weck, L. Albertin, C. L. Barker, G. Battaglia, T. Smart and A. Blanazs, *Faraday Discuss.*, 2007, **139**, 359–368.
117. M. I. Gibson, C. A. Barker, S. G. Spain, L. Albertin and N. R. Cameron, *Biomacromolecules*, 2008, **10**, 328–333.
118. L. Albertin, M. H. Stenzel, C. Barner-Kowollik and T. P. Davis, *Polymer*, 2006, **47**, 1011–1019.
119. L. Albertin and N. R. Cameron, *Macromolecules*, 2007, **40**, 6082–6093.
120. A. M. Granville, D. Quémener, T. P. Davis, C. Barner-Kowollik and M. H. Stenzel, *Macromol. Symp.*, 2007, **255**, 81–89.
121. S. R. S. Ting, A. M. Granville, D. Quemener, T. P. Davis, M. H. Stenzel and C. Barner-Kowollik, *Aust. J. Chem.*, 2007, **60**, 405–409.
122. Y. Miura, T. Ikeda and K. Kobayashi, *Biomacromolecules*, 2003, **4**, 410–415.
123. S. Kitazawa, M. Okumura, K. Kinomura and T. Sakakibara, *Chem. Lett.*, 1990, 1733–1736.
124. X.-L. Sun, D. Grande, S. Baskaran, S. R. Hanson and E. L. Chaikof, *Biomacromolecules*, 2002, **3**, 1065–1070.
125. X.-L. Sun, K. M. Faucher, M. Houston, D. Grande and E. L. Chaikof, *J. Am. Chem. Soc.*, 2002, **124**, 7258–7259.
126. S. Hou, X.-L. Sun, C.-M. Dong and E. L. Chaikof, *Bioconjugate Chem.*, 2004, **15**, 954–959.
127. D. Grande, S. Baskaran, C. Baskaran, Y. Gnanou and E. L. Chaikof, *Macromolecules*, 2000, **33**, 1123–1125.
128. S. Baskaran, D. Grande, X.-L. Sun, A. Yayon and E. L. Chaikof, *Bioconjugate Chem.*, 2002, **13**, 1309–1313.
129. G. Wulff and G. Clarkson, *Macromol. Chem. Phys.*, 1994, **195**, 2603–2610.
130. J. Klein and D. Herzog, *Makromol. Chem. Macromol. Chem. Phys.*, 1987, **188**, 1217–1232.
131. J. Klein and A. H. Begli, *Makromol. Chem. Macromol. Chem. Phys.*, 1989, **190**, 2527–2534.
132. S. R. S. Ting, E. H. Min, P. B. Zetterlund and M. H. Stenzel, *Macromolecules*, 2010, **43**, 5211–5221.
133. J. Klein, A. H. Begli and S. Engelke, *Makromol. Chem. Rapid Commun.*, 1989, **10**, 629–636.
134. H. Kitano, A. Hayashi, H. Takakura, H. Suzuki, N. Kanayama and Y. Saruwatari, *Langmuir*, 2009, **25**, 9361–9368.
135. L. Zhang, J. Bernard, T. P. Davis, C. Barner-Kowollik and M. H. Stenzel, *Macromol. Rapid Commun.*, 2008, **29**, 123–129.
136. J. Bernard, X. J. Hao, T. P. Davis, C. Barner-Kowollik and M. H. Stenzel, *Biomacromolecules*, 2006, 7, 232–238.
137. L. Zhang and M. H. Stenzel, *Aust. J. Chem.*, 2009, **62**, 813–822.

138. A. Miyachi, H. Dohi, P. Neri, H. Mori, H. Uzawa, Y. Seto and Y. Nishida, *Biomacromolecules*, 2009, **10**, 1846–1853.
139. K. Kobayashi, H. Sumitomo and Y. Ina, *Polym. J.*, 1985, **17**, 567–575.
140. W.-J. Zhou, M. E. Wilson, M. J. Kurth, Y.-L. Hsieh, J. M. Krochta and C. F. Shoemaker, *Macromolecules*, 1997, **30**, 7063–7068.
141. W.-J. Zhou, M. J. Kurth, Y.-L. Hsieh and J. M. Krochta, *Macromolecules*, 1999, **32**, 5507–5513.
142. R. Narain and S. P. Armes, *Chem. Commun.*, 2002, 2776–2777.
143. R. Narain and S. P. Armes, *Biomacromolecules*, 2003, **4**, 1746–1758.
144. X. H. Dai and C. M. Dong, *J. Polym. Sci., Part A: Polym. Chem.*, 2008, **46**, 817–829.
145. S. Qiu, H. Huang, X. H. Dai, W. Zhou and C. M. Dong, *J. Polym. Sci., Part A: Polym. Chem.*, 2009, **47**, 2009–2023.
146. X.-H. Dai, H.-D. Zhang and C.-M. Dong, *Polymer*, 2009, **50**, 4626–4634.
147. X.-H. Dai, C.-M. Dong and D. Yan, *J. Phys. Chem. B*, 2008, **112**, 3644–3652.
148. Q. Yang, J. Tian, M.-X. Hu and Z.-K. Xu, *Langmuir*, 2007, **23**, 6684–6690.
149. A. Mateescu, J. Ye, R. Narain and M. Vamvakaki, *Soft Matter*, 2009, **5**, 1621–1629.
150. J. E. Raynor, T. A. Petrie, K. P. Fears, R. A. Latour, A. J. Garcia and D. M. Collard, *Biomacromolecules*, 2009, **10**, 748–755.
151. A. Housni, H. Cai, S. Liu, S. H. Pun and R. Narain, *Langmuir*, 2007, **23**, 5056–5061.
152. K. Kobayashi, H. Sumitomo and Y. Ina, *Polym. J.*, 1985, **17**, 567–575.
153. S.-H. Kim, M. Goto and T. Akaike, *J. Biol. Chem.*, 2001, **276**, 35312–35319.
154. K. Ohno, A. Goto, T. Fukuda, J. Xia and K. Matyjaszewski, *Macromolecules*, 1998, **31**, 2699–2701.
155. R. Narain and S. P. Armes, *Macromolecules*, 2003, **36**, 4675–4678.
156. W. Zhou, X.-H. Dai and C.-M. Dong, *Macromol. Biosci.*, 2008, **8**, 268–278.
157. R. Narain, *React. Funct. Polym.*, 2006, **66**, 1589–1595.
158. K. Mizukami, H. Takakura, T. Matsunaga and H. Kitano, *Colloids Surf., B*, 2008, **66**, 110–118.
159. M. Hetzer, G. Chen, C. Barner-Kowollik and M. H. Stenzel, *Macromol. Biosci.*, 2010, **10**, 119–126.
160. M. Toyoshima and Y. Miura, *J. Polym. Sci., Part A: Polym. Chem.*, 2009, **47**, 1412–1421.
161. K. Yamada, M. Minoda and T. Miyamoto, *J. Polym. Sci., Part A: Polym. Chem.*, 1997, **35**, 751–757.
162. K. Yamada, K. Yamaoka, M. Minoda and T. Miyamoto, *J. Polym. Sci., Part A: Polym. Chem.*, 1997, **35**, 255–261.
163. M. P. Labeau, H. Cramail and A. Deffieux, *Macromol. Chem. Phys.*, 1998, **199**, 335–342.
164. F. D'Agosto, M. T. Charreyre, F. Delolme, G. Dessalces, H. Cramail, A. Deffieux and C. Pichot, *Macromolecules*, 2002, **35**, 7911–7918.
165. F. D'Agosto, M. T. Charreyre, C. Pichot and B. Mandrand, *Macromol. Chem. Phys.*, 2002, **203**, 146–154.

166. K. Aoi, K. Tsutsumiuchi and M. Okada, *Macromolecules*, 1994, **27**, 875–877.
167. M. Okada, K. Aoi and K. Tsutsumiuchi, *Proc. Jpn. Acad., Ser. B*, 1997, **73**, 205–209.
168. K. Aoi, K. Tsutsumiuchi, E. Aoki and M. Okada, *Macromolecules*, 1996, **29**, 4456–4458.
169. K. Tsutsumiuchi, K. Aoi and M. Okada, *Macromolecules*, 1997, **30**, 4013–4017.
170. S.-I. Nishimura, T. Furuike, K. Matsuoka, K. Maruyama, K. Nagata, K. Kurita, N. Nishi and S. Tokura, *Macromolecules*, 1994, **27**, 4876–4880.
171. L. A. Carpino, H. Ringsdorf and H. Ritter, *Makromol. Chem.*, 1976, **177**, 1631–1635.
172. T. Lipinski, P. I. Kitov, A. Szpacenko, E. Paszkiewicz and D. R. Bundle, *Bioconjugate Chem.*, 2011, **22**, 274–281.
173. T. Nagatsuka, H. Uzawa, I. Ohsawa, Y. Seto and Y. Nishida, *ACS Appl. Mater. Interfaces*, 2010, **2**, 1081–1085.
174. T. Nagatsuka, H. Uzawa, K. Sato, I. Ohsawa, Y. Seto and Y. Nishida, *ACS Appl. Mater. Interfaces*, 2012, **4**, 832–837.
175. Y. Nishimura, H. Shudo, H. Seto, Y. Hoshino and Y. Miura, *Bioorg. Med. Chem. Lett.*, 2013, **23**, 6390–6395.
176. Y. Hoshino, M. Nakamoto and Y. Miura, *J. Am. Chem. Soc.*, 2012, **134**, 15209–15212.
177. H. Seto, Y. Ogata, T. Murakami, Y. Hoshino and Y. Miura, *ACS Appl. Mater. Interfaces*, 2012, **4**, 411–417.
178. A. Ghadban, L. Albertin, M. Rinaudo and A. Heyraud, *Biomacromolecules*, 2012, **13**, 3108–3119.
179. K. Matsuoka, H. Yamaguchi, T. Koyama, K. Hatano and D. Terunuma, *Tetrahedron Lett.*, 2010, **51**, 2529–2532.
180. S. Kitazawa, M. Okumura, K. Kinomura and T. Sakakibara, *Chem. Lett.*, 1990, 1733–1736.
181. S. Kimura and M. Imoto, *Makromol. Chem.*, 1961, **50**, 155–160.
182. C. Deng, F. F. Li, J. M. Hackett, S. H. Chaudhry, F. N. Toll, B. Toye, W. Hodge and M. Griffith, *Acta Biomater.*, 2010, **6**, 187–194.
183. M. Obata, T. Kobori, S. Hirohara and M. Tanihara, *Polymer*, 2012, **53**, 4672–4677.
184. M. Obata, M. Shimizu, T. Ohta, A. Matsushige, K. Iwai, S. Hirohara and M. Tanihara, *Polym. Chem.*, 2011, **2**, 651–658.
185. Q. Y. Meng, L. M. Tian and J. X. Wang, *J. Mater. Sci. Mater. Med.*, 2012, **23**, 991–998.
186. C. Zheng, Q. Q. Guo, Z. M. Wu, L. Sun, Z. P. Zhang, C. X. Li and X. G. Zhang, *Eur. J. Pharm. Sci.*, 2013, **49**, 474–482.
187. A. Munoz-Bonilla, G. Marcelo, C. Casado, F. J. Teran and M. Fernandez-Garcia, *J. Polym. Sci., Part A: Polym. Chem.*, 2012, **50**, 5087–5096.
188. A. Munoz-Bonilla, E. Ibarboure, V. Bordege, M. Fernandez-Garcia and J. Rodriguez-Hernandez, *Langmuir*, 2010, **26**, 8552–8558.
189. R. Narain, D. Jhurry and G. Wulff, *Eur. Polym. J.*, 2002, **38**, 273–280.

190. Q. Y. Meng, A. Haque, B. Hexig and T. Akaike, *Biomaterials*, 2012, **33**, 1414–1427.
191. T. Fukuda, Y. Inoue, T. Koga, M. Matsuoka and Y. Miura, *Chem. Lett.*, 2011, **40**, 864–866.
192. L. Albertin, M. H. Stenzel, C. Barner-Kowollik, L. J. R. Foster and T. P. Davis, *Polymer*, 2005, **46**, 2831–2835.
193. S. F. Lou, H. Zhang, G. R. Williams, C. Branford-White, H. L. Nie, J. Quan and L. M. Zhu, *Colloids Surf., B*, 2013, **105**, 180–186.
194. Y. Li, J. Quan, C. Branford-White, G. R. Williams, J. X. Wu and L. M. Zhu, *J. Mol. Catal. B: Enzym.*, 2012, **76**, 15–22.
195. G. Wulff, H. Schmidt and L. M. Zhu, *Macromol. Chem. Phys.*, 1999, **200**, 774–782.
196. A. D. Jenkins, R. G. Jones and G. Moad, *Pure Appl. Chem.*, 2010, **82**, 483–491.
197. A. Ghadban and L. Albertin, *Polymers*, 2013, **5**, 431–526.
198. J. Nicolas, Y. Guillaneuf, C. Lefay, D. Bertin, D. Gigmes and B. Charleux, *Prog. Polym. Sci.*, 2013, **38**, 63–235.
199. K. Babiuch, R. Wyrwa, K. Wagner, T. Seemann, S. Hoeppener, C. R. Becer, R. Linke, M. Gottschaldt, J. Weisser, M. Schnabelrauch and U. S. Schubert, *Biomacromolecules*, 2011, **12**, 681–691.
200. D. Grande, R. Cuerrero and Y. Gnanou, *J. Polym. Sci., Part A: Polym. Chem.*, 2005, **43**, 519–533.
201. S. N. Narla and X. L. Sun, *Org. Biomol. Chem.*, 2011, **9**, 845–850.
202. S. N. Narla and X. L. Sun, *Biomacromolecules*, 2012, **13**, 1675–1682.
203. Z. Lu, L. Mei, X. Zhang, Y. Wang, Y. Zhao and C. Li, *Polym. Chem.*, 2013, **4**, 5743–5750.
204. J. Qu, H. Shao and X. Zhang, *Appl. Mech. Mat.*, 2012, **189**, 3–6.
205. B. B. Ke, L. S. Wan, W. X. Zhang and Z. K. Xu, *Polymer*, 2010, **51**, 2168–2176.
206. O. Leon, A. Munoz-Bonilla, V. Bordege, M. Sanchez-Chaves and M. Fernandez-Garcia, *J. Polym. Sci., Part A: Polym. Chem.*, 2011, **49**, 2627–2635.
207. S. Menon, R. M. Ongungal and S. Das, *Polym. Chem.*, 2013, **4**, 623–628.
208. S. Menon and S. Das, *Polym. Chem.*, 2012, **3**, 2619–2624.
209. M. C. Munisso, A. Mahara, Y. Tachibana, J. H. Kang, S. Obika and T. Yamaoka, in *Biomedical Applications of Smart Technologies*, ed. P. Vincenzini and E. P. Scilingo, Trans Tech Publications, Stafa-Zurich, 2013, pp. 86–91.
210. Y. X. Wang, X. G. Zhang, J. Mu and C. X. Li, *New J. Chem.*, 2013, **37**, 796–803.
211. X. D. Tang, X. Wang, Y. Y. Dou and M. S. Pei, in *Liquid Crystals and Related Materials Ii*, ed. Y. M. Huang, Trans Tech Publications, Stafa-Zurich, 2012, pp. 356–359.
212. N. K. Vishwakarma, A. K. Mishra, A. Mishra, T. Paira, V. K. Patel, C. S. Biswas, T. K. Mandal, P. Maiti and B. Ray, *J. Appl. Polym. Sci.*, 2013, **128**, 1369–1380.
213. Y. X. Wang, X. G. Zhang, P. E. Yu and C. X. Li, *Int. J. Pharm.*, 2013, **441**, 170–180.

214. Y. X. Wang, X. G. Zhang, Y. C. Han, C. Cheng and C. X. Li, *Carbohydr. Poly.*, 2012, **89**, 124–131.
215. H. Kitano, D. Saito, T. Kamada and M. Gemmei-Ide, *Colloids Surf., B*, 2012, **93**, 219–225.
216. R. Fleet, E. T. A. van den Dungen and B. Klumperman, *S. Afr. J. Sci.*, 2011, **107**, 101–111.
217. F. Suriano, O. Coulembier and P. Dubois, *React. Funct. Polym.*, 2010, **70**, 747–754.
218. A. Munoz-Bonilla, J. P. A. Heuts and M. Fernandez-Garcia, *Soft Matter*, 2011, **7**, 2493–2499.
219. F. Suriano, O. Coulembier and P. Dubois, *J. Polym. Sci., Part A: Polym. Chem.*, 2010, **48**, 3271–3280.
220. A. Munoz-Bonilla, O. Leon, V. Bordege, M. Sanchez-Chaves and M. Fernandez-Garcia, *J. Polym. Sci., Part A: Polym. Chem.*, 2013, **51**, 1337–1347.
221. O. Abdelkader, S. Moebs-Sanchez, Y. Queneau, J. Bernard and E. Fleury, *J. Polym. Sci., Part A: Polym. Chem.*, 2011, **49**, 1309–1318.
222. X. P. Chen and N. Ayres, *Macromolecules*, 2010, **43**, 1341–1348.
223. L. Liu, J. C. Zhang, W. H. Lv, Y. Luo and X. J. Wang, *J. Polym. Sci., Part A: Polym. Chem.*, 2010, **48**, 3350–3361.
224. M. Ahmed and R. Narain, *Biomaterials*, 2011, **32**, 5279–5290.
225. A. Pfaff, A. Schallon, T. M. Ruhland, A. P. Majewski, H. Schmalz, R. Freitag and A. H. E. Muller, *Biomacromolecules*, 2011, **12**, 3805–3811.
226. N. Y. Xiao, H. Liang and J. Lu, *Soft Matter*, 2011, **7**, 10834–10840.
227. A. E. Smith, A. Sizovs, G. Grandinetti, L. Xue and T. M. Reineke, *Biomacromolecules*, 2011, **12**, 3015–3022.
228. D. Z. Zhou, C. X. Li, Y. L. Hu, H. Zhou, J. T. Chen, Z. P. Zhang and T. Y. Guo, *Int. J. Biol. Macromol.*, 2012, **50**, 965–973.
229. S. Pearson, W. Scarano and M. H. Stenzel, *Chem. Commun.*, 2012, **48**, 4695–4697.
230. R. J. Mancini, J. Lee and H. D. Maynard, *J. Am. Chem. Soc.*, 2012, **134**, 8474–8479.
231. P. Escale, S. R. S. Ting, A. Khoukh, L. Rubatat, M. Save, M. H. Stenzel and L. Billon, *Macromolecules*, 2011, **44**, 5911–5919.
232. M. Takara, M. Toyoshima, H. Seto, Y. Hoshino and Y. Miura, *Polym. Chem.*, 2014, **5**, 931–939.
233. W. Lu, W. Ma, J. Lu, X. Li, Y. Zhao and G. Chen, *Macromol. Rapid Commun.*, 2014, **35**, 827–833.
234. M. Ahmed, S. Mamba, X. H. Yang, J. Darkwa, P. Kumar and R. Narain, *Bioconjugate Chem.*, 2013, **24**, 979–986.
235. M. Ahmed and R. Narain, *Biomaterials*, 2013, **34**, 4368–4376.
236. D. J. Buckwalter, A. Sizovs, N. P. Ingle and T. M. Reineke, *ACS Macro. Lett.*, 2012, **1**, 609–613.
237. E. H. Song, M. J. Manganiello, Y. H. Chow, B. Ghosn, A. J. Convertine, P. S. Stayton, L. M. Schnapp and D. M. Ratner, *Biomaterials*, 2012, **33**, 6889–6897.

238. W. Wang, D. L. Chance, V. V. Mossine and T. P. Mawhinney, *Glycoconjugate J.*, 2014, **31**, 133–143.
239. L. Albertin, A. Wolnik, A. Ghadban and F. Dubreuil, *Macromol. Chem. Phys.*, 2012, **213**, 1768–1782.
240. T. Xing, X. Z. Yang, L. Y. Fu and L. F. Yan, *Polym. Chem.*, 2013, **4**, 4442–4449.
241. Y. Luo, L. Liu, X. B. Wang, H. T. Shi, W. H. Lv and J. Y. Li, *Soft Matter*, 2012, **8**, 1634–1642.
242. M. Ahmed and R. Narain, *Biomaterials*, 2012, **33**, 3990–4001.
243. Z. K. Wei, X. J. Hao, Z. H. Gan and T. C. Hughes, *J. Polym. Sci., Part A: Polym. Chem.*, 2012, **50**, 2378–2388.
244. J. Lu, W. Zhang, L. Yuan, W. Ma, X. Li, W. Lu, Y. Zhao and G. Chen, *Macromol. Biosci.*, 2013, **14**, 340–346.
245. J. Ishii, M. Chikae, M. Toyoshima, Y. Ukita, Y. Miura and Y. Takamura, *Electrochem. Commun.*, 2011, **13**, 830–833.
246. M. Kaupp, A. P. Vogt, J. C. Natterodt, V. Trouillet, T. Gruendling, T. Hofe, L. Barner and C. Barner-Kowollik, *Polym. Chem.*, 2012, **3**, 2605–2614.
247. A. L. Parry, N. A. Clemson, J. Ellis, S. S. R. Bernhard, B. G. Davis and N. R. Cameron, *J. Am. Chem. Soc.*, 2013, **135**, 9362–9365.
248. E. H. Min, S. R. S. Ting, L. Billon and M. H. Stenzel, *J. Polym. Sci., Part A: Polym. Chem.*, 2010, **48**, 3440–3455.
249. M. Ahmed and R. Narain, *Mol. Pharmaceutics*, 2012, **9**, 3160–3170.
250. S. R. S. Ting, E. H. Min, P. B. Zetterlund and M. H. Stenzel, *Macromolecules*, 2010, **43**, 5211–5221.
251. K. Yamada, M. Minoda and T. Miyamoto, *Macromolecules*, 1999, **32**, 3553–3558.
252. K. Yamada, M. Minoda, T. Fukuda and T. Miyamoto, *J. Polym. Sci., Part A: Polym. Chem.*, 2001, **39**, 459–467.
253. M. C. Schuster, K. H. Mortell, A. D. Hegeman and L. L. Kiessling, *J. Mol. Catal. A: Chem.*, 1997, **116**, 209–216.
254. W. J. Sanders, E. J. Gordon, O. Dwir, P. J. Beck, R. Alon and L. L. Kiessling, *J. Biol. Chem.*, 1999, **274**, 5271–5278.
255. C. Fraser and R. H. Grubbs, *Macromolecules*, 1995, **28**, 7248–7255.
256. K. Nomura and R. R. Schrock, *Macromolecules*, 1996, **29**, 540–545.
257. E. L. Dane, S. L. Chin and M. W. Grinstaff, *ACS Macro. Lett.*, 2013, **2**, 887–890.
258. E. L. Dane and M. W. Grinstaff, *J. Am. Chem. Soc.*, 2012, **134**, 16255–16264.
259. L. Li, Y. Xu, I. Milligan, L. Fu, E. A. Franckowiak and W. Du, *Angew. Chem. Int. Ed.*, 2013, **52**, 13699–13702.
260. M. Tamaki, T. Taguchi, S. Nakabayashi, K. Mori, Y. Kitajyo, R. Sakai, T. Kakuchi and T. Satoh, *Polym. Chem.*, 2010, **1**, 82–92.
261. K. Tsutsumiuchi, K. Aoi and M. Okada, *Macromol. Rapid Commun.*, 1995, **16**, 749–755.
262. K. Aoi, H. Suzuki and M. Okada, *Macromolecules*, 1992, **25**, 7073–7075.

263. A. Takasu and H. Kojima, *J. Polym. Sci., Part A: Polym. Chem.*, 2010, **48**, 5953–5960.
264. F. Zamora, K. Hakkou, A. Alla, M. Rivas, I. Roffé, M. Mancera, S. Muñoz-Guerra and J. A. Galbis, *J. Polym. Sci., Part A: Polym. Chem.*, 2005, **43**, 4570–4577.
265. M. G. García-Martín, E. Benito Hernández, R. Ruiz Pérez, A. Alla, S. Muñoz-Guerra and J. A. Galbis, *Macromolecules*, 2004, **37**, 5550–5556.
266. V. P. Taori, H. Lu and T. M. Reineke, *Biomacromolecules*, 2011, **12**, 2055–2063.
267. P. M. McLendon, K. M. Fichter and T. M. Reineke, *Mol. Pharm.*, 2010, **7**, 738–750.

CHAPTER 3

Synthesis and Characterization of Glycopolypeptides

KAI-STEFFEN KRANNIG[a] AND HELMUT SCHLAAD*[a]

[a]Max Planck Institute of Colloids and Interfaces, Department of Colloid Chemistry, Research Campus Golm, 14424 Potsdam, Germany
*E-mail: helmut.schlaad@mpikg.mpg.de

3.1 Introduction

Glycopolymers, meaning polymers with pendant carbohydrate side chains, are an emerging class of materials that are designed for use in advanced biomedical and life science applications. Although glycopolymers[1–3] with hydrocarbon backbones have been known for over 50 years,[4–6] it took another 35 years until the first glycopolypeptides became available by controlled ring-opening polymerization of amino acid *N*-carboxyanhydrides (NCAs) (also referred to as Leuchs anhydrides).[7] Polypeptides are considered biocompatible and biodegradable and have the intriguing ability to fold into secondary, tertiary, and higher order structures due to multiple non-covalent interactions between the amino acids.

Naturally occurring glycosylated peptides and proteins display a wide range of biological functions, including anti-freezing,[8] mediation of recognition events, proliferation of cells and inflammatory reactions.[9–12] Hence their synthetic counterparts are also expected to have great potential as biomedical materials (*e.g.*, as drug carriers or scaffolds for tissue repair) and

RSC Polymer Chemistry Series No. 15
Glycopolymer Code: Synthesis of Glycopolymers and Their Applications
Edited by C. Remzi Becer and Laura Hartmann

Published by the Royal Society of Chemistry, www.rsc.org

to serve as versatile tools for probing carbohydrate–protein interactions.[13,14] However, the synthesis of well-defined glycopolypeptides in larger quantities is still a challenge and the understanding of their hierarchical assembly and interaction with biological systems is just at the beginning.

This chapter presents a selection of the most recent advances in the synthesis of glycopolypeptides *via* NCA polymerization and their biological applications (mainly studies on interactions of glycopolypeptides with lectins).[15–20] Prior to this, some general aspects and considerations of NCA synthesis and polymerization procedures are briefly discussed.[21–23]

3.2 Polypeptide Synthesis by Ring-Opening Polymerization

3.2.1 Synthesis of Amino Acid *N*-Carboxyanhydrides

Amino acid *N*-carboxyanhydrides are usually prepared by the direct phosgenation of α-amino acids, formally known as the Fuchs–Farthing method.[24,25] Here, the amine reacts with phosgene and forms a chlorocarbonylamine intermediate, which immediately undergoes ring closure to form the anhydride, releasing two equivalents of hydrogen chloride (Scheme 3.1). Gaseous phosgene can be avoided by the use of crystalline bis(trichloromethyl)carbonate (triphosgene). This derivative releases phosgene only at higher temperatures and can be used in smaller excess (0.4 equiv. triphosgene ≡ 1.2 equiv. phosgene), allowing for safer handling. Other synthetic routes include the thermal treatment of *N*-alkoxycarbonyl amino acid chlorides, as originally used by Leuchs,[26] the use of amino acid urethane derivatives,[27] or the direct reaction of α-amine-protected amino acid derivatives with dichloro (methoxy)methane.[28]

Crude NCA products are usually purified by multiple precipitations from heptanes and careful washing under dry conditions, or by extraction with ice-cold water (see below).[29,30] Ultrapure amino acid NCAs can be obtained by column chromatography.[31] Here, the key is the usage of predried silica gel to avoid ring opening of the NCA by water during the interaction with the column material.

Conversion of an amino acid to the respective NCA is typically accompanied by a change in the polarity from hydrophilic to hydrophobic. Hence a

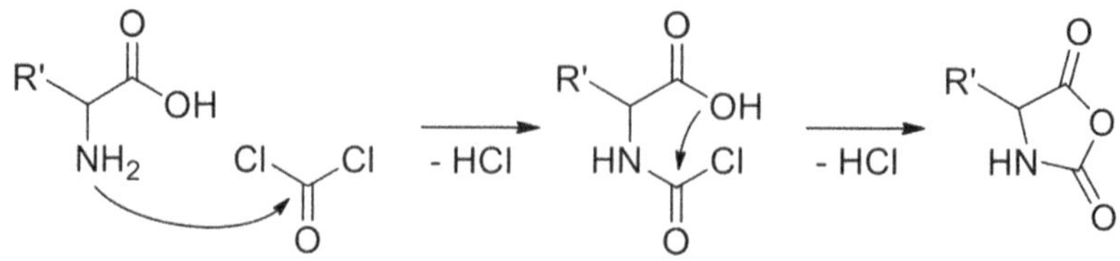

Scheme 3.1 Synthesis of amino acid *N*-carboxyanhydrides using phosgene (Fuchs–Farthing method).

first indicator for the success of the reaction is the solubility of the product in an organic solvent such as tetrahydrofuran (THF) or chloroform. γ-Benzyl-L-glutamate and N^{ε}-Cbz-L-lysine (Cbz = carboxybenzyl), for instance, produce turbid dispersions in THF. After the NCA has formed, the mixture becomes clear (usually within 30–45 min of reaction with triphosgene). Upon purification, the melting-point of the material is a fairly reliable indicator. Many NCAs have melting-points between 90 and 120 °C, whereas those of the respective amino acids are >180 °C.[32]

^{1}H NMR analysis confirms the formation of the NCA when a split signal of the diastereomeric protons adjacent to the α-carbon can be observed, for instance at 2–2.5 ppm for γ-benzyl-L-glutamate NCA in $CDCl_3$. Often a singlet signal for the nitrogen-connected proton appears at relative shifts of >6 ppm. Further proof can be obtained from ^{13}C NMR spectroscopy, where the signals of the carbonyl groups in the anhydride ring appear at relative shifts between 150 and 170 ppm (in $CDCl_3$). In IR spectroscopy, the characteristic bands of the two carbonyl groups of the anhydride ring can be observed at ~1855 and 1780 cm^{-1}.[32]

3.2.2 Polymerization Mechanisms and Protocols

NCAs can be easily polymerized utilizing nucleophilic initiators such as amines, thiols, alcohols, or carbenes; the most commonly used initiators are primary amines such as hexyl- or benzylamine. The reaction follows the normal amine mechanism (NAM) (Scheme 3.2), leading to polymers with fairly narrow molar mass distributions. Silylamines have been demonstrated to be good alternatives, allowing for better control of the polymerization. Tertiary amines have also been utilized to yield polypeptides with broad molar mass distributions, attributable to the competition between the NAM and the activated monomer mechanism (AMM) (Scheme 3.3). Living polymerizations could be achieved using transition metal complexes based on nickel

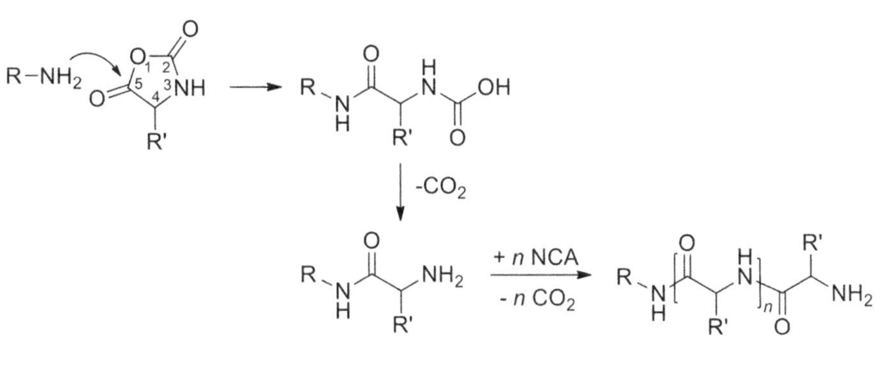

Scheme 3.2 Progress of the polymerization following the normal amine mechanism (NAM). In the initiation step, nucleophilic attack of the primary amine at C-5 leads to ring opening. Release of carbon dioxide recovers a primary amine that continues the polymerization.

or cobalt. The reaction follows a ring-insertion/ring-extension mechanism allowing for the synthesis of high molar mass polypeptides with low dispersity (ratio of weight- to number-average molar mass).[23]

The NAM is the general reaction pathway for all initiators carrying at least one (mobile) hydrogen (base-H) atom such as primary and secondary amines, thiols and alcohols, and also water. In a bimolecular reaction, the nucleophilic base attacks the NCA at C-5, leading to ring opening and formation of an unstable carbamic acid (Scheme 3.2). The carbamic acid decomposes with release of carbon dioxide, recovering an amine with the ability to attack another NCA. Ideally, this process repeats multiple times to yield well-defined polymers with a narrow molar mass distribution (Poisson distribution at best).

The choice of the base plays an important role and only primary or, with reservations, secondary amines guarantee that the initiation step is as fast as (or faster than) the propagation step. Nevertheless, the living and controlled character of the polymerization can be disturbed by several factors, including the carbamic acid–carbon dioxide equilibrium, solvent polarity and the presence of water or other impurities.[23]

Ballard and co-workers showed that for *N*-substituted NCAs, the acidity of the solvent can play an important role.[33,34] They demonstrated that the polymerization of *N*-methylglycine (sarcosine) deviated from pseudo-first-order kinetics when nitrobenzene was used as a solvent instead of DMF, attributing this to a lack of protonation of the carbamic acid in nitrobenzene preventing the release of CO_2 and thus the progress of the reaction. Other groups found that if CO_2 was not removed from the reaction, the formation of the carbamic acid salt was favored, leading also to different kinetics.[35,36] This problem can be overcome by conducting the polymerization under reduced pressure.[37,38] The presence of water as an impurity can influence the course of the polymerization, as it can inherently act as an initiator, leading to broad molar mass distributions or, depending on the amount, only oligomers. However, temperature is a critical factor. Although NCAs can be stored for months in a freezer, they tend to decompose or oligomerize at room or elevated temperatures. Amazingly, Poché *et al.* were able to show that NCAs can be handled and purified in water and aqueous sodium bicarbonate solutions at 0 °C.[30]

The effect of low temperatures during the polymerization was described Vayabouri *et al.* for the polymerization of ε-trifluoroacetyl-L-lysine NCAs.[39] After complete monomer consumption, they found that 99% of the chain ends were still active at 0 °C, decreasing to 22% at room temperature. They reasoned that at low temperature the energy barrier for the propagation step is lower than for side reactions.

Having discussed various side reactions, it should be mentioned that surprisingly neither the ring opening at C-2, leading to termination, nor backbiting in the chain backbone or side chains, particularly expected for poly(γ-benzyl-L-glutamate) carrying a labile ester bond in the side chain, seem to play a role. Nevertheless, Kricheldorf *et al.* described cyclic polypeptides due to end-to-end cyclization using secondary amine initiators such as imidazole or pyridine following a zwitterionic polycondensation mechanism.[40,41]

AMM was first suggested by Ballard and Bamford[42] and later extended by Szwarc[43] to explain ring-opening polymerization using secondary and tertiary amine and alkali metal halide initiators.[23] These initiators act like bases (or basic catalysts) rather than nucleophiles, and the first step is not nucleophilic attack at C-5 but deprotonation of NH in the ring (Scheme 3.3), producing a nucleophilic NCA anion. In the propagation step, the NCA anion attacks another NCA at C-5 to form a tadpole dimer, which then attacks the next NCA. However, depending on the actual structure of the 'catalyst,' particularly when using amines, NAM and AMM mainly occur simultaneously, leading to loss of control of the reaction. Which mechanism dominates depends on steric (bulky amines initiate more slowly) and electronic factors (nucleophilicity and basicity). Moreover, particularly at high conversions, polycondensations may take place when the anionic chain end of one NCA attacks the end-group of another NCA, leading to termination and broad molar mass distributions.[44]

The AMM pathway for primary amines may be suppressed by the use of amine hydrochloride (ammonium chloride) initiators, as suggested by Dimitrov and Schlaad[45] (Scheme 3.4). The amine hydrochloride initiator or chain end dissociates into the amine, which can add to the NCA monomer, and hydrogen chloride, which is used to reprotonate any NCA anion species formed up to this point. After the attack at C-5, the counterion migrates to the amine group of the NCA anion. The reduced amount of free amine leads to a lower degree of deprotonation, and if deprotonation occurs, the NCA is directly reprotonated owing to the small excess of acid in the reaction mixture. The balance of the equilibrium between salt and free amine is a function of temperature, solvent polarity and type of counterion, hence allowing for control by these parameters.[46]

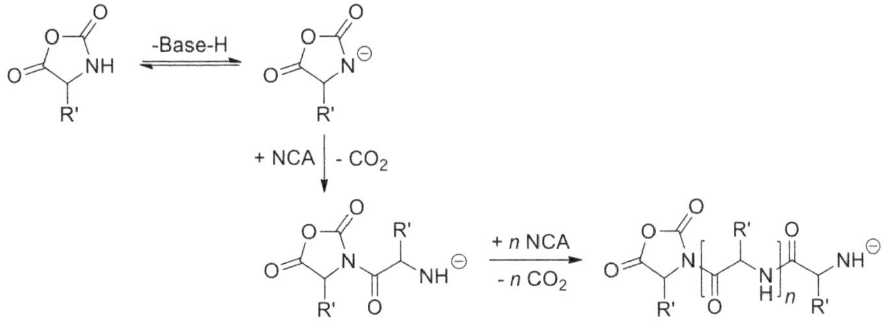

Scheme 3.3 Mechanistic representation of the activated monomer mechanism (AMM) as typically observed for tertiary amines. In addition to nucleophilic attack at C-5 as observed for the NAM, deprotonation of N-3 gives rise to an NCA-based anion that can also initiate or deprotonate another NCA.

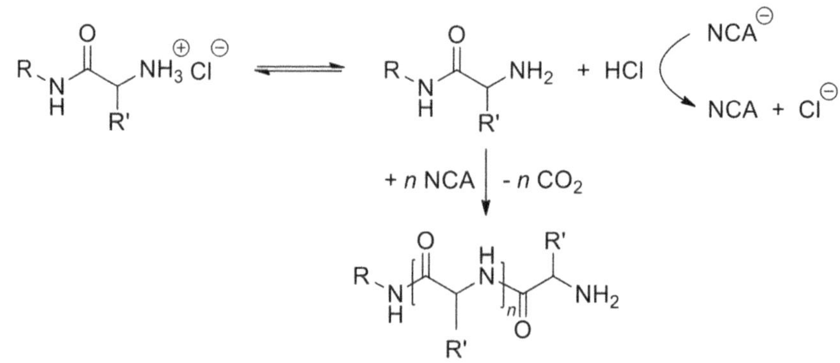

Scheme 3.4 Ammonium chlorides as initiators for controlled polymerization of NCAs *via* the NAM mechanism. Released acid is used to reprotonate any NCA anions, thus avoiding the AMM pathway.

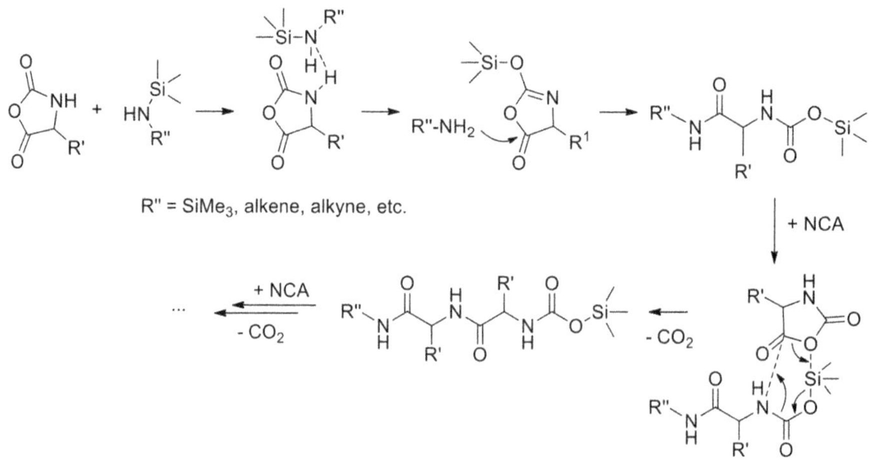

Scheme 3.5 Group transfer polymerization of *N*-unsubstituted NCAs using *N*-trimethylsilylamine initiators to yield polypeptides with defined end-groups.

Lu and Cheng used *N*-trimethylsilylamine-based initiators, for instance hexamethyldisilazane (HMDS), although one could expect that this bulky secondary amine deprotonates the NCA at NH, thus promoting the AMM.[47,48] According to the proposed mechanism, the deprotonation indeed occurs, followed by the transfer of the trimethylsilyl (TMS) group to C-2 (Scheme 3.5). The generated TMS-amine attacks the NCA at C-5 to form a TMS-carbamate. In the propagation step, the non-silyl-bound amine reacts with another NCA with simultaneous transfer of the TMS group to O-2. This technique enables good control of the polymerization and the opportunity to introduce defined end-groups to the chain ends; however, it is restricted to *N*-unsubstituted NCAs.[23]

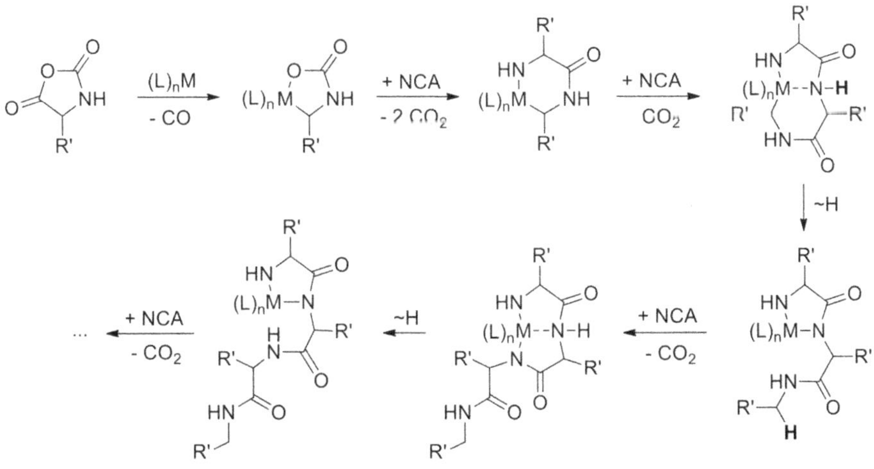

Scheme 3.6 Mechanism of the polymerization of NCAs by transition metal catalysis. In the initiation step, the metal undergoes an oxidative addition to form a metallacycle. The two-step propagation includes ring insertion followed by ring extension. Owing to the need for a proton transfer, the catalysis only works with *N*-unsubstituted NCAs.

In 1997, Deming introduced the use of a zerovalent bipyNi(COD) (bipy = bipyridine, COD = cyclooctadiene) catalyst to perform NCA polymerizations by transition metal complex catalysis. He demonstrated the synthesis of various homopolymers and block copolymers under living conditions with predictable molar masses and narrow molar mass distributions. Moreover, the reactions completed much faster, usually within hours rather than days (as for the conventional amine-initiated polymerizations). Later, he found that cobalt species, such as $(PMe_3)_4Co$, are also efficient initiators for NCA polymerization.[49–52]

The initiation step of the catalysis is the oxidative addition of the metal to the NCA to form a metallacyclic complex (Scheme 3.6). This complex reacts with another NCA, leading to a ring extension to a six-membered amidoalkyl metallacycle. On reacting with a third NCA, carbon dioxide is released, forming a larger metallacycle. Upon migration of the proton from the amide to the metal-bound carbon, a five-membered amidoamidate complex is formed, which is the actual active polymerization intermediate. In the propagation step, the activated amine of this complex attacks C-5 of a another NCA, resulting in a cascade of ring extension, carbon dioxide release, proton migration and finally metal migration to form a new amidoamidate complex at the chain end.[23]

The limits of this reaction type are the need for transition metals, which are always a concern for biomedical materials, and that it can only be used with *N*-unsubstituted NCAs as N-3 proton migration is necessary.[53] Furthermore, the monomers have to be highly pure as otherwise the catalyst is poisoned and deactivated.[31]

3.3 Glycopolypeptide Synthesis and Characterization[20]

Two different pathways for the controlled synthesis of glycosylated polypeptides have been described. The first route starts with the synthesis of a glycosylated NCA, referred to as glyco-NCA, which can be polymerized in a controlled manner to yield high molar mass glycopolypeptides. The main drawback is the preparation and purification of the glyco-NCA, which requires a high level of synthetic expertise. Alternatively, a suitable polypeptide precursor chain is synthesized, onto which the carbohydrate residues are grafted by post-polymerization modification (preferably applying 'click' chemistry).[54–56] This second approach is modular and much less elaborate than the first route, without compromising the degree of glycosylation. However, glycopolypeptides obtained by post-polymerization modification may have considerably lower molar masses than those produced by glyco-NCA polymerization.

3.3.1 Synthesis

3.3.1.1 *Polymerization of Glycosylated Amino Acid N-Carboxyanhydrides*

Okada and co-workers described the first synthesis of a glycopeptide through amine-initiated ring-opening polymerization of a glyco-NCA, namely *O*-(tetra-*O*-acetyl-β-D-glucopyranosyl)-L-serine.[7] The glyco-NCA was obtained by coupling tetra-*O*-acetylated glucose to serine, as described by Rüde *et al.*,[57] and subsequent reaction with phosgene. Primary amine-initiated polymerization produced well-defined glycopeptides with number-average molar masses of ~10 kg mol^{-1} and a dispersity of 1.1. Block copolypeptides with polyalanine as the second block have also been prepared.

The synthesis of higher molar mass glycopolypeptides requires glyco-NCAs in high purity. Kramer and Deming described the coupling of peracetylated sugars (glucose, galactose or mannose) to L-lysine in four steps and conversion into NCAs by treatment with dichloro(methoxy)methane (Scheme 3.7); the glyco-NCAs were purified by anhydrous flash chromatography.[31,58] Polymerization using $(PMe_3)_4Co$ as a transition metal catalyst (*cf.* Scheme 3.6) yielded homo and block copolypeptides with high molar masses ($M_n \approx$ 160 kg mol^{-1}) and low dispersity, which after deacetylation could be readily dispersed in water.[59]

Another procedure towards glyco-NCAs based on L-lysine was reported by Sen Gupta and co-workers (Scheme 3.8).[60–63] A fully protected lysine was coupled with the 1,2-orthoesters of either glucose or mannose in the presence of a gold catalyst. After removal of the acid protecting group by hydrogenation, the amino acid was converted into the NCA using triphosgene and α-pinene as hydrogen chloride scavenger. These lysine derived glyco-NCAs were purified by conventional recrystallization. Polymerizations of the

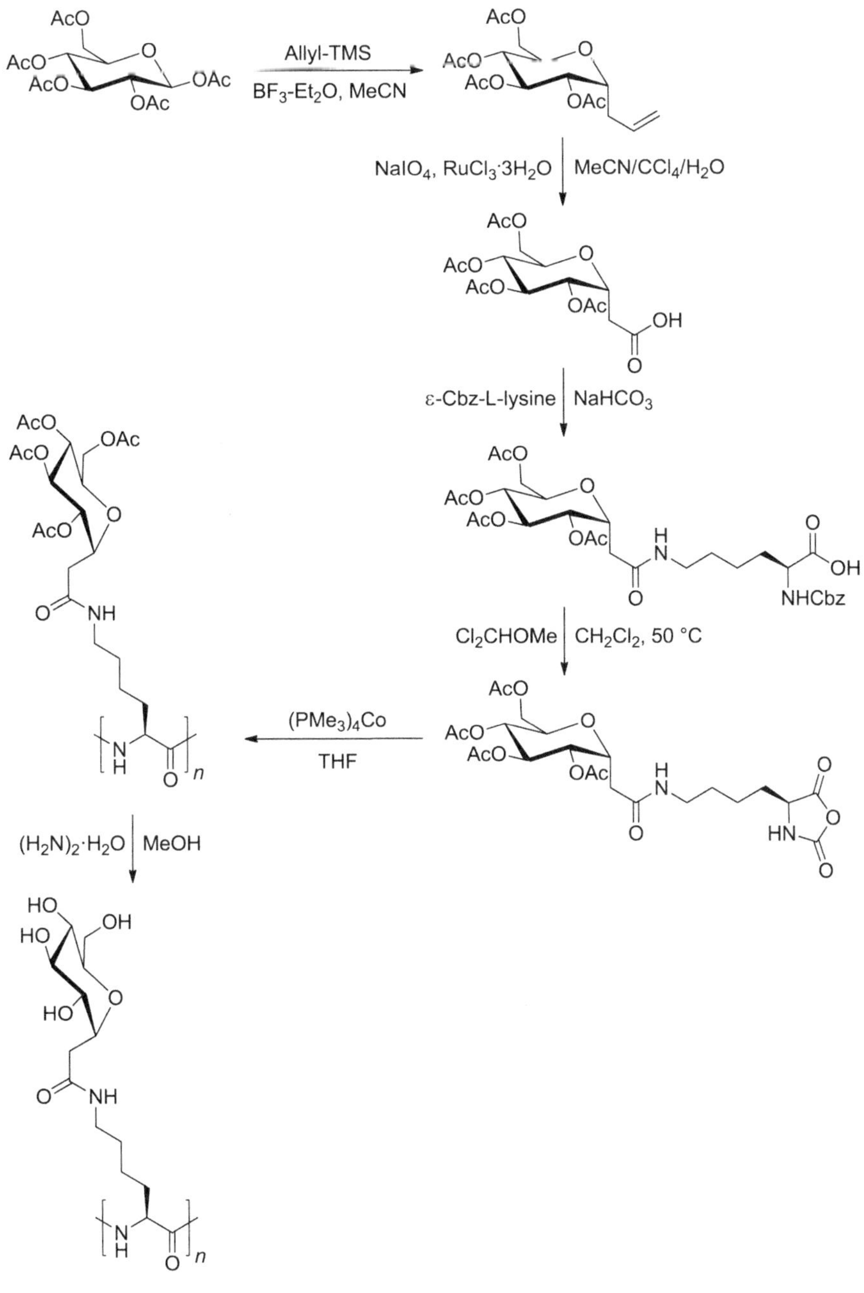

Scheme 3.7 Synthesis and polymerization of glycosylated lysine-NCAs according to Kramer and Deming.

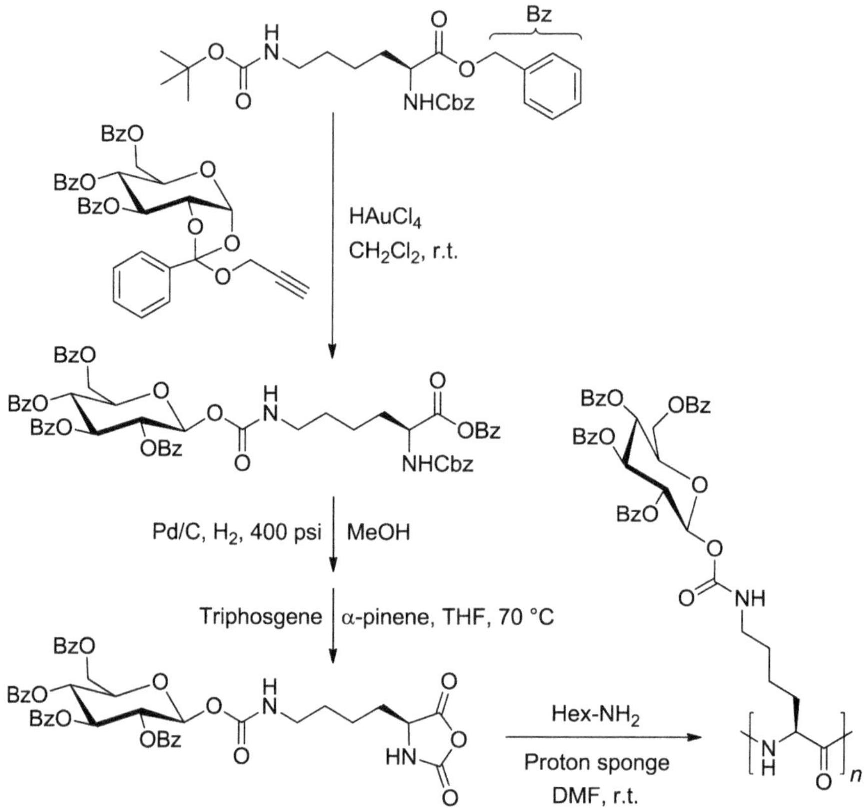

Scheme 3.8 Synthesis and polymerization of glycosylated lysine-NCAs according to Sen Gupta and co-workers.

glyco-NCAs initiated by 1-hexylamine (*cf.* Scheme 3.2) were only successful in the presence of 1,8-bis(dimethylamino)naphthalene as a 'proton sponge.' The proton sponge is thought to neutralize residual acid impurities, but its function is not yet fully understood. The homo- and block copolypeptides obtained exhibited low dispersity but usually much higher than expected molar masses.

As described by Wenz and co-workers, glycosylated lysine-NCA can also be obtained by glycosylation of 2-(2-isothiocyanatoethoxy)ethanol with 1-trichloroacetimidate-2,3,4,6-tetra-*O*-acetylglycopyranose, coupling of the isothiocyanate group to the ε-amino group of lysine, and treatment with dichloromethyl(methoxy)methane.[64] The glyco-NCA was then copolymerized with N^{ε}-substituted lysine-NCAs using triethylamine or (bipy)Ni(COD) (*cf.* Schemes 3.3 and 3.6).

Glyco-NCAs were also prepared by Kramer and Deming from L-cysteine by photochemical addition of the amino acid thiol to 1-allyl-2,3,4,6-tetra-*O*-acetyl-α-D-glucose or -galactose, employing 2,2-dimethoxy-2-phenylacetophenone as a photoinitiator and UV irradiation at 365 nm, and subsequent

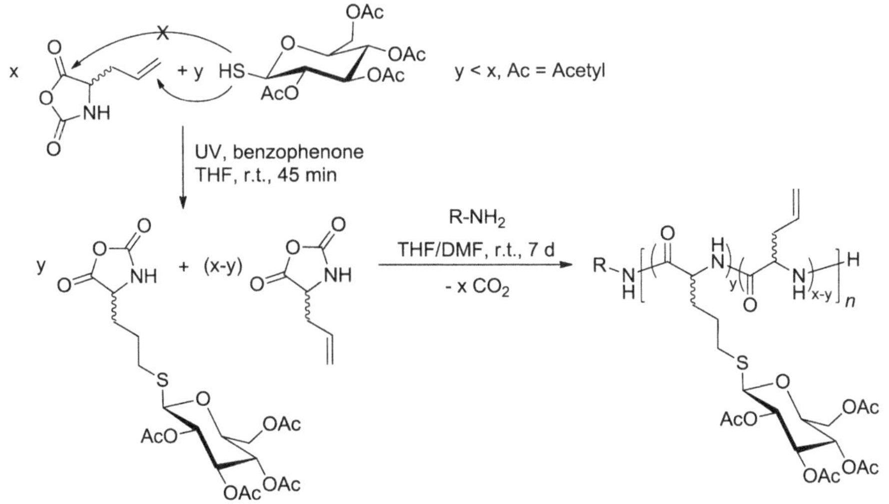

Scheme 3.9 *In situ* glycosylation and polymerization of allylglycine NCA according to Schlaad and co-workers.

reaction with dichloro(methoxy)methane.[65] Purification was effected by column chromatography or by aqueous workup and repeated precipitation. The glyco-NCAs could be polymerized with $(PMe_3)_4Co$ in THF solution at room temperature to yield protected glycopolypeptides with molar masses up to 90 kg mol^{-1} and dispersities <1.1.

Schlaad and co-workers demonstrated that allylglycine NCA can be (partially) glycosylated by photochemical addition of 1-thio-β-D-glucopyranose-2,3,4,6-tetraacetate (using benzophenone as a photoinitiator and UV light from an energy-saving bulb) and *in situ* polymerized using 1-hexylamine as the initiator at room temperature (Scheme 3.9).[66] This one-pot modular strategy permits the preparation of glycopolypeptides with molar masses up to 20 kg mol^{-1} (dispersity ~1.2) without the need to apply tedious synthesis and purification protocols.

Cameron and co-workers described the syntheses of *O*- and *S*-linked glyco-NCAs based on L-threonine, L-serine and L-cysteine.[67] The acetobromo derivatives of glucose, galactose or lactose were coupled to the respective amino acids in the presence of iodine as Lewis acid promoter and subsequently reacted to give the NCA using triphosgene and α-pinene in ethyl acetate. However, polymerizations of these NCAs have not yet been reported.

3.3.1.2 Post-Polymerization Glycosylation of Synthetic Polypeptides

Tang and Zhang synthesized a γ-chloropropyl-L-glutamate-NCA by monoesterification of glutamic acid with 3-chloropropyl alcohol in the presence of chlorotrimethylsilane and subsequent treatment with triphosgene, and polymerized it using an HDMS initiator (*cf.* Scheme 3.5).[68,69] The polypeptides obtained were reacted with sodium azide to give poly(γ-azidopropyl-L-glutamate) with M_n = 5–28 kg mol^{-1} (dispersity ~1.2), on to which 1-propargyl-α-D-mannose could be added quantitatively using copper-catalyzed azide–alkyne [3 + 2]-cycloaddition (CuAAC). Alternatively, poly(γ-propargyl-L-glutamate) was prepared by primary amine-initiated ring-opening polymerization of the corresponding amino acid-NCA and then reacted with sugar-azides applying CuAAC (Scheme 3.10).[70]

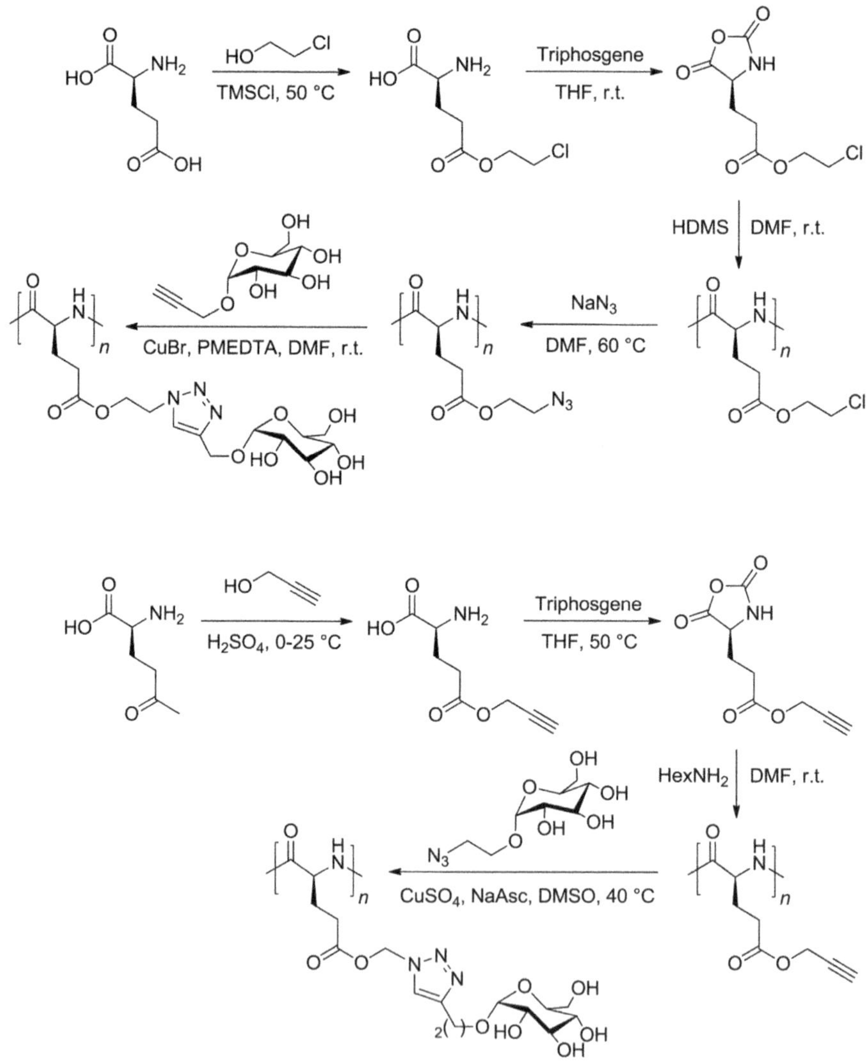

Scheme 3.10 Synthesis of glycopolypeptides using glutamate-based NCAs and CuAAC click chemistry.

The CuAAC approach was also applied by Heise and co-workers to polypropargylglycine, avoiding any hydrolytically unstable ester linkage in the polypeptide side chains.[71–73] Polypropargylglycine homo- and copolymers were synthesized by primary amine-initiated ring-opening polymerization of propargylglycine-NCA (readily obtained by the reaction of the commercial amino acid with triphosgene) and were glycosylated with 1-azido-β-D-galactose in the presence of triethylamine and catalytic amounts of $(PPh_3)_3CuBr$ in DMSO solution. Oligosaccharide chains, *i.e.* dextran or hyaluronan, were also grafted onto poly(γ-benzyl-L-glutamate)-*block*-polypropargylglycine to yield 'tree-like' copolypeptides.[73]

Polyallylglycines (and related polydepsipeptides) have been introduced as another clickable polypeptide by thiol–ene chemistry, connecting functional groups to the polypeptide chains with a thioether linker.[74,75] Polyallylglycines are poorly soluble in common solvents, making the synthesis of higher molar mass polypeptides and quantitative modification difficult or even impossible. Polyallylglycine with a low molar mass of M_n < 2 kg mol^{-1} could only be modified quantitatively with 1-thio-β-D-glucopyranose when the reaction was run in trifluoroacetic acid over 2 days. However, trifluoroacetic acid is a hazardous solvent and may not be applicable to sensitive biological substrates. Low molar mass poly(*N*-allylglycine) with M_n < 2 kg mol^{-1}, a structural isomer of polyallylglycine, however, is soluble in water and could be reacted quantitatively with 1-thio-β-D-glucopyranose within less than 1 day.[76]

Statistical copolymers of allyl- or propargylglycine and γ-benzyl-L-glutamate, exhibiting much better solubility, have been synthesized and glycosylated in organic media or, after debenzylation of the glutamate units, in aqueous acetate buffer (Scheme 3.11).[77,78] The single addition of the thiosugar to allylglycine units proceeded smoothly, but double addition of sugar to propargylglycine was not quantitative (<1.5 equiv.). Importantly, the modification reaction occurs in water using unprotected sugars and may well be applicable to other thiol-bearing biological materials.

Cationic glycopolypeptides have been synthesized by the quantitative alkylation of poly(L-methionine)s with alkyl halides or triflates (Scheme 3.12).[79] Alkylation of methionine with 2-(2,3,4,6-tetra-*O*-acetyl-α-D-galactopyranosyl)ethyl triflate was conducted in a solvent mixture of dichloromethane–acetonitrile at 20 °C and produced the corresponding *S*-linked glycopolypeptide in 97% yield. Alkylation is reversible and dealkylation of methionine units can be induced when desired.[80]

N-Linked glycopolypeptides were obtained by condensation of β-glycosylamines to poly(L-glutamic acid)[81] or by aqueous amide coupling of D-glucosamine hydrochloride to poly(L-glutamic acid) (M_n = 9–180 kg mol^{-1}) using 4-(4,6-dimethoxy-1,3,5-triazin-2-yl)-4-methylmorpholinium chloride (DMT-MM) as coupling agent (Scheme 3.13).[82] The latter reaction was performed without any organic solvents, additives or buffers, and the degree of substitution could be controlled by the molar ratio of DMT-MM to glutamate to about 80%. A similar approach has been used for the glycosylation of a protein sequence containing glutamic acid residues with β-D-galactosamine

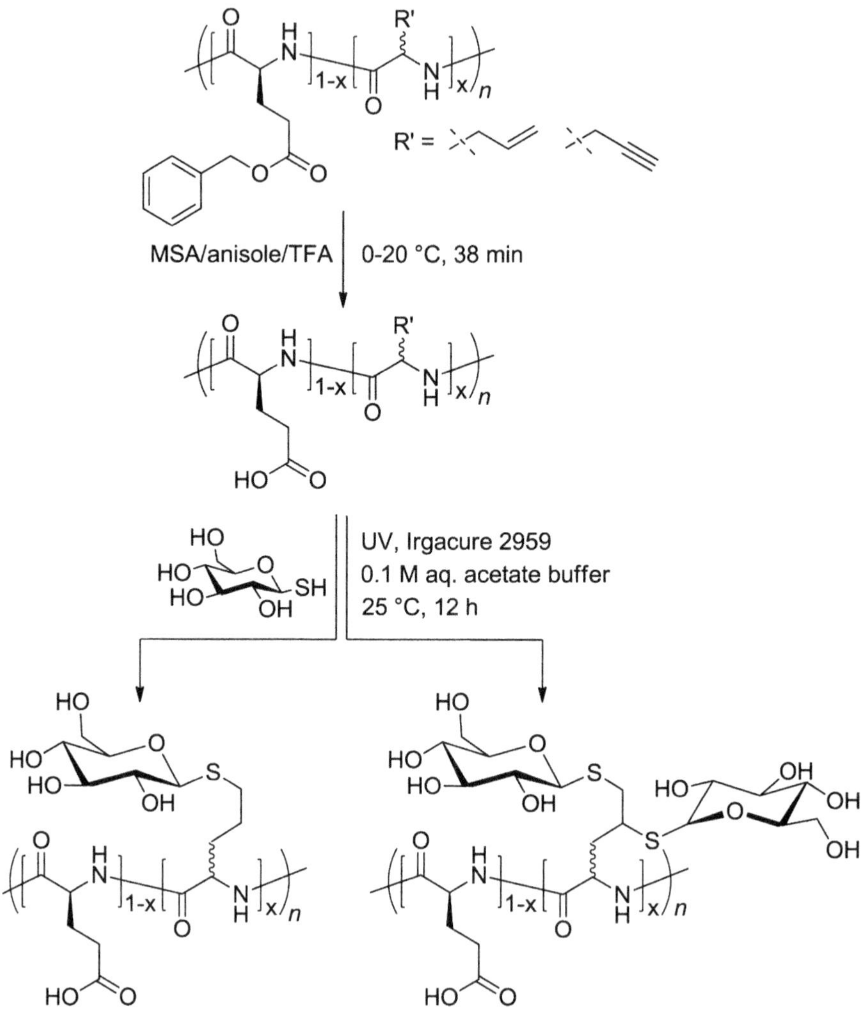

Scheme 3.11 Glycosylation of copolymers of allyl- or propargylglycine and (γ-benzyl-) L-glutamate by thiol–ene/yne photochemistry.

using *N*,*N*,*N*′,*N*′-tetramethyl-*O*-(1*H*-benzotriazol-1-yl)uranium hexafluorophosphate (HBTU) as coupling agent under basic conditions.[83] Alternatively, poly(L-lysine) was glycosylated with D-gluconolactone in the presence of diisopropylethylamine in DMSO–water at room temperature.[84]

3.3.2 Glycopolypeptide–Lectin Interactions

Recognition is the first step in numerous biological processes based on cell–cell interactions. In many cases, these recognition events are based on specific carbohydrate–protein (lectin) interactions that occur on the surface of

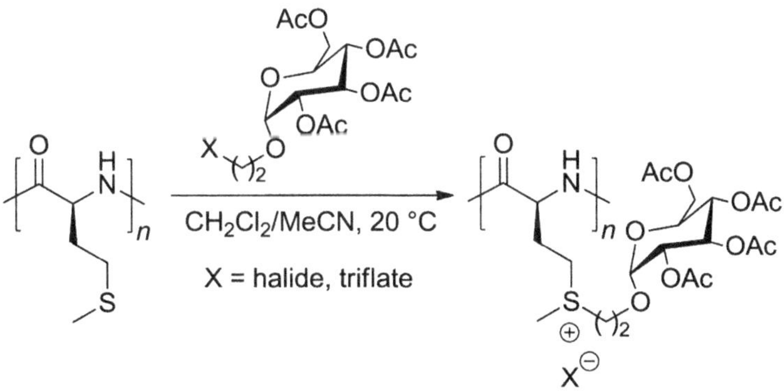

Scheme 3.12 Glycosylation of poly(L-methionine) to yield cationic glycopolypeptides.

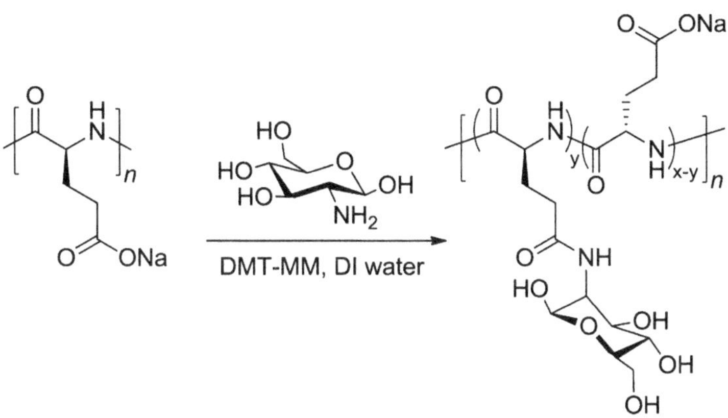

Scheme 3.13 Glycosylation of poly(L-glutamic acid) with β-glycosylamines.

cells.[85] Lectins bind specifically but weakly to carbohydrates, and the binding can be enhanced by combining several carbohydrates in the same molecule (oligosaccharide or glycopolymer) or aggregate, allowing for multiple binding events at the same time (cluster glycoside effect).[1,86] However, the mechanisms and influence of ligand structure on the multivalent binding interactions and the glyco code are largely still unknown.

Several studies have been reported that demonstrate that the synthetic glycopolypeptides can indeed recognize and specifically interact with lectins, usually using turbidity assays,[59,61,62,70,72,77,81] surface plasmon resonance (SPR)[87] and isothermal titration calorimetry (ITC).[61] The lectins used were usually concanavalin A (ConA) for the selective binding of glucose and mannose and *Ricinus communis* agglutinin (RCA_{120}) for galactose. Interestingly, it has been suggested that the epitope density (degree of glycosylation)[13] affects the kinetics of lectin binding[70] and that the secondary structure of the glycopolypeptide has no impact on the binding affinity.[61]

Wenz and co-workers showed that galactosylated fluorescent poly(L-lysine)s could be specifically incorporated in human T lymphocytes at 37 °C, and are therefore potentially useful for selective staining of cells for targeted drug delivery.[64]

3.4 Conclusion

The state-of-the-art in the synthesis of polypeptide materials by ring-opening polymerization of amino acid *N*-carboxyanhydrides (NCAs) has been described, including monomer synthesis and purification and the main polymerization mechanisms and protocols. The developed techniques have been applied and extended to the synthesis of glycosylated NCAs and polypeptides. Two pathways for the controlled synthesis of glycopolypeptides have been used: (1) synthesis of highly pure glycosylated NCAs and subsequent controlled polymerization initiated by primary amines or by transition metal catalysis and (2) controlled synthesis of a suitable polypeptide precursor followed by post-polymerization glycosylation, preferably using click chemistry (CuAAC or thiol–ene/yne photochemistry).

The glycosylated polypeptides are designed for use in advanced biomedical and life science applications, especially for molecular recognition of biological species. First studies suggest that these synthetic glycopolypeptides are indeed suitable for this purpose as they are able specifically to recognize proteins/lectins. However, research in this direction is still in its infancy – but the prospects are very promising.

References

1. V. Ladmiral, E. Melia and D. M. Haddleton, *Eur. Polym. J.*, 2004, **40**, 431–449.
2. S. R. S. Ting, G. Chen and M. H. Stenzel, *Polym. Chem.*, 2010, **1**, 1392–1412.
3. A. M. Eissa and N. R. Cameron, *Adv. Polym. Sci.*, 2013, **253**, 71–114.
4. H. P. Panzer and R. L. Whistler, *Chem. Eng. News*, 1959, **37**, 41.
5. T. P. Bird, W. A. P. Black, E. T. Dewar and D. Rutherford, *Chem. Ind.*, 1960, 1331–1332.
6. S. Kimura and M. Imoto, *Makromol. Chem.*, 1961, **50**, 155–160.
7. K. Aoi, K. Tsutsumiuchi and M. Okada, *Macromolecules*, 1994, **27**, 875–877.
8. Y. Yeh and R. E. Feeney, *Chem. Rev.*, 1996, **96**, 601–618.
9. C. R. Bertozzi and L. L. Kiessling, *Science*, 2001, **291**, 2357–2364.
10. R. A. Dwek, *Chem. Rev.*, 1996, **96**, 683–720.
11. M. R. Pratt and C. R. Bertozzi, *Chem. Soc. Rev.*, 2005, **34**, 58–68.
12. T. Buskas, S. Ingale and G.-J. Boons, *Glycobiology*, 2006, **16**, 113R–136R.
13. C. W. Cairo, J. E. Gestwicki, M. Kanai and L. L. Kiessling, *J. Am. Chem. Soc.*, 2002, **124**, 1615–1619.
14. J. E. Gestwicki, C. W. Cairo, L. E. Strong, K. A. Oetjen and L. L. Kiessling, *J. Am. Chem. Soc.*, 2002, **124**, 14922–14933.

15. J. Huang and A. Heise, *Chem. Soc. Rev.*, 2013, **42**, 7373–7390.
16. C. Bonduelle and S. Lecommandoux, *Biomacromolecules*, 2013, **14**, 2973–2983.
17. J. R. Kramer and T. J. Deming, *Polym. Chem.*, 2014, **5**, 671–682.
18. M. A. Quadir, M. Martin and P. T. Hammond, *Chem. Mater.*, 2014, **26**, 461–476.
19. H. Lu, J. Wang, Z. Song, L. Yin, Y. Zhang, H. Tang, C. Tu, Y. Lin and J. Cheng, *Chem. Commun.*, 2014, **50**, 139–155.
20. K.-S. Krannig and H. Schlaad, *Soft Matter*, 2014, **10**, 4228–4235.
21. T. J. Deming, *Adv. Mater.*, 1997, **9**, 299–311.
22. T. J. Deming, *Adv. Polym. Sci.*, 2006, **202**, 1–18.
23. N. Hadjichristidis, H. Iatrou, M. Pitsikalis and G. Sakellariou, *Chem. Rev.*, 2009, **109**, 5528–5578.
24. A. C. Farthing and R. J. W. Reynolds, *Nature*, 1950, **165**, 647–647.
25. F. Fuchs, *Ber. Dtsch. Chem. Ges.*, 1922, **55**, 2943–2943.
26. H. Leuchs, *Ber. Dtsch. Chem. Ges.*, 1906, **39**, 857–861.
27. K. Koga, A. Sudo, H. Nishida and T. Endo, *J. Polym. Sci. Part A: Polym. Chem.*, 2009, **47**, 3839–3844.
28. K. Poduska and H. Gross, *Chem. Ber.*, 1961, **94**, 527–537.
29. W. H. Daly and D. S. Poché, *Tetrahedron Lett.*, 1988, **29**, 5859–5862.
30. D. S. Poché, M. J. Moore and J. L. Bowles, *Synth. Commun.*, 1999, **29**, 843–854.
31. J. R. Kramer and T. J. Deming, *Biomacromolecules*, 2010, **11**, 3668–3672.
32. H. R. Kricheldorf, *α-Aminoacid-N-Carboxyanhydrides and Related Heterocycles*, Springer, Berlin, 1987.
33. D. G. H. Ballard and C. H. Bamford, *Proc. R. Soc. Lond. A*, 1954, **223**, 495–520.
34. D. G. H. Ballard, C. H. Bamford and F. J. Weymouth, *Proc. R. Soc. Lond. A*, 1955, **227**, 155–183.
35. D. Thunig, J. Semen and H.-G. Elias, *Makromol. Chem.*, 1977, **178**, 603–607.
36. M. Goodman and J. Hutchison, *J. Am. Chem. Soc.*, 1965, **87**, 3524–3525.
37. T. Aliferis, H. Iatrou and N. Hadjichristidis, *Biomacromolecules*, 2004, **5**, 1653–1656.
38. G. J. M. Habraken, K. Wilsens, C. E. Koning and A. Heise, *Polym. Chem.*, 2011, **2**, 1322–1330.
39. W. Vayaboury, O. Giani, H. Cottet, A. Deratani and F. Schué, *Macromol. Rapid Commun.*, 2004, **25**, 1221–1224.
40. H. R. Kricheldorf, *Angew. Chem., Int. Ed.*, 2006, **45**, 5752–5784.
41. H. R. Kricheldorf, C. Von Lossow and G. Schwarz, *J. Polym. Sci. Part A: Polym. Chem.*, 2006, **44**, 4680–4695.
42. D. G. H. Ballard and C. H. Bamford, *J. Chem. Soc.*, 1956, 381–387.
43. M. Szwarc, *The Kinetics and Mechanism of N-Carboxy-α-Amino-Acid Anhydride (NCA) Polymerisation to Poly-Amino Acids*, Springer, Berlin, 1965.
44. T. J. Deming, *Peptide Hybrid Polymers. Advances in Polymer Science*, Vol. 202, Springer, Berlin, 2006, pp. 1–18.
45. I. Dimitrov and H. Schlaad, *Chem. Commun.*, 2003, 2944–2945.

46. M. Meyer and H. Schlaad, *Macromolecules*, 2006, **39**, 3967–3970.
47. H. Lu and J. Cheng, *J. Am. Chem. Soc.*, 2007, **129**, 14114–14115.
48. H. Lu and J. Cheng, *J. Am. Chem. Soc.*, 2008, **130**, 12562–12563.
49. T. J. Deming, *Macromolecules*, 1999, **32**, 4500–4502.
50. T. J. Deming, *Adv. Drug Delivery Rev.*, 2002, **54**, 1145–1155.
51. T. J. Deming, *J. Am. Chem. Soc.*, 1998, **120**, 4240–4241.
52. T. J. Deming, *Nature*, 1997, **390**, 386–389.
53. T. J. Deming and S. A. Curtin, *J. Am. Chem. Soc.*, 2000, **122**, 5710–5717.
54. H. C. Kolb, M. G. Finn and K. B. Sharpless, *Angew. Chem., Int. Ed.*, 2001, **40**, 2004–2021.
55. J.-F. Lutz and H. Schlaad, *Polymer*, 2008, **49**, 817–824.
56. C. Barner-Kowollik, F. E. Du Prez, P. Espeel, C. J. Hawker, T. Junkers, H. Schlaad and W. Van Camp, *Angew. Chem., Int. Ed.*, 2011, **50**, 60–62.
57. E. Rüde, O. Westphal, E. Hurwitz, S. Fuchs and M. Sela, *Immunochemistry*, 1966, **3**, 137–151.
58. J. R. Kramer and T. J. Deming, *J. Am. Chem. Soc.*, 2010, **132**, 15068–15071.
59. J. R. Kramer, A. R. Rodriguez, U.-J. Choe, D. T. Kamei and T. J. Deming, *Soft Matter*, 2013, **9**, 3389–3395.
60. D. Pati, A. Y. Shaikh, S. Hotha and S. Sen Gupta, *Polym. Chem.*, 2011, **2**, 805–811.
61. D. Pati, A. Y. Shaikh, S. Das, P. K. Nareddy, M. J. Swamy, S. Hotha and S. Sen Gupta, *Biomacromolecules*, 2012, **13**, 1287–1295.
62. S. Das, D. Pati, N. Tiwari, A. Nisal and S. Sen Gupta, *Biomacromolecules*, 2012, **13**, 3695–3702.
63. D. Pati, N. Kalva, S. Das, G. Kumaraswamy, S. Sen Gupta and A. V. Ambade, *J. Am. Chem. Soc.*, 2012, **134**, 7796–7802.
64. T. Stöhr, A. R. Blaudszun, U. Steinfeld and G. Wenz, *Polym. Chem.*, 2011, **2**, 2239–2248.
65. J. R. Kramer and T. J. Deming, *J. Am. Chem. Soc.*, 2012, **134**, 4112–4115.
66. K.-S. Krannig, A. Doriti and H. Schlaad, *Macromolecules*, 2014, **47**, 2536–2539.
67. M. I. Gibson, G. J. Hunt and N. R. Cameron, *Org. Biomol. Chem.*, 2007, **5**, 2756–2757.
68. H. Tang and D. Zhang, *Biomacromolecules*, 2010, **11**, 1585–1592.
69. H. Tang and D. Zhang, *Polym. Chem.*, 2011, **2**, 1542–1551.
70. C. Xiao, C. Zhao, P. He, Z. Tang, X. Chen and X. Jing, *Macromol. Rapid Commun.*, 2010, **31**, 991–997.
71. J. Huang, G. Habraken, F. Audouin and A. Heise, *Macromolecules*, 2010, **43**, 6050–6057.
72. J. Huang, C. Bonduelle, J. Thévenot, S. Lecommandoux and A. Heise, *J. Am. Chem. Soc.*, 2012, **134**, 119–122.
73. C. Bonduelle, J. Huang, E. Ibarboure, A. Heise and S. Lecommandoux, *Chem. Commun.*, 2012, **48**, 8353–8355.
74. J. Sun and H. Schlaad, *Macromolecules*, 2010, **43**, 4445–4448.
75. N. Franz and H.-A. Klok, *Macromol. Chem. Phys.*, 2010, **211**, 809–820.
76. J. W. Robinson and H. Schlaad, *Chem. Commun.*, 2012, **48**, 7835–7837.

77. K.-S. Krannig and H. Schlaad, *J. Am. Chem. Soc.*, 2012, **134**, 18542–18545.
78. K.-S. Krannig, J. Huang, A. Heise and H. Schlaad, *Polym. Chem.*, 2013, **4**, 3981–3986.
79. J. R. Kramer and T. J. Deming, *Biomacromolecules*, 2012, **13**, 1719–1723.
80. J. R. Kramer and T. J. Deming, *Chem. Commun.*, 2013, **49**, 5144–5146.
81. X. X. Zeng, T. Murata, H. Kawagishi, T. Usui and K. Kobayashi, *Biosci. Biotechnol. Biochem.*, 1998, **62**, 1171–1178.
82. R. Mildner and H. Menzel, *J. Polym. Sci. Part A: Polym. Chem.*, 2013, **51**, 3925–3931.
83. Y. Wang and K. L. Kiick, *J. Am. Chem. Soc.*, 2005, **127**, 16392–16393.
84. Z. Tian, M. Wang, A. Y. Zhang and Z. G. Feng, *Polymer*, 2008, **49**, 446–454.
85. M. Ambrosi, N. R. Cameron and B. G. Davis, *Org. Biomol. Chem.*, 2005, **3**, 1593–1608.
86. C. R. Becer, *Macromol. Rapid Commun.*, 2012, **33**, 742–752.
87. X. X. Zeng, T. Murata, H. Kawagishi, T. Usui and K. Kobayashi, *Carbohydr. Res.*, 1998, **312**, 209–217.

CHAPTER 4

Multivalent Glycocalixarenes: Synthesis and Molecular Recognition

ALESSANDRO DONDONI*[a] AND ALBERTO MARRA*[b]

[a]Interdisciplinary Center for the Study of Inflammation, Università di Ferrara, Via Borsari 46, 44100 Ferrara, Italy; [b]Institut des Biomolécules Max Mousseron (IBMM), UMR 5247, Université de Montpellier, Ecole Nationale Supérieure de Chimie de Montpellier, 8 Rue de l'Ecole Normale, 34296 Montpellier cedex 5, France
*E-mail: adn@unife.it, alberto.marra@univ-montp2.fr

4.1 Introduction

Specific, reversible interactions between carbohydrates at the cell surface (glycolipids and glycoproteins) and their protein receptors (lectins) play a crucial role in a multitude of biological and pathological processes, including fertilization, intercellular communication, viral and bacterial infection, inflammation, tumor cell metastasis and immune response.[1] Hence the analysis of carbohydrate–protein interactions and the development of inhibitors/probes of these interactions are at the forefront of modern glycobiology. In this context, it is important to take into account that lectins typically bind monovalent carbohydrate ligands with low affinity and poor selectivity.[2] In Nature, this low intrinsic affinity is compensated by the architecture of the lectin

RSC Polymer Chemistry Series No. 15
Glycopolymer Code: Synthesis of Glycopolymers and Their Applications
Edited by C. Remzi Becer and Laura Hartmann

Published by the Royal Society of Chemistry, www.rsc.org

itself, bearing clustered binding sites, and by the presence of multiple copies of the lectin on the same surface (*e.g.* the viral envelope).[3] Moreover, the carbohydrate ligands are also present as clusters on the bacterial or cell membrane. The resulting multivalent complexes have more affinity, called avidity, and selectivity than the combination of individual binding events, that is, the whole interaction is greater than the sum of its parts. This effect is known as the glycoside cluster effect.[4] Therefore, multivalency has been a leading feature in the design of inhibitors or probes of lectins.[5] Consequently, researchers have been investigating synthetic oligosaccharides and their conjugation to a variety of multivalent scaffolds, including cyclopeptides,[6] glycoproteins,[7] dendrimers and dendrons,[8] polymers,[9] polymeric and gold nanoparticles,[10] fullerenes,[11] calixarenes,[12] cyclodextrins,[12a,13] DNA[14] and silsesquioxanes.[15] Each multivalent platform displays glycans differently, with varying spacing, orientation, density, flexibility and overall architecture. Along with the broad window of possible structural representations come the challenges of identifying the scaffold that can provide the optimal arrangement for a particular glycan–lectin interaction, particularly when the spacing and geometries of the lectin binding sites are not known. Therefore, the construction of glycoclusters with a known number of glycans that are confined to the surface of new scaffolds and the evaluation of their affinity for specific lectins appear to be an empirical yet viable approach towards the discovery of new inhibitors.

Based on these premises, we report in this chapter on the synthesis of multivalent glycocalixarenes and their binding affinities to specific lectins. The choice of the calixarene as scaffold on which install a precise number of carbohydrate ligands relies on several factors, including their chemical stability and defined conformation and also the selective functionalization of either the lower or upper rim or both (Figure 4.1).[16] Although different families of calix[*n*]arenes are known depending on the number of phenol residues, calix[4]arenes, *i.e.* compounds constituted of four phenol residues connected through methylene bridges, have been employed as the platform for sugar clustering because their formation can be easily carried out by adopting suitable phenol–formaldehyde condensation conditions and their conformational mobility can be controlled by placing appropriate substituents in the phenyl rings. Thus, a fixed cone conformation of calix[4]

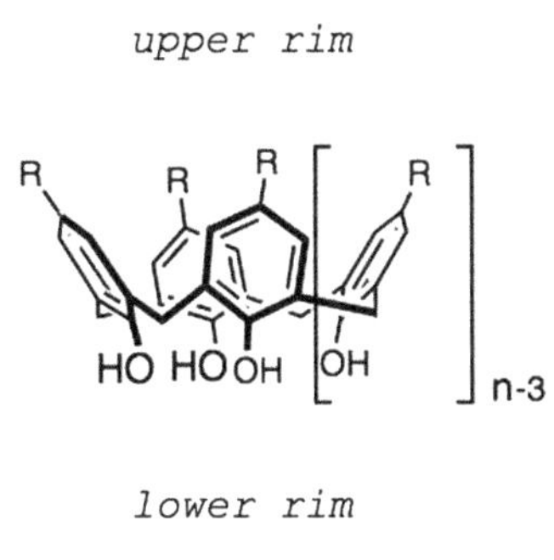

Figure 4.1 Calix[*n*]arenes in cone conformation.

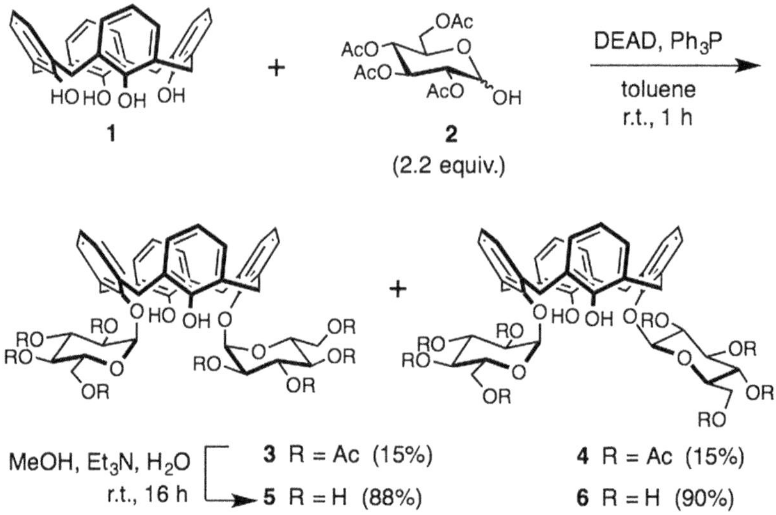

Scheme 4.1

arenes appeared to be the most suitable for the preparation of clusters with a well-defined arrangement around the macrocycle.

4.2 Calixsugar Assembly *via* Mitsunobu Reaction

The first synthesis of glycocalixarenes was pioneered in the mid-1990s by the groups led by Dondoni and Ungaro.[17] The method employed for the glycosylation of calix[4]arene **1** at the lower rim consisted of the Mitsunobu reaction with tetra-*O*-acetyl-α, β-D-glucopyranose **2** (2.2 equiv.) (Scheme 4.1). The reaction afforded a mixture of α, α-bis-glucoside **3** and α, β-bis-glucoside **4** in a *ca.* 1 : 1 ratio. Both products were isolated in pure form by HPLC in modest yield (15% each). The ^{1}H and ^{13}C NMR spectra clearly indicated the assigned configuration of the anomeric carbons and the cone conformation of the macrocycle.[18] It was also shown that upon deacetylation with sodium methoxide, both compounds could be transformed into the corresponding unprotected calix[4]arene glucosides **5** and **6**. Unfortunately, these compounds showed minimal water solubility, preventing their use in molecular recognition processes.

In a subsequent and more extensive report, Dondoni *et al.* described improved conditions targeting calix[4]arene *O*-glycosides whose synthesis was inefficient or failed in earlier work.[19] Thus, mono- and bis-glycosylation at the lower rim of calix[4]arene **1** was successfully performed *via* Mitsunobu reaction with the configurationally stable α-D-mannofuranose diacetonide **7** to give the calix[4]arene mannosides **8** and **9** in good yields (Scheme 4.2). The mannosyl residues of these compounds showed a β-D-linkage, in agreement with the well-known inversion of configuration of secondary alcohols *via* Mitsunobu reaction. Interestingly, both calixsugars **8** and **9** appeared to be fixed in the cone conformation, as demonstrated by NMR analysis. It was

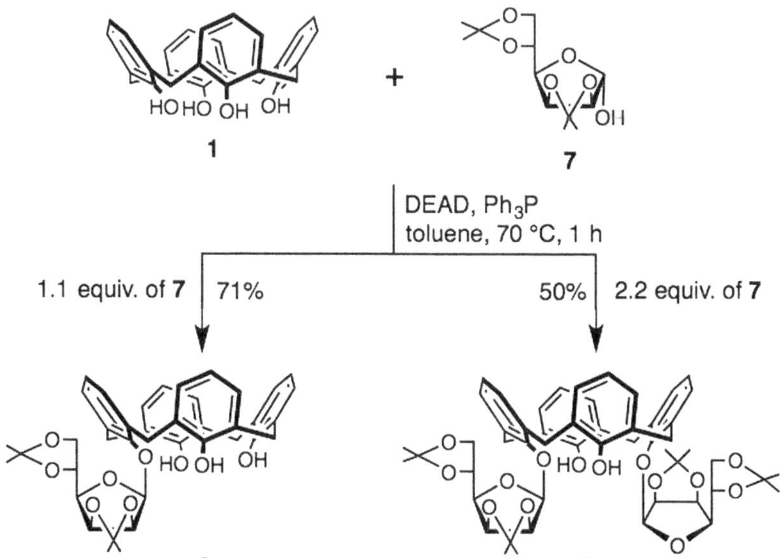

Scheme 4.2

demonstrated that this structural arrangement was maintained over a wide range of temperature (from −80 to 160 °C). Attempted deacetonization of these glycosides was unsuccessful and therefore unprotected compounds could not be prepared.

4.3 Calixsugar Assembly *via* *O*-Glycosylation

Dondoni and Ungaro and their co-workers reported[17] on the efficient synthesis of the upper rim tetra-*O*-galactosylcalix[4]arene **12** (Scheme 4.3) *via* glycosylation of the tetrahydroxymethyl-substituted calix[4]arene **10** that was rigidified in the cone conformation by the four *O*-propyl groups at the lower rim.[20] Ethyl tetra-*O*-benzoyl-1-thio-β-D-galactoside **11** was used as a suitable glycosyl donor because the presence of participating benzoyl group at C-2 ensured a totally β-D-stereoselective glycosylation reaction. Copper(II) triflate [Cu(OTf)$_2$]-promoted coupling between **11** (6 equiv.) and **10** occurred readily in acetonitrile at room temperature. Then, removal of the benzoyl protective groups by treatment with sodium methoxide afforded the expected glycocalix[4]arene **12** in 60% yield. Compound **12** represented the first tetravalent glycocluster anchored to a structurally well-defined calixarene platform. Interestingly, this calixarene turned out to be water soluble up to 5 mM and therefore was employed for binding studies with various organic compounds. Unfortunately, no tests were carried out on the recognition of galectins by this compound.[21]

The same group investigated the reactions of calix[4]arene **10** with other thioglycosides. Side products were isolated and characterized as

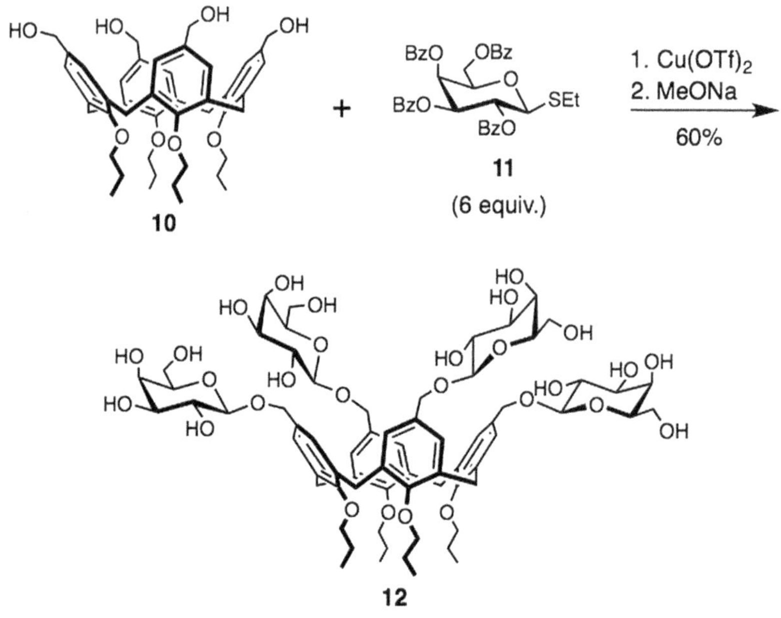

Scheme 4.3

ether-bridged calix[4]arenes derived from acid-catalyzed intramolecular coupling of two distal hydroxymethyl groups.[19]

The effect of the alkyl chain length on the coupling of calix[4]arene polyols with Lewis acid-activated glycosyl donors was investigated some years later by the same group.[22] The aim of this study was to overcome the problem of ether-bridged side-product formation and provide new structures to probe the glycoside cluster effect. Thus, the difficulty of achieving extensive glycosylation of calix[4]arene tetrol **10** bearing short hydroxymethyl arms (see Scheme 4.3) by using *O*-benzoyl-protected thioethyl gluco- and mannopyranosides was first confirmed. On the other hand, calix[4]arene tetrol **13** bearing long spacer arms (hydroxypropyl groups) turned out to be effectively glycosylated by *galacto-*, *gluco-* and *manno*-thioglycosides to give the corresponding *O*-glycosides in very good isolated yields (70–80%) and without the formation of ether-bridged side products (Scheme 4.4).

A special kind of calix[4]arene *O*-glycoside featuring ketosyl residues linked to the macrocycle platform was prepared by Dondoni and co-workers.[23] This work stemmed from the ready access to thiazolylketoses *via* thiazole-based chemistry[24] and the high reactivity of their anomeric acetates as glycosyl donors.[25] Thus, it was demonstrated that excess of thiazolyl galactoketose acetate **16** under Lewis acid catalysis (trimethylsilyl triflate, TMSOTF) gave rise to highly stereoselective glycosylation of the model hydroxymethyl-substituted calix[4]arenes **15** and **10** to give exclusively the α-D-linked di- and tetra-*O*-galactosides **17** and **18**, respectively (Scheme 4.5). The low yield (11%) of isolated tetravalent calixsugar **18** was very likely due to the increasing congestion of

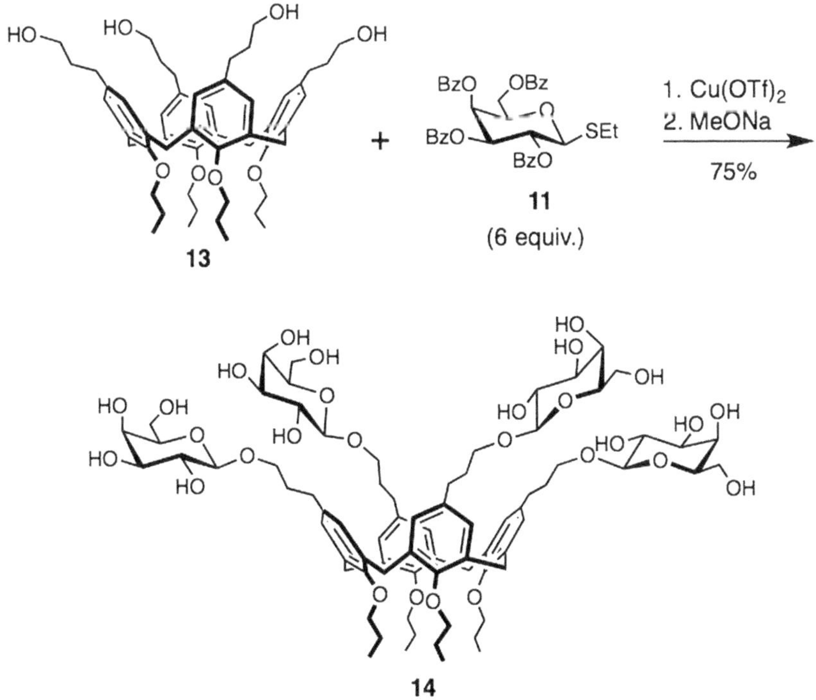

Scheme 4.4

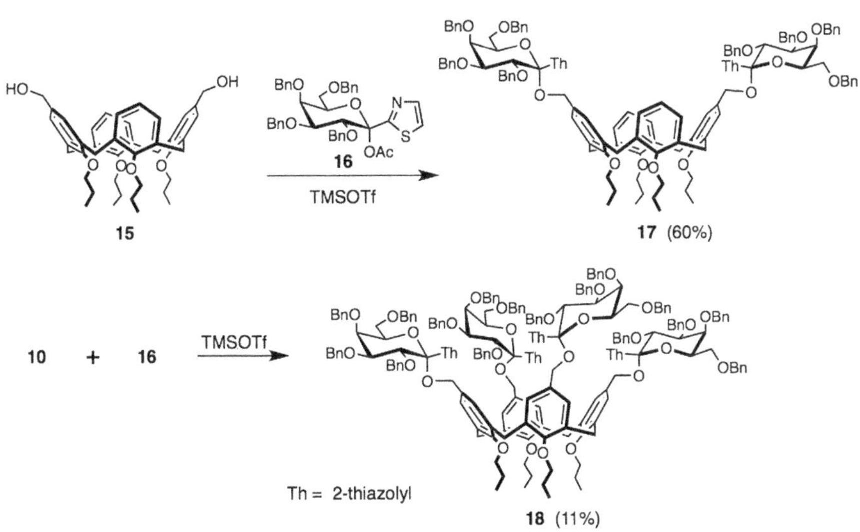

Scheme 4.5

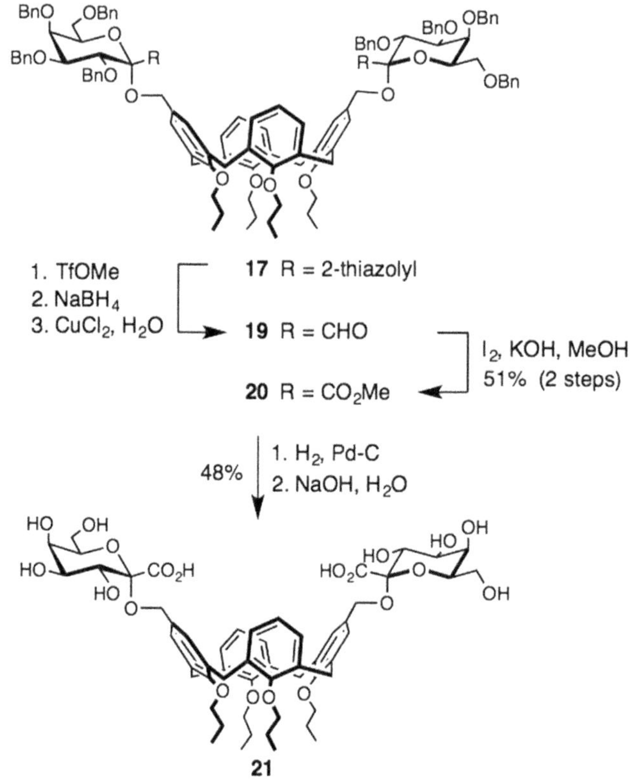

Scheme 4.6

the system by the sequential insertion of the bulky ketosyl fragments. Accordingly, di- and triadducts were also formed. Nevertheless, the multiple glycosylation of diol and tetrol calix[4]arenes **15** and **10** by **16** confirmed the high and unexpected reactivity of thiazolylketose acetates as glycosyl donors.[25] Moreover, the exclusive formation of the α-glycosidic linkage was noteworthy. This stereochemical outcome was interpreted as due to a chair-like transition state derived from an axial attack of the hydroxyl group of the acceptor to the less hindered face of the sugar oxycarbenium ion intermediate.

The facile transformation of the thiazole ring into formyl group[26] was conveniently exploited for the conversion of **17** into the dialdehyde **19** that in turn was oxidized to the diester **20** (Scheme 4.6). Debenzylation of the sugar fragments and saponification of the ester group afforded the totally unprotected glycoconjugate **21**, a calix[4]arene bis-ulosonide.

4.4 Calixsugar Assembly *via S*-Alkylation

The first example of calix[4]arene *S*-glycoside synthesis was reported as early as 1996 by Meunier and Roy in a brief communication.[27] The *p*-*tert*-butylcalix[4]arene tetraamine **22** was converted into the tetra-*N*-chloroacetylated

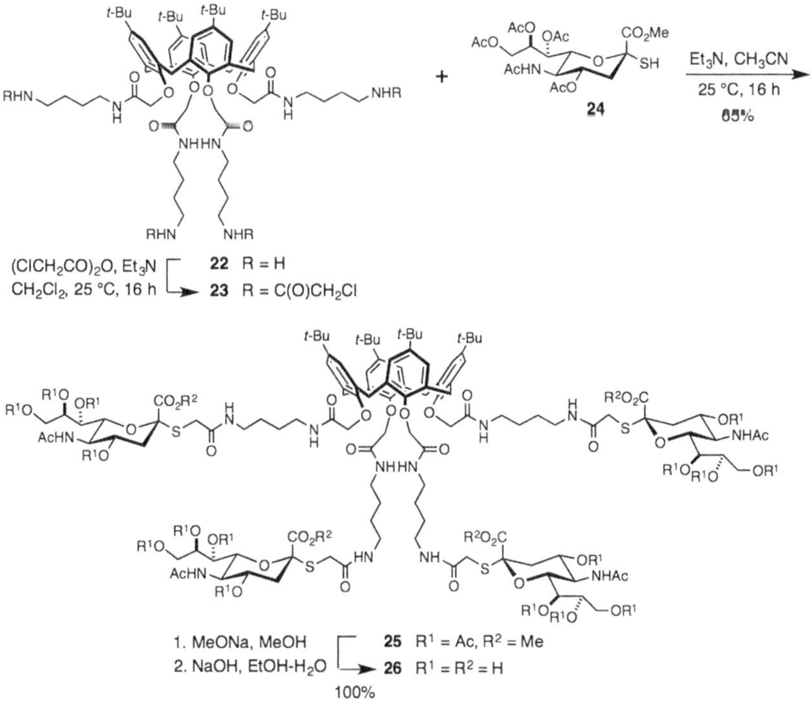

Scheme 4.7

calix[4]arene **23** and this was treated with an excess of sialyl thiol **24** in the presence of triethylamine to give the tetrasialylated calix[4]arene **25** in 65% yield (Scheme 4.7). This compound upon basic hydrolysis gave the water-soluble (~1 mM) deprotected calix[4]arene *S*-glycoside **26**, suitable for lectin binding studies. Standard turbidimetric analysis revealed the formation of crosslinking aggregates between **26** and wheat germ agglutinin (WGA) that was known to bind sialosides. The lattice thus formed was disrupted by addition of a monovalent ligand, *i.e.* phenyl α-D-*S*-sialoside. As the methyl ester of **26**, also water-soluble, formed stable aggregates with WGA, it was concluded that simple electrostatic binding interactions were absent.

4.5 Calixsugar Assembly *via* Amide Bond Formation

Roy and Kim identified *p*-*tert*-butylcalix[4]arene glycoconjugates as amphiphilic systems that assembled on a hydrophobic surface through the lipophilic *tert*-butyl groups could mimic the cell saccharide surface through the hydrophilic carbohydrate ligands.[28] Having chosen the α-linked *N*-acetyl-D-galactosamine (α-GalNAc) residue as ligand, they prepared the lower rim tetravalent calix[4]arene-based glycocluster **29** (Scheme 4.8). The synthesis involved amidation of the calix[4]arene-based tetraacyl chloride **27** by the free amine derived from the peracetylated *O*-glycoside **28** followed by de-*O*-acetylation.

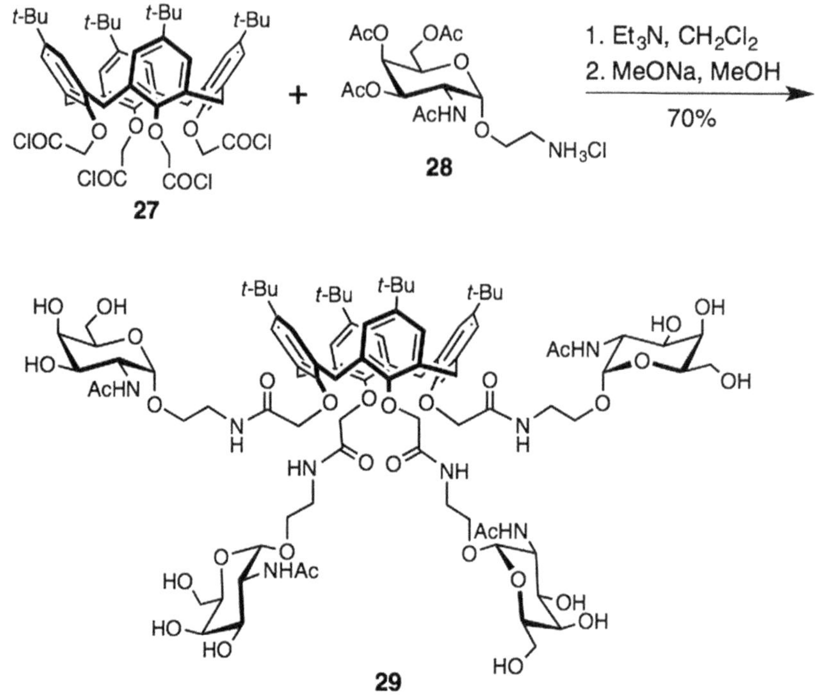

Scheme 4.8

Roy and Kim also prepared the octavalent calix[4]arene-based glycocluster **31** featuring long spacer arms between the macrocyle and the GalNAc residues. In this case, the calixarene-based tetraamine **22** was allowed to react with 10 equiv. of the GalNAc-bearing bromide **30** in the presence of *N*,*N*-diisopropylethylamine (DIPEA) and the reaction product was deacetylated to give the target *O*-glycoside **31** in very good yield (Scheme 4.9). This synthetic approach was expanded in a remarkable manner by reacting 20 equiv. of **30** with a calix[4]arene-based octameric amine to give the hexadecameric glycocalix[4]arene **32**, a true dendrimer (Figure 4.2).

Interestingly, compounds **29**, **31** and **32** were water soluble and therefore their binding properties against the plant lectin *Vicia villosa* agglutinin (VVA) were evaluated.[28] Turbidimetric analysis revealed that all three glycoclusters exhibited direct binding abilities to VVA while the use of allyl 2-acetamido-2-deoxy-α-D-galactopyranoside (allyl α-D-GalNAc) as an inhibitor for the crosslinking of octavalent **31** with VVA demonstrated the sugar specificity of the binding. The interaction between **31** and VVA was so strong that a 250-fold molar excess of the above inhibitor was necessary to disrupt the crosslinking interaction. The efficiency of the glycoclusters to inhibit the binding of asialoglycophorin to VVA was also measured by enzyme-linked lectin assay. The most efficient inhibitor was the hexadecavalent conjugate **32** (IC_{50} = 13.4 μM), which in fact was 12-fold more potent than allyl α-D-GalNAc (IC_{50} = 158.3 μM). It was also demonstrated,

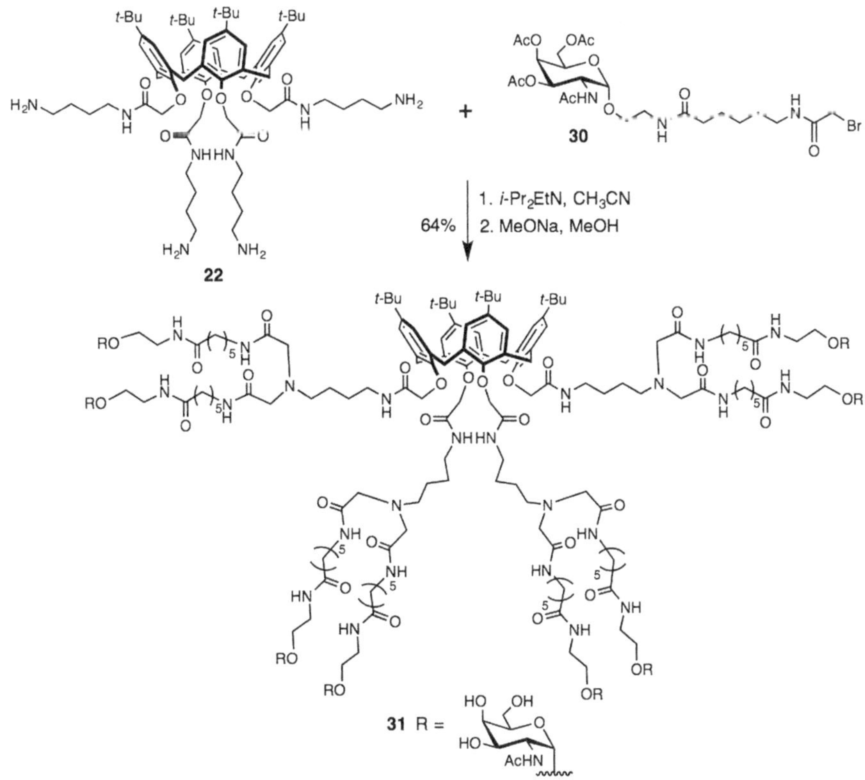

Scheme 4.9

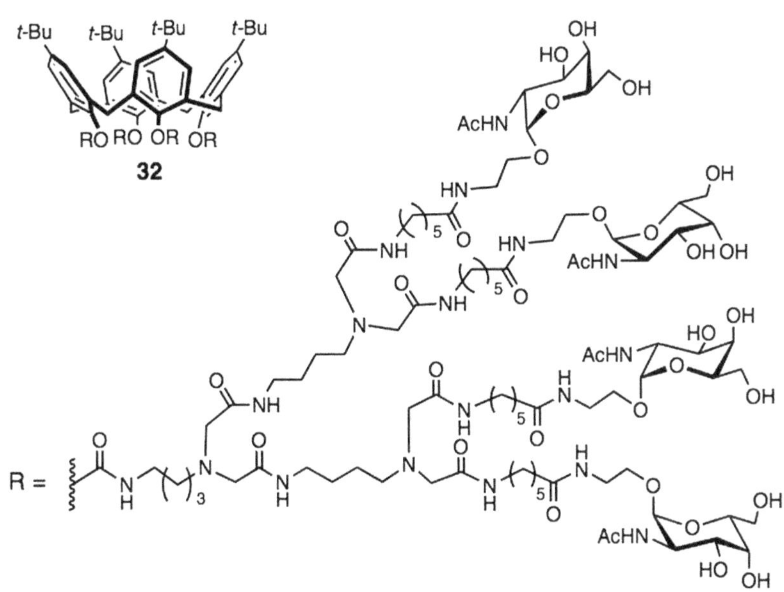

Figure 4.2 Hexadecaglycosylcalix[4]arene prepared by Roy and Kim.[28]

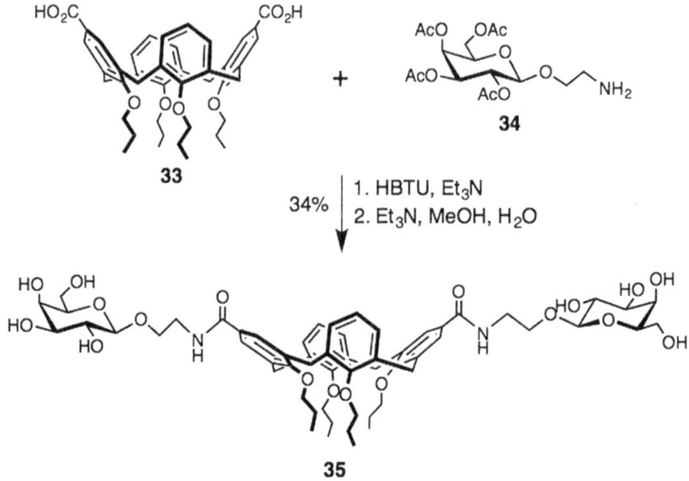

Scheme 4.10

as probably the most important finding, that the GalNAc-containing glycoclusters **31** and **32** have the ability to form monolayer assemblies on polystyrene microtiter plates. This hydrophobic interaction was inhibited by an excess of allyl α-D-GalNAc.

Calix[4]arene *O*-glycosides assembled through amide bonds have been the target of synthetic efforts by Ungaro's group in the mid-2000s.[29] The contribution of the amide tether to molecular recognition processes by binding to either acidic or basic substrates was emphasized. Initial attempts to introduce carbohydrate fragments by coupling the tetrapropoxycalix[4]arene dicarboxylic acid **33** (Scheme 4.10) with 2,3,4,6-tetra-*O*-acetylgalactosamine or -glucosamine in the presence of a benzotriazole-based activating agent [*O*-(benzotriazol-1-yl)-*N*,*N*,*N*′,*N*′-tetramethyluronium hexafluorophosphate, HBTU] and a base (Et_3N) were unsuccessful. Also, the use of the dichloride of the diacid **33** and a base did not produce any desired glycoconjugate. As the low reactivity of the glycosylamines appeared to be responsible for the unsuccessful coupling reactions, the use of *O*-ethylaminoglycosides was considered. In the event, successful coupling took place between the known galactose-derived amine **34** and the diacid **33** in the presence of HBTU and Et_3N to give, after deacetylation of the carbohydrate residues, the target upper rim calix[4]arene *O*-glycoside **35**, adopting an open flattened cone conformation, in 34% overall yield (Scheme 4.10). Comparable results were obtained using a D-*gluco* analogue of **34**.

The cholera toxin, a protein featuring five sugar-binding sites on a single face, selectively binds to the cell-surface ganglioside GM1 (**36**) (Figure 4.3) through specific interactions with the terminal galactose and sialic acid residues.[30] In earlier reports, Bernardi and co-workers described the synthesis of the pseudotrisaccharide **37** (Figure 4.3), in both monovalent[31] and multivalent form,[32] as a mimic of the oligosaccharidic portion of **36**.

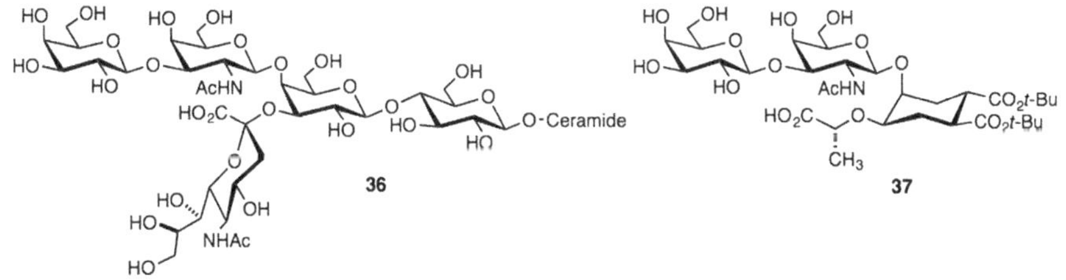

Figure 4.3 Ganglioside GM1 (**36**) and one of its mimics, the pseudotrisaccharide **37**.

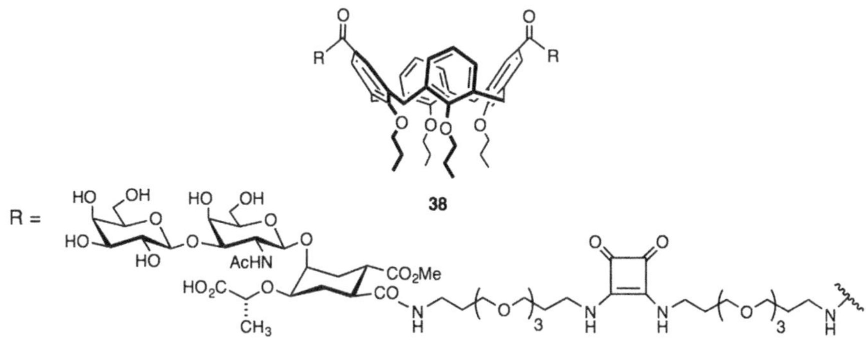

Figure 4.4 Calixsugar bearing two ganglioside GM1 mimic units.

Aiming to mimic the external surface of the cell where the GM1 pentasaccharide units protrude from the lipophilic cell membrane, Bernardi and co-workers prepared the bis-glycosylated calix[4]arene **38** (Figure 4.4).[33] The interaction of the latter compound with the cholera toxin was studied using fluorescence titration and ELISA experiments. Unfortunately, the adsorption of **38** at the GM1-coated ELISA plates that occurred at high concentrations (>0.2 mM) of ligand prevented an extensive study of its inhibition properties. On the other hand, the fluorescence titration data indicated that the affinity of **38** for the cholera toxin was 2000-fold (per pseudotrisaccharide unit) higher than the affinity observed for the monovalent mimic **37**. This result was surprising since a dendrimer-based divalent analogue of **37** showed only a ninefold (per ligand unit) affinity enhancement.[32] Therefore, it was concluded that a more detailed thermodynamic analysis based on other multivalent ligands was required to explain these controversial results.

The natural Tn antigen (2-acetamido-α-D-galactosyl-L-serine or -L-threonine) is a well-known carbohydrate antigen, present as di- to pentavalent clusters, overexposed on the surface of the tumor cells. To enhance their immunogenicity, the antigen epitopes have been conjugated to carrier proteins or macrolipids.[34] Moreover, in order to mimic the mucine surface, multivalent vaccine candidates have been prepared.[35] In 2008, Geraci *et al.*[36] described the synthesis and immunological properties of calixsugar **41** (Scheme 4.11). This compound featured four 2-acetamido-α-D-galactosyl-L-cysteine units

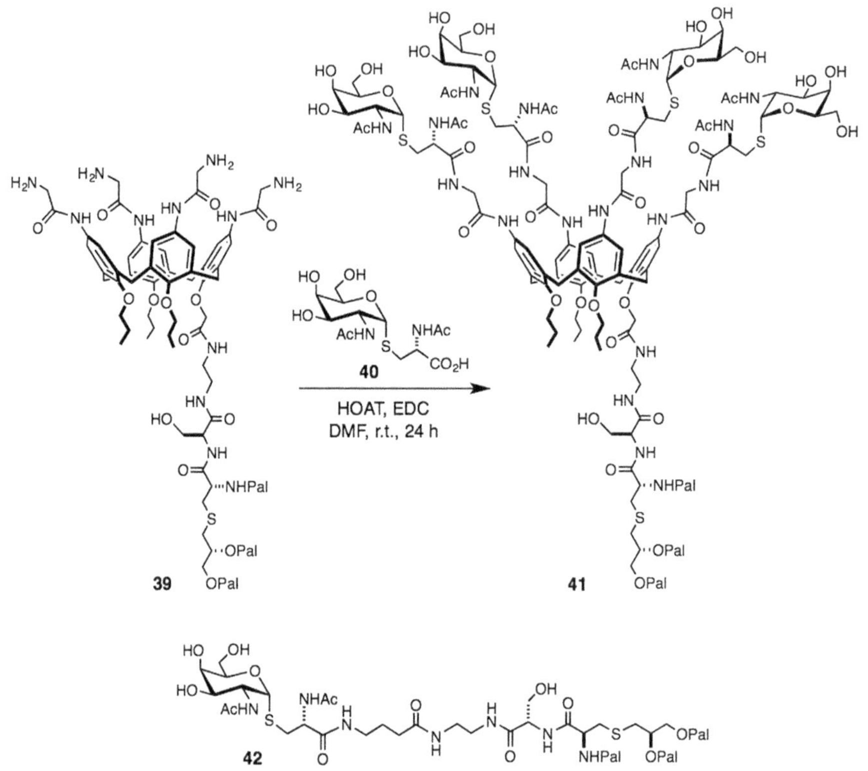

Scheme 4.11

linked to the upper rim of the macrocycle through four glycine residues and a lipopeptide as immunoadjuvant moiety at the lower rim. The latter derived from the N-terminal sequence of the principal lipoprotein of *Escherichia coli*. The *S*-glycosylamino acid **40** was used as a hydrolytically stable mimic of the natural Tn antigen, which is an *O*-glycosylamino acid. The tetravalent cluster **41** showed, by ELISA experiments, a fourfold increase of the end-point titer compared with that of the previously reported[37] monovalent analogue **42** at the same molar concentration.

Aiming to discover a novel class of L-fucose-based multivalent systems endowed with anti-biofilm properties, Consoli *et al.* described the synthesis of the tetravalent α-L-*C*-fucosylcalix[4]arene derivative **46** (Scheme 4.12) and its inhibitory activity against the formation of biofilm derived by the pathogen *Pseudomonas aeruginosa*.[38] The target compound was prepared in good yield from the tetraaminocalixarene **43** by coupling with *N*-Boc-glycine, amine deprotection to afford **44** and amide bond formation with the peracetylated α-L-*C*-fucopyranosylacetic acid **45** to give, after quantitative deacetylation, the calixsugar **46** in a fixed cone conformation. The anti-biofilm activity was investigated by minimum biofilm inhibitory concentration (MBIC) assay. The experiments demonstrated that the activity of **46** was dose

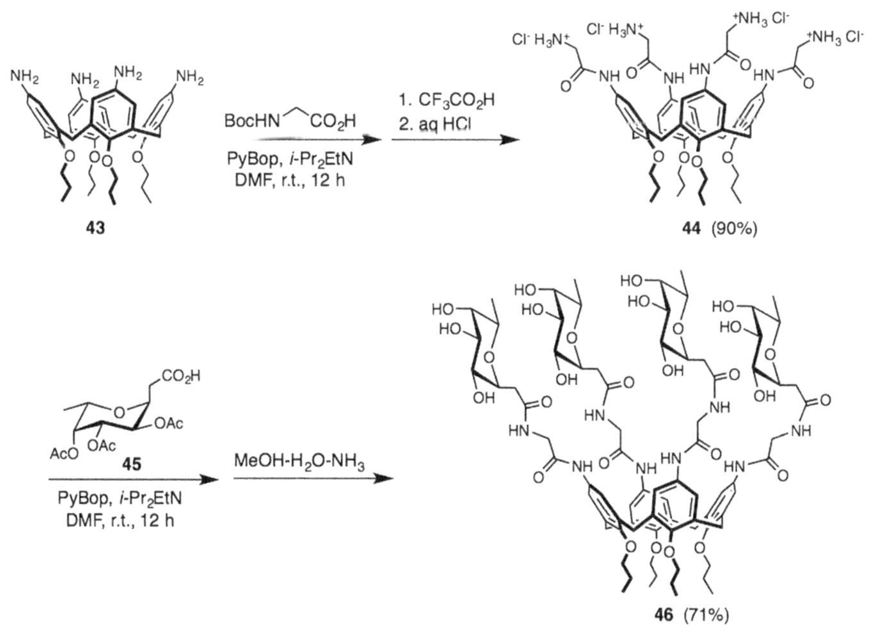

Scheme 4.12

dependent and even at 12.5 μM some biofilm inhibition was observed (35%), whereas the tetraammoniumcalixarene derivative **44** used as a control was able to inhibit the formation of biofilm only in the concentration range 200–50 μM.

4.6 Calixsugar Assembly *via* Palladium-Catalyzed Coupling Reactions

Another interesting route to glycocalix[4]arene was reported by Parrot-Lopez and co-workers.[39] It consisted of Suzuki-type coupling in the presence of Pd(0) between a *p*-bromophenyl glycoside and the boronic acid derivative of a calix[4]arene. The work aimed both to increase the size of the macrocycle cavity for a better binding of potential guest molecules and to incorporate saccharide antennae for specificity of transport and recognition properties. Thus, the $Pd(PPh_3)_4$-catalyzed cross-coupling of diboronic acid calix[4]arene **47** with 2 equiv. of peracetylated *p*-bromophenyl galactoside **48** or glucoside **49** afforded the corresponding upper rim bis-glycosylated calix[4]arenes **50** and **51**, respectively, in fair yields (Scheme 4.13). Then, the acetyl protecting groups were removed using sodium methoxide. It was reported, however, that the unprotected products were difficult to purify owing to their amphiphilic character so that they were isolated in very low yields (~10%). The same group also developed a one-pot procedure, again *via* Suzuki coupling, leading to a calix[4]arene bearing a single phenyl *O*-maltoside fragment at the upper rim.

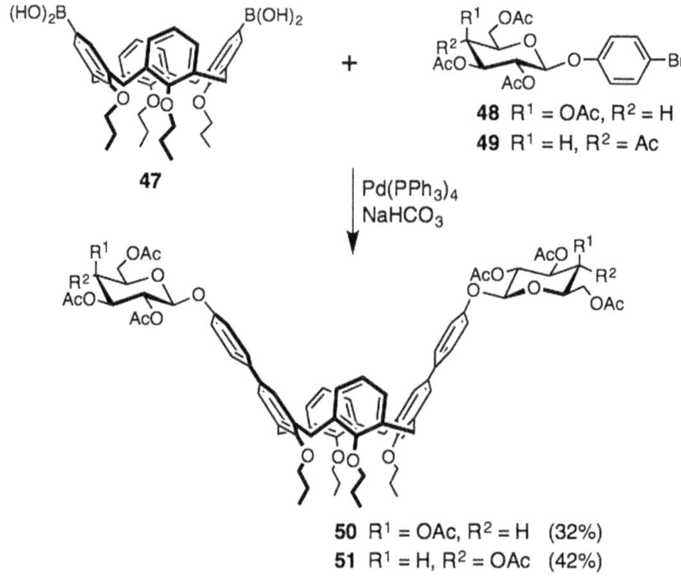

Scheme 4.13

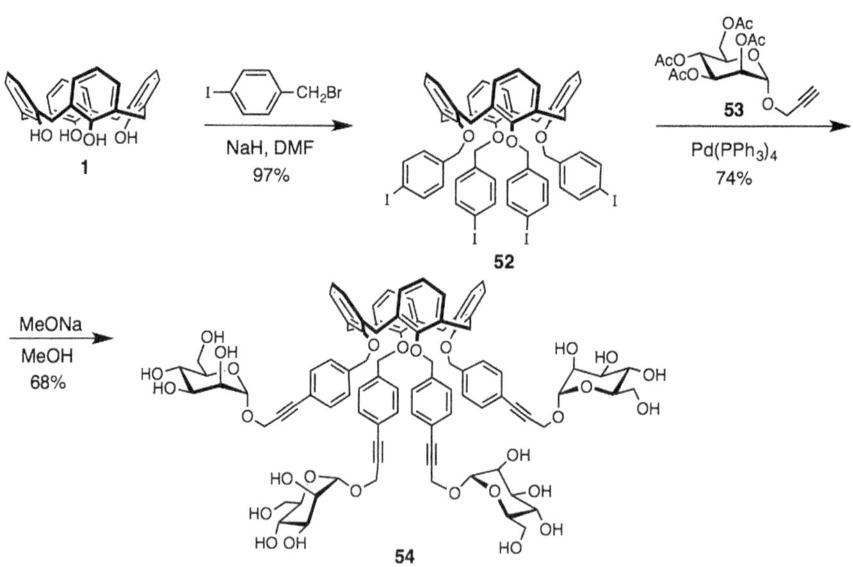

Scheme 4.14

Pérez-Balderas and Santoyo-González reported the use of the Pd(0)-catalyzed Sonogashira reaction for the glycosylation of a calix[4]arene at the lower rim.[40] A deeper lipophilic cavity was constructed by this approach because the first step introduced four iodophenyl rings at the lower rim by ether formation between calix[4]arene **1** and *p*-iodobenzyl bromide (Scheme 4.14).

The functionalized calix[4]arene **52**, obtained in excellent yield, was then reacted with the peracetylated propargyl α-D-mannopyranoside **53** using catalytic $Pd(PPh_3)_4$. The four carbohydrate residues of this Sonogashira product were deacetylated by transesterification to give a tetravalent *O*-glycocluster grafted onto the calix[4]arene platform through phenylethynyl tethers (**54**). The upper rim tetra-*tert*-butyl derivative analogue of **54** was also efficiently prepared by the same procedure. Unfortunately, the low water solubility of the *tert*-butyl derivative prevented its use for biological testing. On the other hand, the recognition properties of water-soluble **54** toward the plant lectin concanavalin A were determined by enzyme-linked lectin assay (ELLA) using methyl α-D-mannopyranoside as a reference standard. Unfortunately, no glycoside cluster effect turned out to be operative because **54** showed lower inhibitor properties than the reference monosaccharide.

4.7 Calixsugar Assembly *via* Wittig Olefination

It is well known that *O*-glycosides are easily degraded *via* chemical or enzymatic hydrolysis. The substitution of the labile exocyclic carbon–oxygen bond with a more robust carbon–carbon bond to give *C*-glycosides provides higher stability.[41] This issue was addressed within the context of calix[4]arene glycosidation by Dondoni, Marra and co-workers since the early stages of calixsugar development. The first approach to an all-carbon tethered calixsugar dated back to 1997 and consisted of multiple Wittig olefination of a tetraformylcalix[4]arene with sugar phosphoranes followed by reduction of the double bonds of the alkene-linked calix[4]arene glycoconjugate intermediates.[42] The products thus formed, however, could not be considered as *C*-glycosides because the carbohydrate fragments were linked to the alkyl chain by their non-reducing end. The first genuine calix[4]arene *C*-glycoside, **57**, was prepared some years later[22] *via* the same Wittig-based approach by using a sugar phosphorane generated from tetra-*O*-benzylgalactosylmethylphosphonium iodide **56** and the long-arm calix[4]arene dialdehyde **55** (Scheme 4.15). Unfortunately, the entry to a tetravalent *C*-glycocluster was precluded by the failure to prepare the required long-arm calix[4]arene tetraaldehyde.

4.8 Calixsugar Assembly *via* Cu(I)-Catalyzed Azide–Alkyne Cycloaddition

The Cu(I)-catalyzed azide–alkyne cycloaddition (CuAAC),[43] the quintessential click reaction,[44] leading to the regiospecific formation of 1,4-disubstituted triazoles attracted attention as a formidable tool for calixarene decoration with biologically active molecules. Dondoni and Marra reported in 2006 on the first use of this ligation tool for multiple carbohydrate anchoring to calix[4]arene scaffolds.[45] In these products, the glycosyl residues featured an anomeric carbon–carbon bond and were anchored to the calixarene platform through 1,4-disubstituted triazole rings. Thus, di- and tetravalent

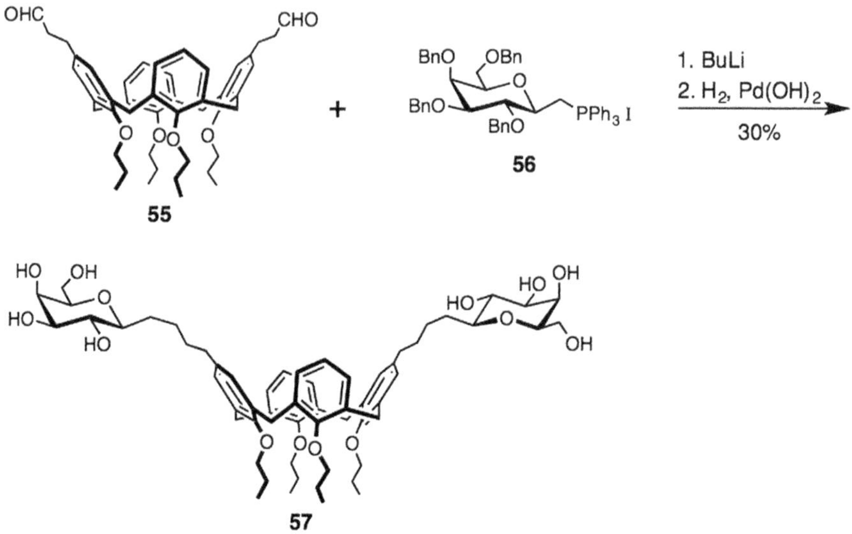

Scheme 4.15

carbon-linked glycoclusters were anchored to the conformationally rigidified tetrapropoxycalixarene scaffold using two complementary reagent combinations. The first was based on the coupling of an ethynyl *C*-glycoside, for example the peracetylated *gluco* derivative **58**, with the tetraazidopropyl calix[4]arene **59** in the presence of catalytic CuI to afford the corresponding calix[4]arene-based tetravalent *C*-glycoclusters **60** (86% yield) featuring the triazole ring as a tether (Scheme 4.16). The high yields of isolated products indicated that each cycloaddition occurred in a nearly quantitative manner. Removal of the *O*-acetyl group by transesterification was key to the transformation of calixsugar **60** into compound **61** suitable for bioassays.

The second approach consisted of the coupling of an ethynyl-functionalized calix[4]arene with a sugar azide. With this aim, a glycosylmethyl azide was used in order to produce a glycoconjugate with a stable methylene bridge between the sugar and triazole residue. This approach is exemplified in Scheme 4.17, illustrating the CuI-catalyzed cycloaddition between the galactosylmethyl azide **63** and the tetraethynylcalix[4]arene **62** to give the calix[4]arene-based tetravalent glycocluster **64** in excellent yield.[45] This compound differs from the tetravalent calixsugar **60** in the position of the carbohydrate and calixarene fragments attached to each triazole ring. The 1,4-disubstitution pattern in the triazole rings was confirmed by NMR analysis on the basis of the large and positive $\Delta(\delta_{C4} - \delta_{C5})$ values (~20 ppm), in agreement with earlier findings on simple molecules.[46]

A special study was carried out on the synthesis of the *galacto* analogue of the tetraglucosylcalixarene **60** *via* CuAAC in unconventional solvents such as ionic liquids.[47] The use of three different ionic liquids combined with microwave dielectric heating confirmed the high efficiency (yields up to 90%) of the

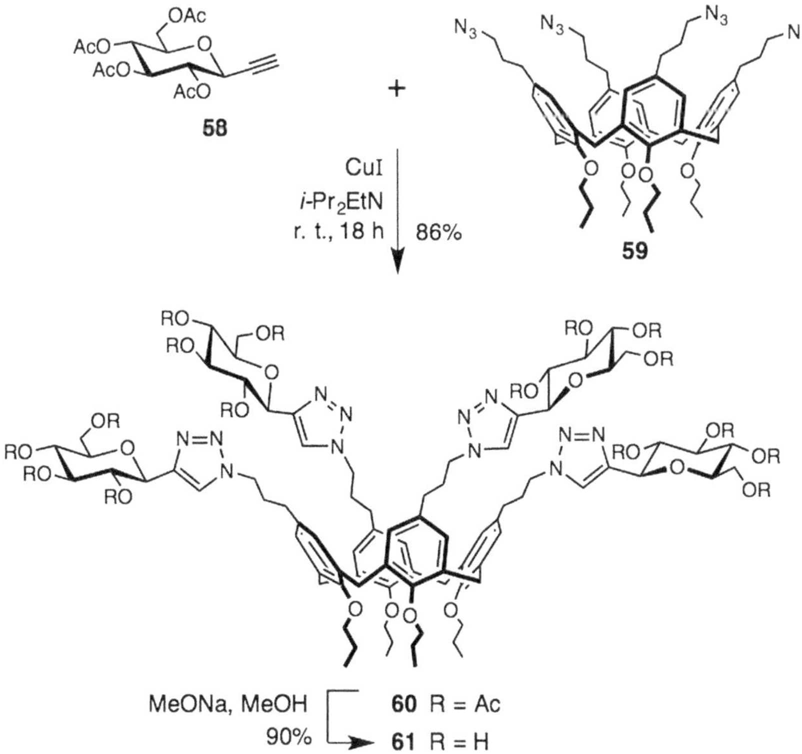

Scheme 4.16

reaction described in Scheme 4.16. The excellent yields of isolated tetravalent glycocluster and the lack of products arising from partial coupling of the sugar acetylene with tetraazidocalix[4]arene **59** led to the suggestion that once the first triazole ring was formed, subsequent coupling reactions occurred much readily owing to the formation of Cu(I)–triazolide–ethynyl complexes.

In 2007, Bew *et al.* described the synthesis of two triazole-linked calix[4]arene *O*-glycosides *via* CuAAC[48] (Figure 4.5). In both cases, peracetylated carbohydrate residues were introduced at the upper rim of the macrocycle, namely two glucosyl and two lactosyl moieties, as shown in **66** and **67**, respectively.[49] A tetra-*O*-lactosylcalix[4]arene[49] was also prepared and this was mono- an bis-sialylated using a *trans*-sialidase. No bioassays were carried out with these compounds, however.

Twelve years after the Meunier and Roy report,[27] Dondoni, Marra and co-workers considered sialyl clustering on calix[4]arene scaffold an important issue to be addressed because of the potential applications in glycobiology. First, synthetic efforts were made towards the preparation of polyvalent calix[4]arene *S*-sialosides with different structures and compositions[50] by the use of the CuAAC. The key reagent in this approach was the propargyl thiosialoside **68**, which served as cycloaddition partner with azidocalix[4]arenes.

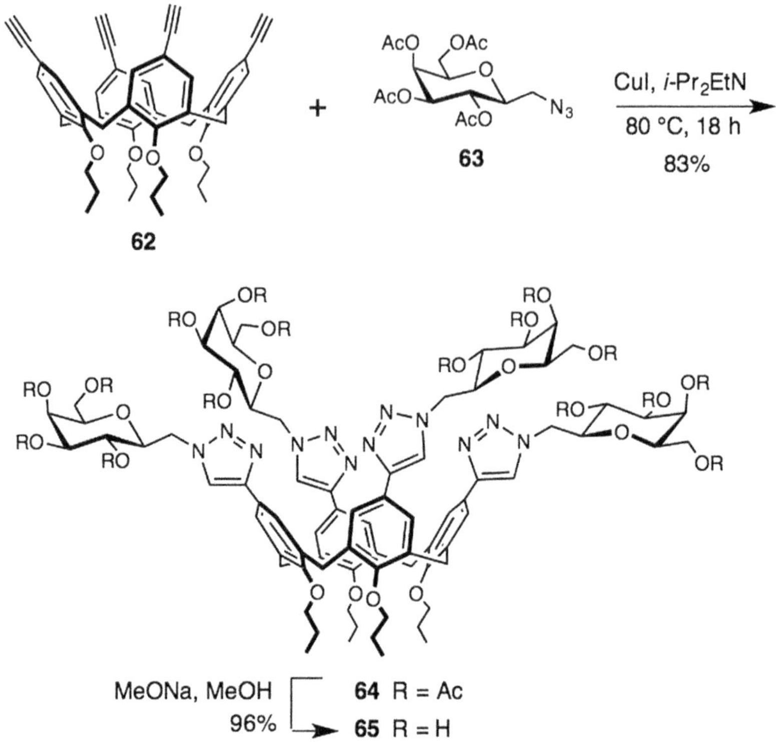

Scheme 4.17

An excess of this sugar alkyne was allowed to react with the calix[4]arene tetraazide **59** in the presence of CuI and *i*-Pr_2EtN to give, after removal of the ester protecting groups, the upper rim tetravalent calix[4]arene *S*-sialoside **70** in good isolated yield (Scheme 4.18). Thus, the sialyl residues in **70** formed a tetravalent cluster with all sugar moieties anchored to the macrocycle platform through an *N*-propyltriazole tether that ensured enough flexibility to the whole system. The calix[4]arene *S*-sialoside **71**, displaying shorter connecting arms, was prepared in a similar way by reacting **68** with the calix[4] arene tetraazide **69**.

In a second instance, sialyl groups were introduced at the lower rim of the calix[4]arene platform. This new structure was aimed at positioning the sialyl residues closer to one another than in the upper rim functionalized compounds **70** and **71**. The synthesis of the new target calix[4]arene *S*-glycoside **73** was also performed by CuAAC, which in this case involved the tetraazide **72** and the propargyl thiosialoside **68** (Scheme 4.19).

As a final endeavor, the introduction of sialyl residues at both the upper and lower rims of a calix[4]arene was considered. It was assumed that such a densely sialylated molecule could bind simultaneously to a pair of hemagglutinin trimers located on a single virion or to two distinct viral particles. The coupling of the sugar alkyne **68** with the calix[4]arene octaazide in the

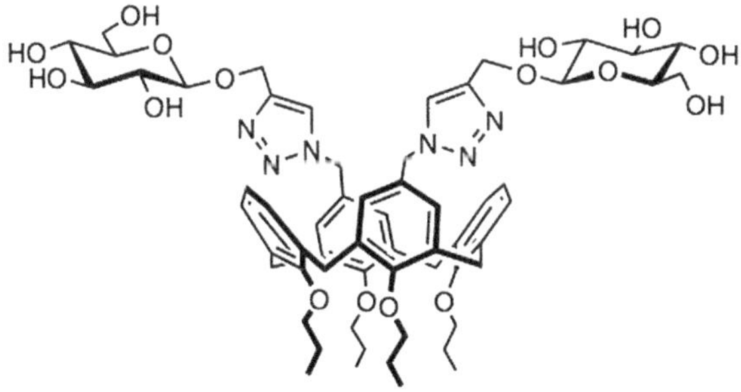

66

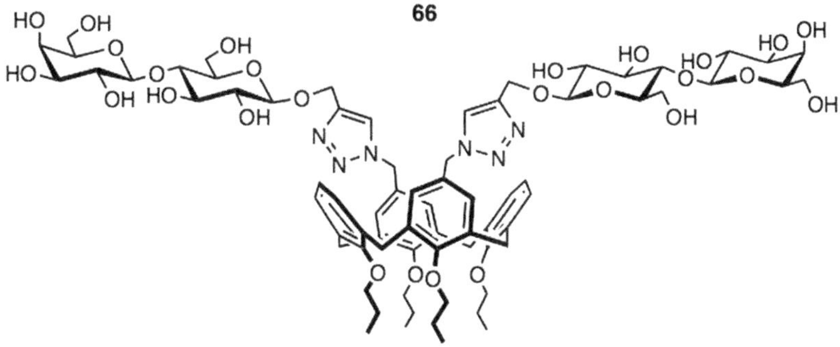

67

Figure 4.5 Bis-*O*-glycosylated calix[4]arenes prepared *via* CuAAC by Bew *et al.*[48]

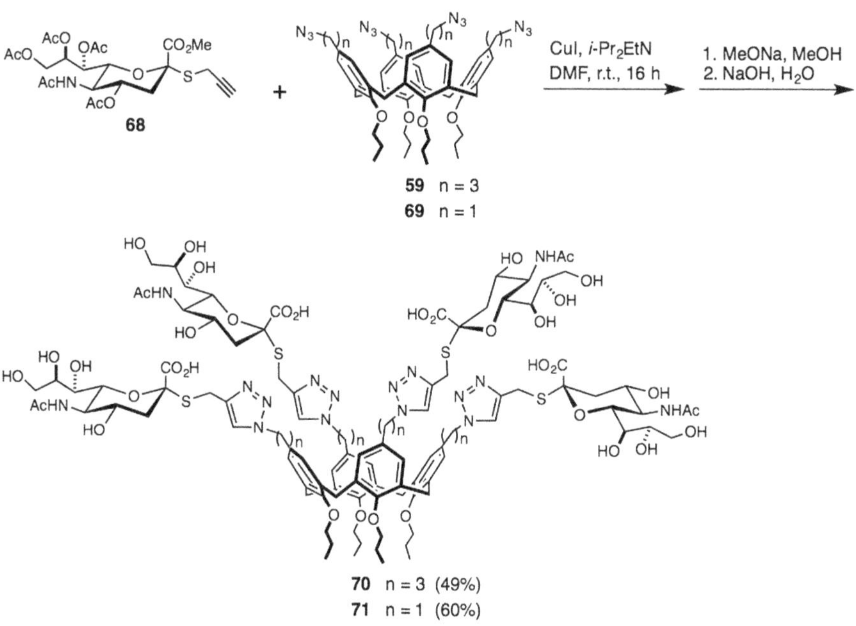

Scheme 4.18

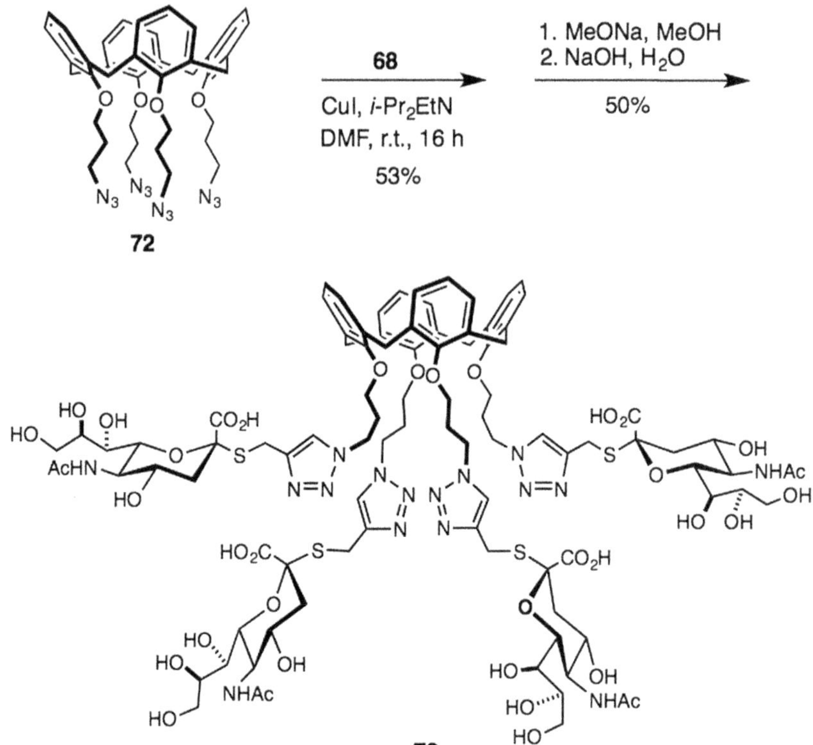

Scheme 4.19

presence of CuI and i-Pr$_2$EtN proceeded smoothly to give, after deprotection, the octavalent sialocluster **74** (Figure 4.6).

The capability of calix[4]arenes *S*-sialosides **70**, **71**, **73** and **74** to bind BK virus was tested using the hemagglutination inhibition (HI) assay and compared with that of the monovalent analogue **75** (Figure 4.7), which featured a similar aglycone structure. All compounds were able to inactivate BKV at submillimolar concentrations, although each multivalent sialoside was less active than **75** when the actual concentration of sialic acid moieties was taken into consideration. The HI assay demonstrated also that the sialoclusters **70**, **73** and **74** inhibited influenza A virus-induced hemagglutination, whereas **71** failed to show HI activity at concentrations up to 50 mM and the monovalent sialoside **75** was a very weak inhibitor. The data indicated that, in this case, a moderate glycoside cluster effect was operative since, considering the inhibition activity per sialic acid unit, the multivalent sialosides were 50–83 times more active than the monovalent derivative **75**. Finally, the HI activity of the octasialoside **74** was close to that of the tetrasialosides **70** and **73**. This indicated that only one of the two sets of sialic acid units linked at both calixarene rims was involved in the interaction with the influenza A hemagglutinin. Moreover, experiments of viral infectivity proved that the above sialoclusters inhibited the adsorption of

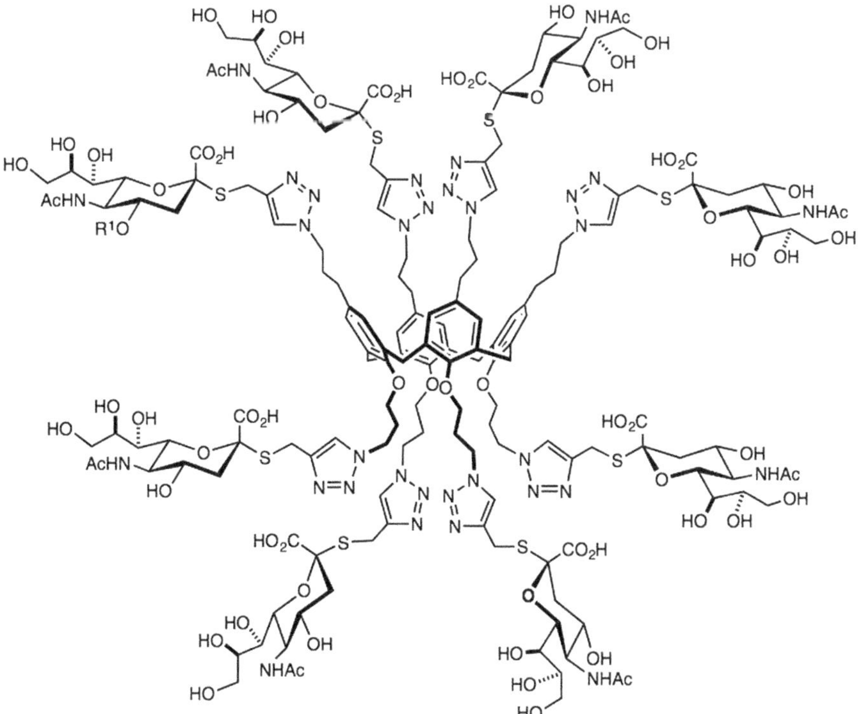

74

Figure 4.6 Octavalent sialyl calix[4]arene prepared by Dondoni, Marra and co-workers.[50]

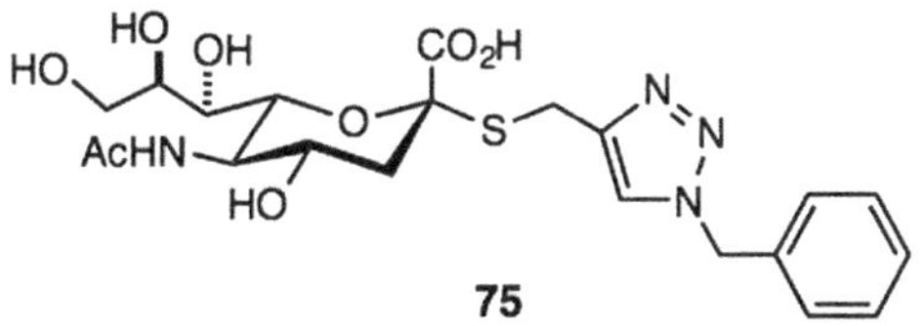

75

Figure 4.7 Monovalent sialoside ligand of viral hemagglutinin.

both BK and influenza A (H3N2 strain) virions to the host cells. Therefore, the use of the non-natural calixarene scaffold and triazole linker did not prevent the molecular recognition properties of the Neu5Ac ligand.

Another issue that was addressed was the immobilization of triazole-tethered calix[4]arene *O*-glycosides on TiO_2 nanoparticles.[51] This represented the first example of TiO_2 nanoparticle–calixsugar hybrid materials. As reported earlier by de la Fuentes and Penadés,[52] metal and metal oxide nanoparticles coated by biologically relevant oligosaccharides, named glyconanoparticles, constitute a good biomimetic model of carbohydrate presentation at the cell surface. While the direct immobilization of glycoside units on the TiO_2 nanoparticle surface represented a difficult task, the use of a suitable functionalized calix[4]

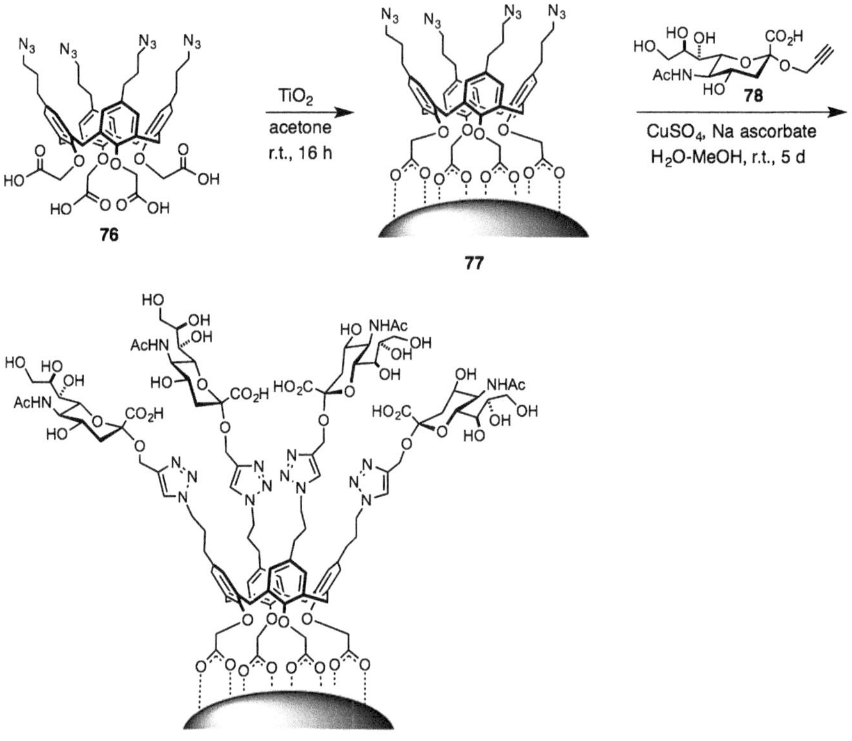

Scheme 4.20

arene as a platform was considered. Thus, given the well-known ability of carboxylate groups to form strong bonds with TiO_2 surfaces,[53] the orthogonally fuctionalized calix[4]arene **76**, with azido and carboxyl groups at the upper and lower rim, respectively, was grafted on TiO_2 nanoparticles (Scheme 4.20). In this way, the calix[4]arene **76** acted as a tetrapodal ligand using the four carboxylate groups while the azido groups were preserved for the execution of click reaction with alkyne-functionalized carbohydrates. In the event, treatment of the calixarene-coated TiO_2 nanoparticles **77** thus obtained with free hydroxypropargyl *O*-sialoside **78** in the presence of $CuSO_4$ and sodium ascorbate gave rise to multiple triazole forming reactions on each calix[4]arene platform anchored to the TiO_2 surface. This resulted in the formation of the target glyconanoparticles **79**. The occurrence of the azide–alkyne cycloaddition was monitored by infrared analysis while thermogravimetric analysis (TGA) was used to evaluate the molar concentration of the glycocluster per gram of TiO_2 and the number of organic molecules on the surface of the nanoparticles.

The CuAAC reaction was also employed for the preparation of calix[4]arene-based *C*-glycosides **80** and **81**, featuring a single azido group attached to the lower rim through suitable tethers[54] (Figure 4.8). The synthesis of these compounds was carried out in a program of an Italian–French consortium

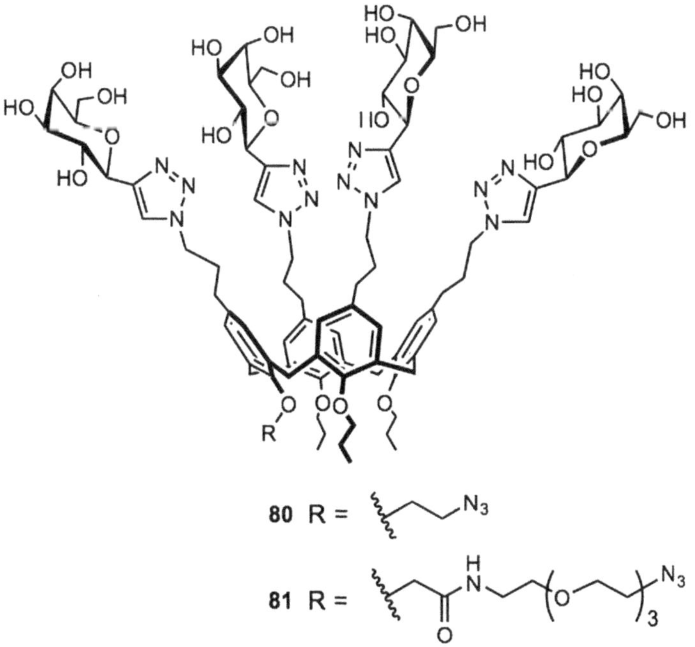

Figure 4.8 *C*-Glycosyl calix[4]arenes bearing an azido group at the lower rim.

directed towards the construction of multivalent glycocluster–oligonucleotide hybrids **82** and **83** and evaluation of their affinity to galactose-recognizing lectins (Figure 4.9). Compounds **80** and **81** were coupled *via* CuAAC to oligonucleotides bearing the fluorescent dye Cy3 and one or two propargyl groups to give **82** and **83**, respectively.

The glycocluster-oligonucleotide hybrids **82** and **83** were immobilized on DNA chips constituted of complementary oligonucleotide sequences covalently linked to 52-well glass slides. Next, the *Pseudomonas aeruginosa* lectin (PA-IL) and *Ricinus communis* agglutinin (RCA 120), labeled with specific fluorescent dyes, were deposited in the wells and the fluorescent signals of the conjugates were measured. The results indicated that the hybrids **82** and **83** had no affinity to the PA-IL lectin whereas they displayed good affinity to the RCA 120 lectin. The level of affinity of the latter was similar to that observed for both a linear conjugate bearing three galactose moieties and a cluster bearing ten galactose residues in an antenna spatial arrangement. Moreover, IC_{50} values, *i.e.* the concentration of lactose required for the removal of 50% of RCA 120 bound to the immobilized glycoconjugates, were measured and compared with those determined for linear monovalent and trivalent galactoside conjugates. From these assays, it appeared that the glycoconjugates **82** and **83** displayed a modest glycoside cluster effect as the affinity per galactose moiety was only five and seven times higher than that displayed by the monovalent conjugate. Hence it was concluded that for the attainment of multivalency, the

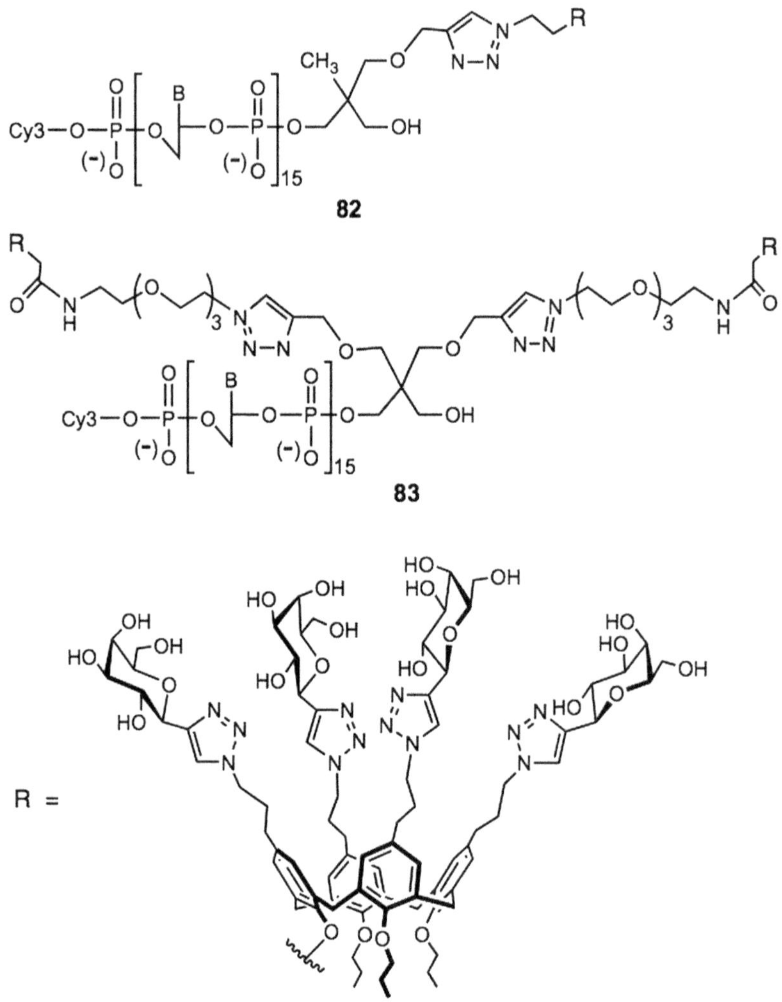

Figure 4.9 Multivalent glycocluster–oligonucleotide hybrids.

spatial arrangement was more important than the number of sugar residues because the linear trivalent cluster was capable of binding to lectins more efficiently than calixarene and antenna systems bearing four, eight and ten galactose moieties.

A series of partially and fully *O*-galactosylated calix[4]arene clusters in cone, partial cone and 1,3-alternate conformations were prepared by Imberty and co-workers *via* CuAAC.[55] The di- (**84** and **85**), tri- (**86**) and tetrapropargyl (**87**) *tert*-butylcalix[4]arenes, locked in the cone conformation, were allowed to react with the 8-azidodioxaoctyl galactopyranoside **88** in the presence of catalytic CuI and *N*,*N*-diisopropylethylamine under microwave irradiation to give, after deacetylation, the corresponding triazole-linked lower rim glycoclusters **89–92** in 50–92% overall yield (Scheme 4.21). The diastereomeric

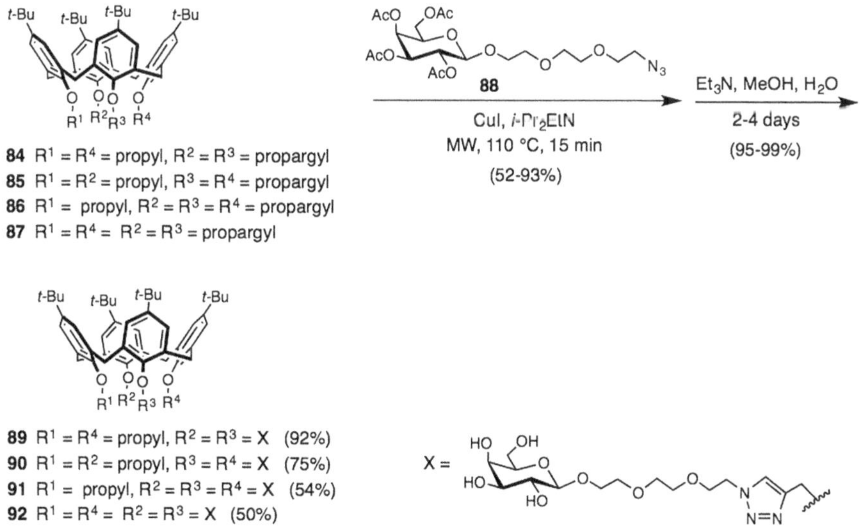

Scheme 4.21

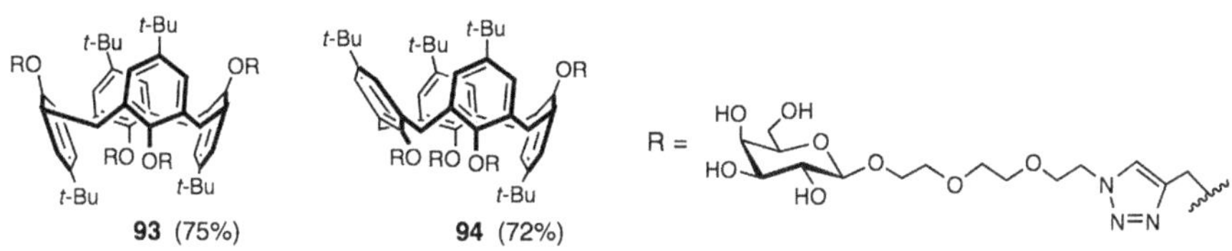

Figure 4.10 Tetra-*O*-glycosylated calix[4]arenes in 1,3-alternate (**93**) or partial cone conformation (**94**) prepared *via* CuAAC.[55]

tetragalactosyl calixarenes displaying a 1,3-alternate (**93**) or partial cone conformation (**94**) were also prepared in good yields by the same click azide–alkyne cycloaddition (Figure 4.10).

The affinity of these glycoclusters towards *Pseudomonas aeruginosa* lectin PA-IL, which selectively recognizes D-galactosides, was determined by isothermal titration calorimetry (ITC) and compared with the affinity of a monovalent ligand. Unfortunately, the monogalactosylated calixarene **95** (Figure 4.11) could not be used in this study because of its poor solubility in H_2O and H_2O–DMSO mixtures. Therefore, the *O*-glycoside **96** (Figure 4.11), although devoid of the hydrophobic macrocyclic scaffold, was used as monovalent ligand.

Titration experiments showed that the bis-galactoside clusters **89** and **90** did not bind to PA-IL, whereas the trivalent glycocluster **91** and also tetravalent clusters (**92–94**) displayed an enhancement in binding up to 200-fold larger (per D-Gal unit) than that observed for **96**. Docking calculations established that the best ligand of the series, the tetragalactosyl cluster **93**, adopting a fixed 1,3-alternate conformation, was well suited for the topology of the rectangular-shaped PA-IL tetramer. The modeling study indicated that two galactose units located on the opposite sides of **93** were linked to the adjacent

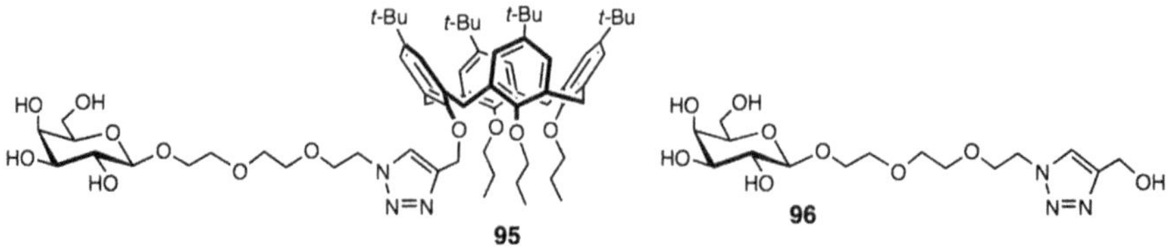

Figure 4.11 Monovalent ligands of *Pseudomonas aeruginosa* lectin PA-IL.

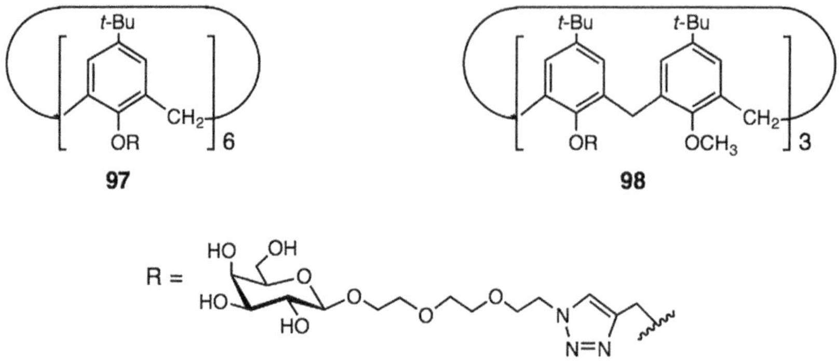

Figure 4.12 Hexavalent and trivalent galactosyl calix[6]arenes prepared by Vidal and co-workers.[57]

binding sites on the small face of one PA-IL tetramer, whereas the other two galactose residues were available for binding to another PA-IL tetramer. It was found that the two lectin tetramers face each other without steric conflict. Glycoclusters **93** and **94** were also efficient inhibitors of the adhesion of PA-IL to a biotinylated polygalactosylated surface onto a streptavidin-coated chip, as proved by surface plasmon resonance analysis. Therefore, it was concluded that these anti-adhesive molecules could find potential applications as antimicrobials.

The binding of the tetragalactosylcalixarene in a 1,3-alternate conformation (**93**, Figure 4.10) to the tetrameric PA-IL was also investigated by means of atomic force microscopy (AFM).[56] The self-assembly of this lectin in the presence of **93** afforded filaments showing linear segments (length 90–500 nm, width ~36 nm), constituted of 10–50 PA-IL tetramers, interrupted by bifurcations and branching points. This study revealed that two galactose units of the calixsugar chelate two adjacent binding sites of the lectin whereas the two other galactose moieties interact with another tetrameric lectin.

In 2011, Vidal and co-workers[57] reported the synthesis of a series of multivalent galactosylated glycoconjugates, including hexa- (**97**) and trivalent (**98**) *tert*-butylcalix[6]arene derivatives (Figure 4.12), tetravalent linear and cyclic β-peptoids and a tetravalent porphyrin derivative, all obtained by copper-mediated cycloaddition of the azidogalactoside **88** (see Scheme 4.21) with the corresponding propargylated scaffold followed by deacetylation. Two galactose-binding lectins, PA-IL from *Pseudomonas aeruginosa* and *Erythrina* *cristagalli* agglutinin (ECA), were selected to evaluate the glycoside cluster effect shown by the above-mentioned glycoconjugates and also the tetragalactosylcalix[4]arenes in a cone (**92**, Scheme 4.21) or 1,3-alternate conformation (**93**, Figure 4.10) previously described by the same group.[55] However, the trivalent calix[6]arene **98** could not be included in this study because of its low solubility in water and H_2O–DMSO. The rabbit red cell hemagglutination induced by PA-IL or ECA was inhibited to various extents by the calixsugars. The glycoconjugates **92**, **93** and **97** displayed poor affinity towards ECA, whereas a strong inhibition effect on PA-IL was observed for the three compounds, in particular in the case of the hexavalent derivative **97**, compared with that of the galactoside **96** (see Figure 4.11) used as a monovalent reference. These results were confirmed by surface plasmon resonance (SPR) experiments, which clearly indicated that calixsugar **97** was the best inhibitor for both lectins (IC_{50} in the micromolar range). Finally, isothermal titration calorimetry (ITC) showed a dissociation constant (K_d) of 136 nM for **97**, *i.e.* the lowest K_d value of the ligands of PA-IL under investigation.

One year later, the same group[58] prepared tetralactosylated calix[4]arenes in fixed cone (**99**), 1,3-alternate (**100**) and partial cone (**101**) conformations (Figure 4.13) using the synthetic approach already reported for the preparation of the corresponding galactosyl derivatives **92**–**94** (see Scheme 4.21 and Figure 4.10). These glycoconjugates were evaluated as ligands of a plant lectin, *Erythrina cristagalli* agglutinin (ECA), and a human lectin, recombinant galectin-1, by hemagglutination inhibition assay (HIA), enzyme-linked lectin assay (ELLA) and surface plasmon resonance (SPR). The 1,3-alternate calixsugar **100** showed the highest affinity towards the ECA lectin in the ELLA experiments, although no glycoside cluster effect was observed. On the other hand, the cone calixsugar **99** displayed significant inhibition of the rabbit erythrocyte hemagglutination mediated by galectin-1 in comparison with the monovalent reference (156-fold increase per lactose unit). With the latter lectin, the peroxidase-based ELLA experiments could not be performed, so SPR was used to complement the HIA measurements. However, the HIA data were not confirmed by the SPR experiments, probably because the glycocluster–lectin interactions at the surface of erythrocytes are different from those occurring on a gold surface coated with lactosylated polyacrylamide. It was concluded that the binding properties of the synthesized clusters towards galectin-1 should be evaluated using other analytical techniques, such as isothermal titration calorimetry (ITC).

In 2011, the trypanocidal activity of the above-mentioned[59] lower rim lactosylated calix[4]arenes in cone (**99**) and partial cone (**101**) conformations (Figure 4.13), the lower rim lactosyl calix[4]arene **102** and the upper rim lactosyl calix[4]arenes **103** and **104** (Figure 4.14) was evaluated by Field and co-workers.[60] Benznidazole, the drug currently used for the treatment of Chagas disease, a tropical parasitic disease caused by the protozoan *Trypanosoma cruzi*, was employed as a positive control. Although D-lactose and its derivatives inhibit[61] the *T. cruzi trans*-sialidase, an enzyme present in multiple copies on the parasite surface that plays an essential role in the infection process, monovalent lactose derivatives are endowed with poor inhibitory

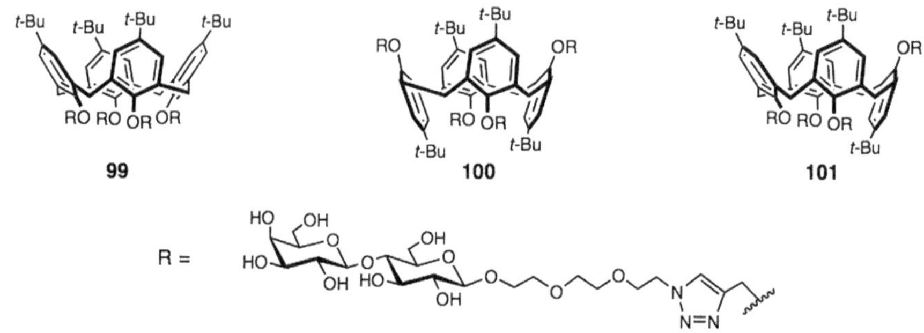

Figure 4.13 Tetravalent lactosyl calix[4]arenes in cone, 1,3-alternate and partial cone conformations.

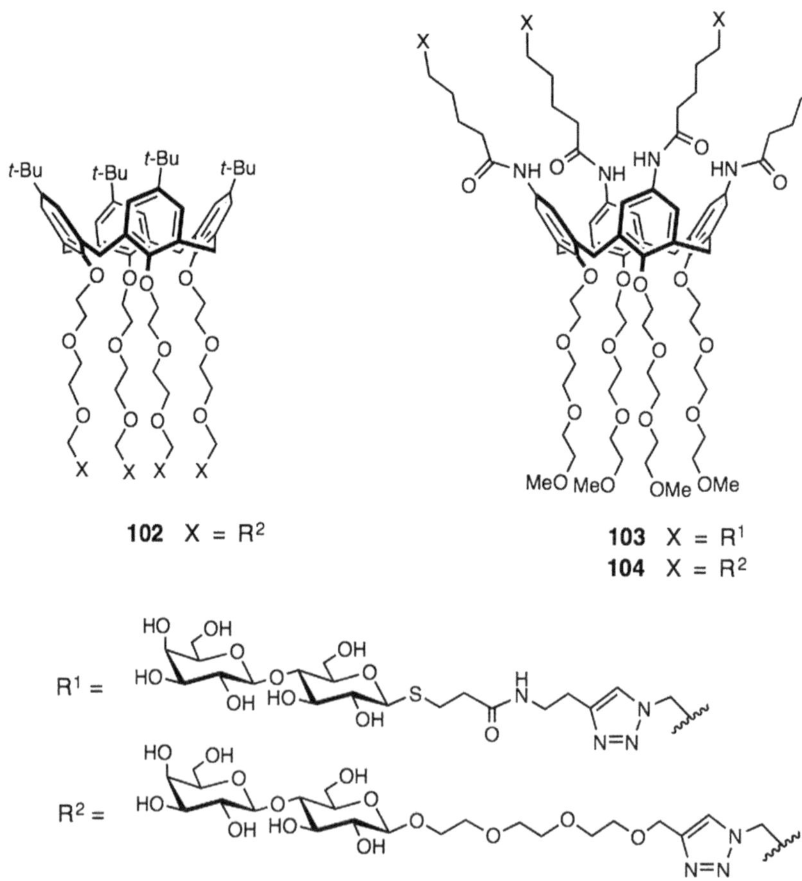

Figure 4.14 Tetravalent lactosylcalix[4]arenes prepared by Field and co-workers.[60]

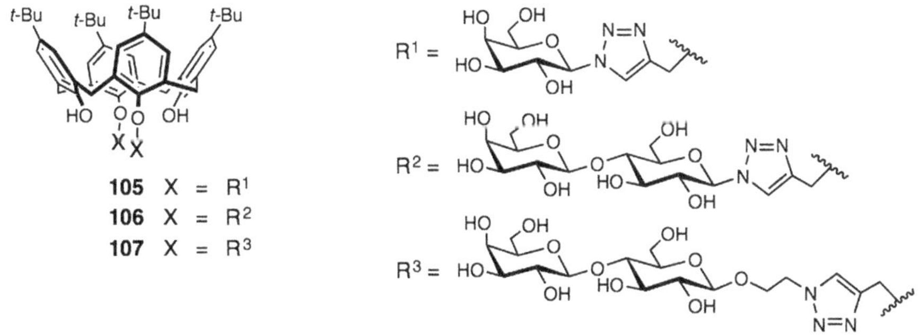

Figure 4.15 Lower rim divalent calix[4]arenes as ligands for jacalin.

activity. Therefore, the authors envisaged the tetravalent lactosylcalix[4]arenes adopting a well-defined three-dimensional arrangement as stronger sialidase inhibitors. Actually, all calixsugars in the cone conformation showed potent trypanocidal activity, the glycocalixarene **99** being the most efficient inhibitor (**99**, IC_{50} = 62 mM; benznidazole, IC_{50} = 67 mM).

Three divalent glycocalix[4]arenes (**105–107**, Figure 4.15) were synthesized by Chinta and Rao as probes for jacalin, a galactose-binding plant lectin.[62] The absorption titration of the bis-galactoside derivative **105** with L-tryptophan, L-tyrosine and L-phenylalanine in methanol was carried out in order to study the interactions between the aromatic moiety of these natural amino acids and the calixsugar. Actually, the decrease in the absorbance of the 213 and 280 nm bands shown by **105** upon gradual addition of L-Trp indicated an interaction of the triazole ring with the aromatic amino acid, which was confirmed by ^{1}H NMR titration and MALDI-TOF mass spectrometry experiments. Then, the binding of the glycocalixarenes **105–107** with jacalin was studied by fluorescence spectroscopy. The 340 nm emission band of the lectin was quenched by the addition of increasing amounts of calixsugars in methanolic buffer solution. The galactose- (**105**) and the lactose-based calixarenes (**106** and **107**) produced similar quenching; however, circular dichroism analysis showed that only **107**, bearing a spacer at the anomeric carbon, led to secondary structural changes of jacalin.

In 2013, Mancini, Sansone and co-workers reported the synthesis of a single glucocalix[4]arene in a 1,3-alternate conformation (**108**, Figure 4.16) and its incorporation into the liposome bilayer.[63] In fact, calixsugar **108** displayed two hydrophilic regions separated by a long lipophilic spacer and therefore it was able to span the bilayer of liposomes constituted by phosphatidylcholines. It was found by fluorescence experiments that **108** rigidified the bilayers of unsaturated lipids, avoiding the release of calcein from liposomes. Moreover, the binding of the calixsugar-containing liposomes to the plant lectin concanavalin A was evaluated by steady-state fluorescence experiments and turbidimetric measurements. Taking advantage of the selective affinity of multivalent glycoconjugates towards lectins, these supramolecular structures could be used as site-specific drug-delivery systems.

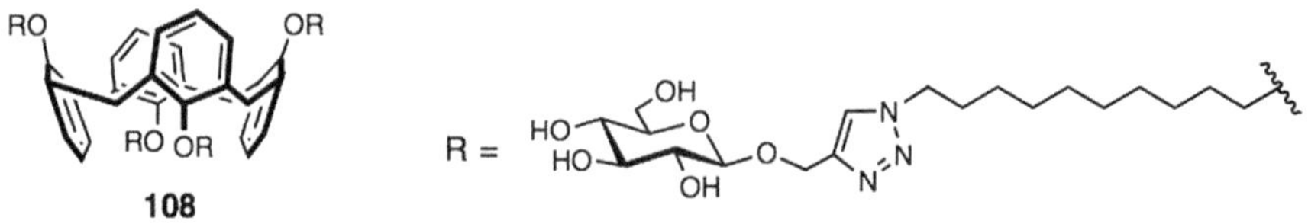

Figure 4.16 Tetraglucosyl 1,3-alternate calix[4]arene that was included in liposomes.

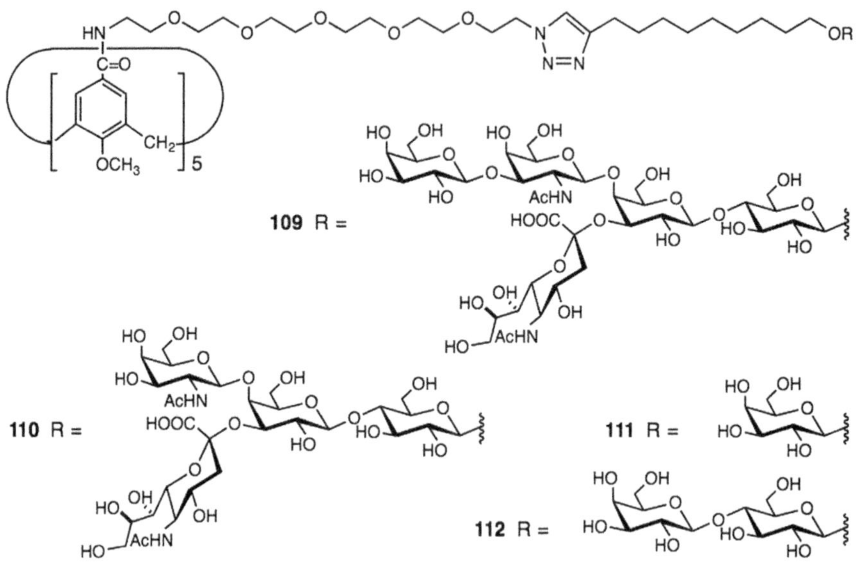

Figure 4.17 Pentavalent glycosylcalix[5]arenes prepared by Casnati, Zuilhof and co-workers.[64]

The cholera toxin, produced by the *Vibrio cholera* bacterium, contains a pentameric binding domain that allows the recognition and binding to the cell membranes. The natural ligand of the toxin is the glycosphingolipid ganglioside GM1 that is present on the cell surface. Inhibitors of this recognition process can prevent the binding and thus the endocytosis of the cholera toxin. In 2013, Casnati and co-workers[64] synthesized the pentavalent ganglioside GM1 (**109**) and GM2 oligosaccharide calix[5]arene (**110**) by CuAAC reaction starting from the a pentaazidocalix[5]arene derivative and the corresponding, chemo-enzymatically produced alkynylpenta- and -tetrasaccharides (Figure 4.17). The inhibitory potency of the glycoclusters was evaluated by ELISA experiments. Calixsugar **109** showed an extremely low IC_{50} value (0.45 nM), which, compared with that of the monovalent reference compound (44 μM), corresponds to a *ca.* 100 000-fold increase in inhibition activity on a molar basis. The much higher IC_{50} value (9 μM) found for **110** indicated that the GM2-based cluster did not show any glycoside cluster effect towards the cholera toxin. On the other hand, the replacement of the sialylated oligosaccharide moieties with simple galactose (**111**) or lactose units (**112**) led to a total loss of inhibition potency (IC_{50} > 1 mM).

4.9 Calixsugar Assembly *via* Nucleophilic Substitution in Tosyltetrazoles

Owing to stability towards both acids and bases and also oxidizing and reducing reagents, the tetrazole ring was also considered by Dondoni and Marra to be a useful system for establishing a robust connection between carbon-linked carbohydrate fragments and a calix[4]arene platform.[65] They envisaged that tetrazole, similarly to triazole, being a multinitrogenated heterocyclic compound, could actively participate in hydrogen bonding processes and therefore provide additional recognition properties to the system in which it was embedded. A simple approach to tetrazole formation was the thermal azide–nitrile 1,3-dipolar cycloaddition, which was known to afford 1,5-disubstituted tetrazole derivatives.[66] This reaction, however, is of preparative value only with nitriles activated by electron-withdrawing groups. Therefore, in order to gain access to a wide range of disubstituted tetrazoles, Demko and Sharpless developed a simple and expedient two-step sequence.[67] This involved the coupling of an organic azide with *p*-toluenesulfonyl cyanide and then the displacement of the tosyl group in the 5-sulfonyltetrazole thus formed by a suitable nucleophile. This approach was successfully used to introduce four ribofuranose, galactopyranose and glucopyranose units in the calix[4]arene tetrol **13** (Scheme 4.22). First, 1-glycosylmethyl-5-tosyltetrazoles

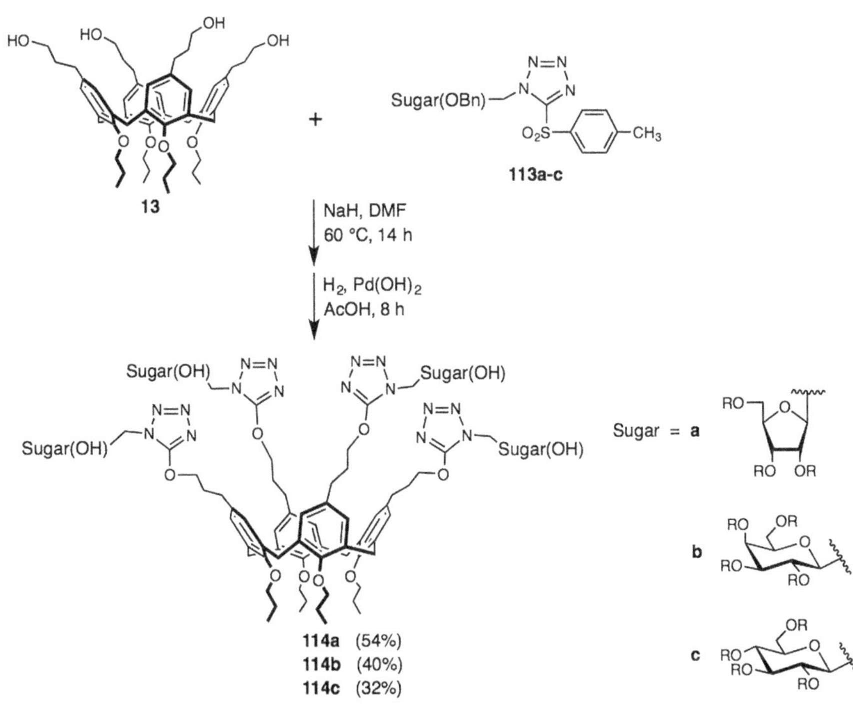

Scheme 4.22

113a–c were prepared in good to excellent yields (62–96%) by coupling *O*-benzylated glycosylmethyl azides with *p*-toluenesulfonyl cyanide. Then, each of these compounds (6 equiv.) was allowed to react under basic conditions with the polydentate alcohol **13** to give, after removal of the benzyl protective groups from the sugar fragments, the corresponding calix[4]arene *C*-glycosides **114a–c**. These compounds all featured the 1,5-disubstituted tetrazole ring (the substituents are adjacent) in the carbohydrate-to-platform tether. Hence these tetravalent carbon-linked glycoclusters showed a connectivity pattern around the heterocycle ring that was different to that established in triazole-tethered glycoclusters in which the substituents were distant.

4.10 Calixsugar Assembly *via* Isothiocyanate–Amine Addition

Ungaro and co-workers realized the potential of the classical isothiocyanate–amine coupling for introducing amino acids or small peptides and carbohydrate fragments into calix[4]arenes through a thiourea spacer.[68] They envisaged the strong hydrogen bonding donor thiourea group as a point of interaction in molecular recognition processes. Moreover, the convenient choice of this ligation tool resided on the easy access of carbohydrates and calix[4]arenes functionalized with isothiocyanate and amino groups.

The first report on calix[4]arene *N*-glycoside synthesis *via* reaction of glycosyl isothiocyanates with lower rim dialkylamino calix[4]arene **115** was reported in 1999 by Santoyo-González and co-workers.[69] Scheme 4.23 exemplifies this approach by showing the reaction of **115** with peracetylated lactosyl isothiocyanate **116** to give the calix[4]arene *N*-glycoside **117**. The same approach was successfully applied to *gluco*-, *galacto*- and *manno*-configured isothiocyanates. Interestingly, NMR analysis revealed that these calix[4]arene glycoconjugates adopted a 1,2-alternate conformation as shown for **117** and form dimers through intermolecular hydrogen bondings. This structural arrangement precluded the use of these compounds for probing the multivalent effect.

The thioureido-tethered calix[4]arene *N*-glycoside **120** assembled in a well-defined cone conformation was prepared by Ungaro's group in the early 2000s[70] (Scheme 4.24). With this aim, the conformationally rigidified tetrapropoxycalix[4]arene diamine **118** was allowed to react with the peracetylated glucosyl isothiocyanate **119** and the product was deacetylated to give **120** in good overall yield. The close flattened cone conformation for this compound was supported by NMR data. This preferential structure was attributed to hydrogen bonding interactions involving the thiourea group of one tether and the glycoside endocyclic oxygen in the other chain. Ungaro and co-workers also prepared the tetrafunctionalized calix[4]arene *N*-glycoside **121** by addition of glucosyl isothiocyanate **119** to the calix[4]arene tetraamine **43** (Scheme 4.25). In this case, the NMR spectra showed the expected C_4 symmetry and no particular conformational feature connected with the calix[4]arene ring mobility. It was pointed out, however,

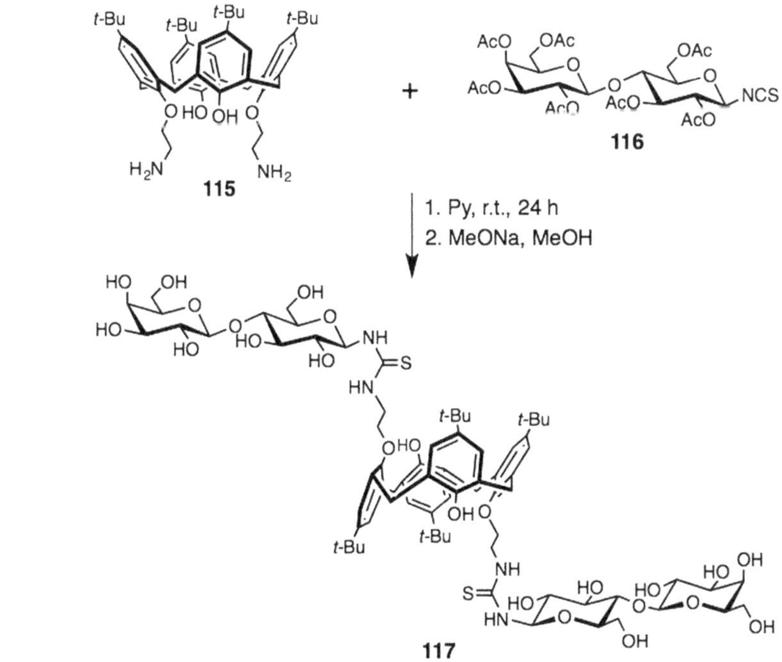

Scheme 4.23

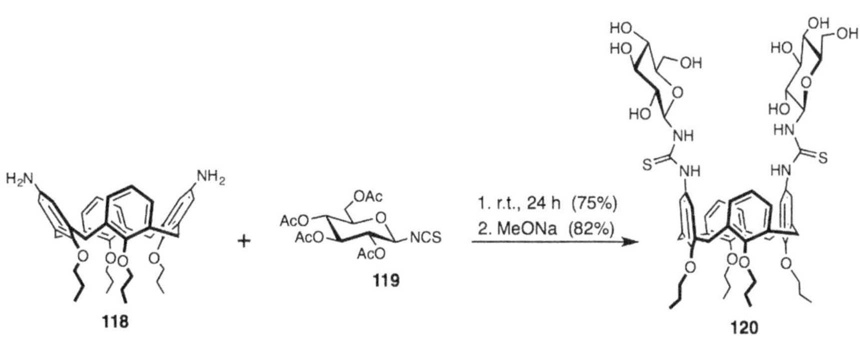

Scheme 4.24

that the NMR spectra of calixsugars **120** and **121** at room temperature showed some broad signals, in particular those corresponding to the NHs and the aromatic and anomeric protons. This was attributed to rotamers involving the thiourea groups. Accordingly, sharpening of the signals was observed at higher temperatures.

Ungaro and co-workers[70] focused their attention on the interaction of concanavalin A (ConA), which binds selectively α-D-glucoside and α-D-mannoside residues, and peanut lectin (PNA), which binds β-D-galactosides, with the tetravalent glucocluster **121** and the galactocluster **146** (see Figure 4.20)

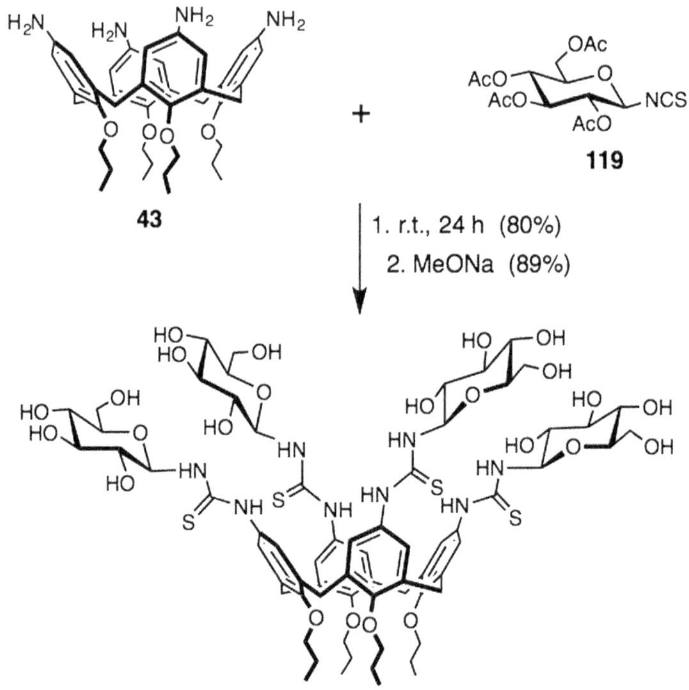

Scheme 4.25

by turbidimetric analysis. The absorbance increase for a mixture of ConA and calix[4]arene *N*-glucoside **121** in water at various time intervals, followed by an absorbance decrease upon addition of a large excess of D-glucose, revealed the specific involvement of the glucose residues of **121** in the interaction with the lectin. On the other hand, no turbidity was observed on mixing PNA with **121**.

In the same paper,[70] Ungaro and co-workers reported the viability of calix[4]arene *N*-glycosides by a complementary approach to that shown in Scheme 4.24. Specifically, the synthesis of the *N*-lactoside **124** involved the coupling of the peracetylated lactosylamine **123** with the diisothiocyanate calix[4]arene **122** followed by deacetylation (Scheme 4.26). The yield of **124**, however, was much lower than those of **120** and **121** and this approach was therefore abandoned.

The calixsugars **120** and **121** and also the methylated analogue of the latter, *i.e.* the tetraglucosyl calix[4]arene bearing four methoxy groups instead of four propoxy groups at the lower rim, were subjected to complexation assays with a series of simple mono- and dicarboxylates.[71] The selective complexation of the anions was studied by electrospray ionization Fourier transform ion cyclotron resonance (ESI-FTICR) mass spectrometry (MS) in the negative ion mode. The MS experiments established that the selectivity increases

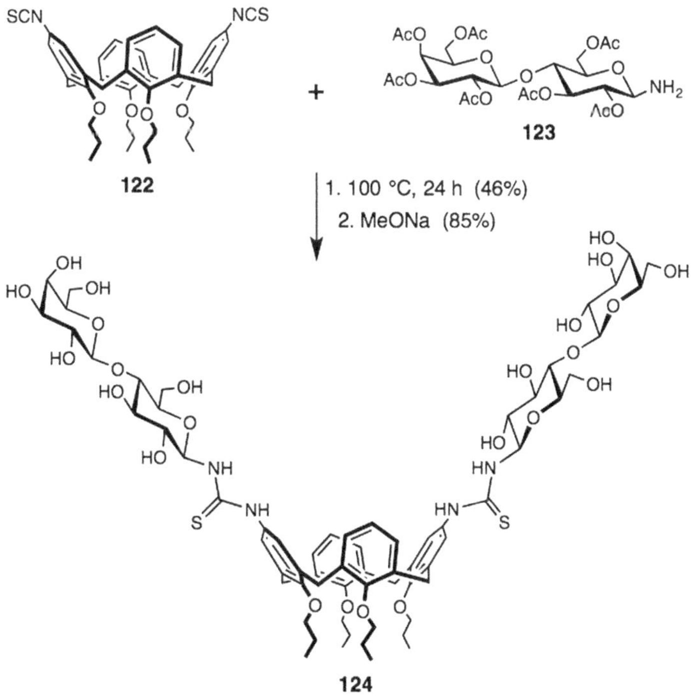

Scheme 4.26

along with the valency of the calixsugar and its rigidity. On the other hand, no chiral discrimination was observed when the two tetravalent glucosylcalixarenes were complexed with the enantiomers of 2-phenylpropionate and 2-hydroxy-3-phenylpropionate.

Thioureido tethering for calix[4]arene glycoclustering was also used by Consoli *et al.*,[72] who focused on the use of a calix[8]arene platform instead of the traditional calix[4]arene. They speculated that the large number of carbohydrate residues, the larger size of the hydrophobic cavity and the flexibility of the macrocycle would be beneficial for intense molecular recognition processes. Thus, the octaaminocalix[8]arene **125**, which was prepared for the first time, was allowed to react with peracetylated glycosyl isothiocyanates **119** and **126–130** (D-glucosyl, 2-acetamido-2-deoxy-D-glucosyl, D-galactosyl, L-fucosyl, D-maltosyl and D-cellobiosyl isothiocyanate, respectively) to give the corresponding thioureido-tethered calix[8]arene *N*-glycosides in very good yields (Scheme 4.27). The removal of the *O*-acetyl protecting groups by transesterification afforded the target octavalent calix[8]arene-based glycoclusters **131a–f**, all showing good solubility in pure water. Complexation experiments followed by NMR titration revealed the formation of complexes between D-glucosamine hydrochloride and octagalactosyl (**131c**) and octacellobiosyl derivatives (**131f**) with association constants of 934 ± 90 and 764 ± 65 M^{-1}, respectively. Hence there appeared to be a higher affinity for the guest by

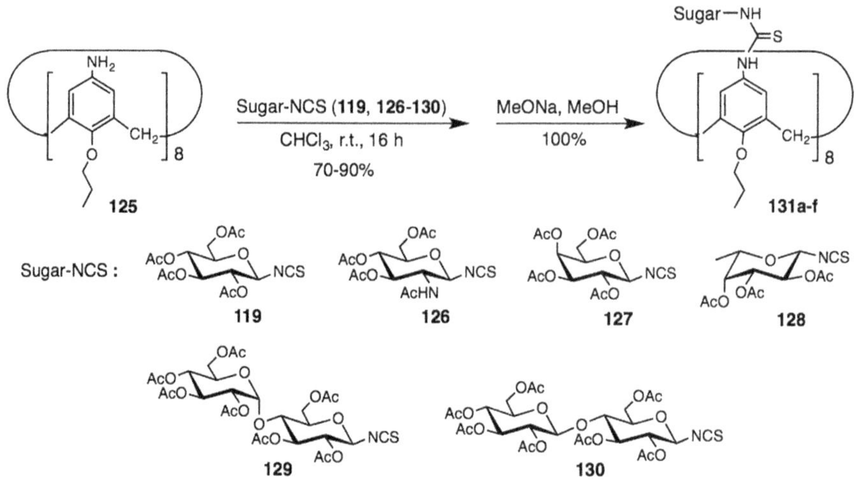

Scheme 4.27

the monosaccharide-bearing receptor with respect to the disaccharide-based receptor. This demonstrated that, in contrast to the above premises, increasing the number of sugar hydroxy groups in the host does not necessarily imply an increase in affinity. Sortino and co-workers[73] found that the glycocluster **131b**, exposing eight GlcNAc residues,[72] was able to bind strongly to the β-1,4-galactosyltransferase V, a cell surface enzyme responsible for the invasion, growth and survival of glioma tumor cells. Owing to this efficient recognition process, **131b** inhibited rat C6 glioma cell migration and proliferation, although the latter effect did not depend on the presence of sugar units but was related to the multiple ureido functions.

Consoli *et al.* also carried out the synthesis of *N*-glycosylated calix[4]arenes and calix[8]arenes exposing *N*-acetyl-D-glucosamine (GlcNAc) clusters at the upper rim.[74] The carbohydrate fragments were grafted on the macrocycle platform through a rather complex and chiral tether constituted of amino acid and thioureido groups. The selected *N*-Boc-protected amino acid residues (*e.g.* L-glycine) were introduced in the calixarene platform by multiple amidation of the calix[4]arene tetraamine **43** or calix[4]arene octaamine **125** (Scheme 4.28). Then the *N*-Boc groups in **132** and **133** were removed by acid treatment and the polyamines thus formed were reacted with 2-acetamido-2-deoxy-D-glucosyl isothiocyanate **126** to give, after de-*O*-acetylation, the target thioureido-amino acid-bridged calix[*n*]arenes *N*-glycosides **134** ($n = 4$) and **135** ($n = 8$) in very good isolated yields. The NMR spectra confirmed the usual cone conformation for **134** and a flexible arrangement for **135**.

Interaction of tetra- (**134, 138**) and octavalent (**131b, 135**) glycoclusters with wheat germ agglutinin (WGA), a lectin known to be GlcNAc specific, was examined by turbidimetric analysis and hemagglutination inhibition tests.[74] The WGA–glycocluster crosslinking was evidenced by turbidity formation and the disruption of such GlcNAc–protein interaction was demonstrated by addition

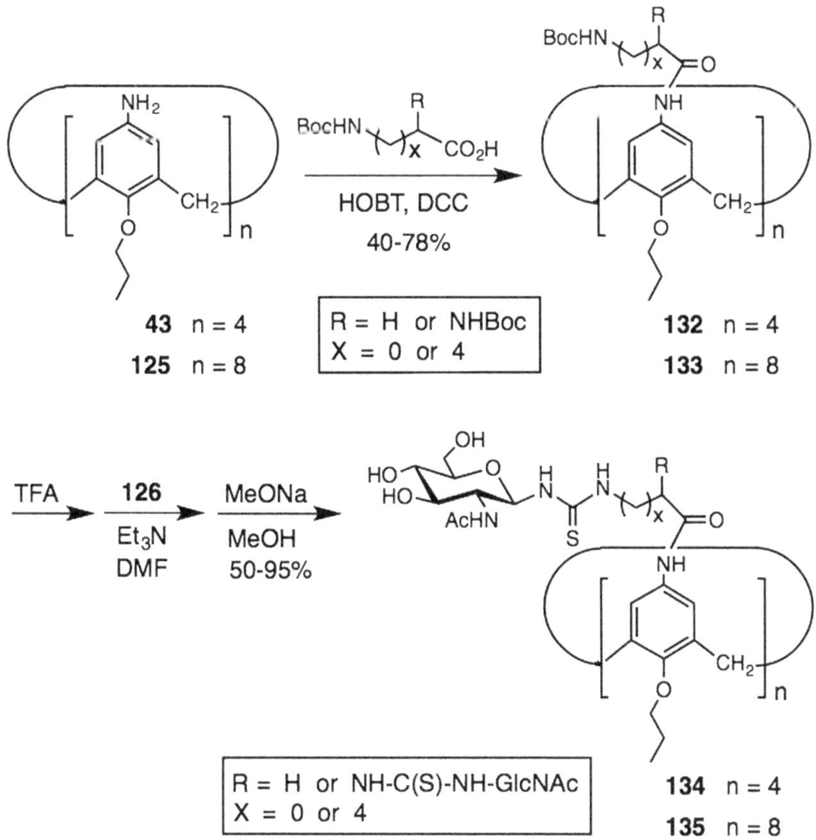

Scheme 4.28

of a large excess of D-GlcNAc. In support of the recognition specificity of WGA by GlcNAc residues were the observations that D-galactose did not induce any disruption of the observed aggregation and that PNA lectin, specific for β-D-galactosides, did not bind to any of the glycoclusters **131b, 134, 135** and **138**.

The occurrence of the glycoside cluster effect was demonstrated by the ability of **131b**, **134**, **135** and **138** to inhibit the WGA-dependent hemagglutination at 0.1–1 mM concentrations. All the prepared glycoclusters turned out to be more potent than D-GlcNAc (up to 312-fold in the case of **134** having R = H, X = 4). Two other important observations were made about the three-dimensional arrangement of the scaffold and the linker holding the cluster domain. The tetravalent compound **134** blocked in the cone conformation and bearing both the amino acid and thioureido groups as spacers was shown to be more potent than the tetravalent and octavalent derivatives **138** and **131b**, respectively, featuring only the thioureido group as a spacer. Therefore, the fine combination of special structural features appeared to be crucial for an effective glycoside cluster effect to operate. Moreover, it was also demonstrated that a higher valency glycocluster such as a hexadecavalent

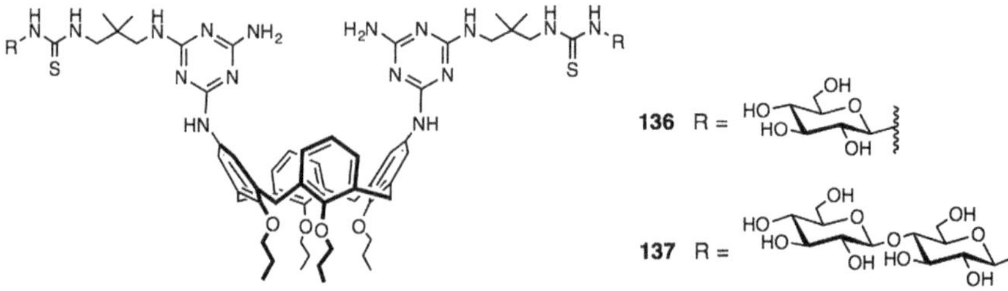

Figure 4.18 Bisglycosylated calix[4]arenes prepared by Crego-Calama, Reinhoudt and co-workers.[75]

derivative displayed lower inhibitory activity than the tetra- and octavalent derivatives. It was suggested that the crosslinking with crowded glycoclusters was prevented because of an improper spacing or geometry of the carbohydrate fragments.

Targeting natural receptor mimics, Crego-Calama and co-workers[75] reported on the synthesis of calix[4]arenes diametrically substituted at the upper rim with two melamine units bearing simple achiral groups (alkyl, pyridyl, *etc.*) or biologically relevant moieties such as the *N*-glycosylated calixarenes **136** and **137** (Figure 4.18). The self-assembly of these calixsugars with barbituric or cyanuric acid derivatives into hydrogen-bonded, double-rosette nanostructures was studied by NMR and circular dichroism spectroscopy.

The preparation of the *N*-acetyl-D-glucosamine (GlcNAc)-containing calix[4]arenes **138, 139** and **140** (Figure 4.19) featuring only the thioureido group as a tether was achieved by Krenek *et al. via* the usual isothiocyanate-to-amine coupling reaction.[76] The geometric arrangements of these compounds shown in Figure 4.19 were supported by the method of synthesis and NMR data. The authors considered the binding activity of their GlcNAc-based calixsugars to C-type lectin-like activation receptors NKR-P1 (CD161) expressed by rat natural killer (NK) cells and CD69, expressed by human NK cells. Although ManNAc and GalNAc have the highest affinities for NKR-P1, the convenient choice of calix[4]arene-based GlcNAc clusters relied on their easy preparation and the low cost of the monosaccharide. Moreover, the CD69 receptor displayed higher affinity for GlcNAc than for GalNAc. Based on the results obtained from a set of experiments, it was concluded that di- and tetravalent glycoclusters **140** and **138** (see Figure 4.19) were better inhibitors for NKR-P1 than **139** featuring a blocked 1,3-alternate conformation. Parallel experiments demonstrated that the tetraglycosylated calixarene **138** was the best ligand for the CD69 receptor. Moreover, the latter glycocluster was also tested for its immunomodulatory effect in a functional cytotoxicity assay *in vitro*. It was found that **138** was more active, at picomolar concentrations, than a known[77] octavalent polyamidoamine (PAMAM) GlcNAc dendrimer used as a reference compound.[78] Moreover, unlike the above-mentioned dendrimer, the glycocluster **138** was able to induce the apoptosis of tumor cells.

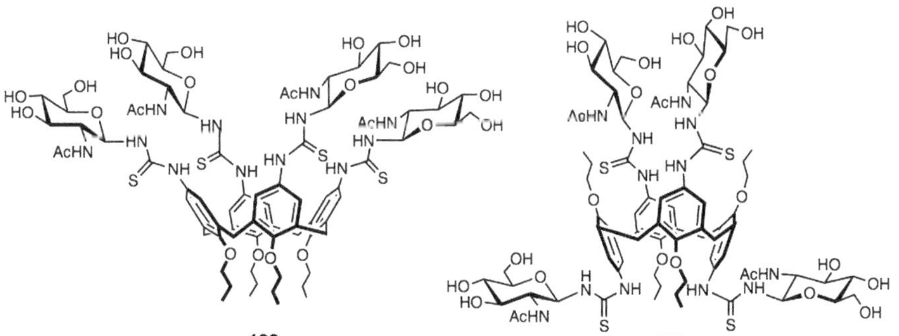
138 139 140

Figure 4.19 Thioureido-tethered glycocalix[4]arene prepared by Krenek *et al.*[76]

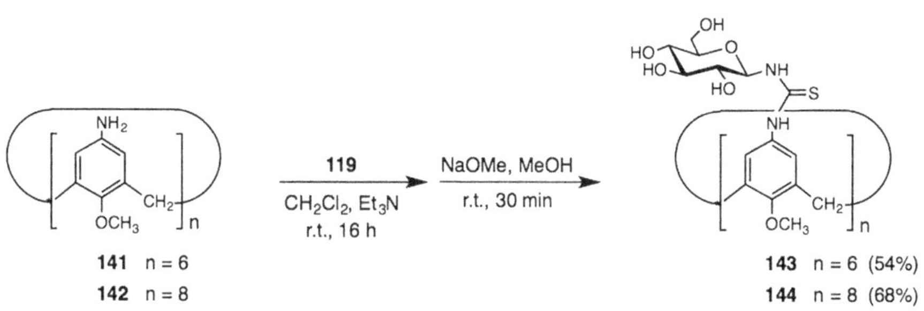

Scheme 4.29

Casnati and co-workers[79] described the synthesis of the glucosylthioureidocalix[6]arene **143** and calix[8]arene derivatives **144** starting from the corresponding hexa- and octaaminocalixarenes **141** and **142** (Scheme 4.29), respectively, the latter being the methylated analogue of **125** (see Scheme 4.27). The isothiocyanate-to-amine coupling, followed by transesterification under basic conditions, gave the hexa- and octavalent *N*-glucoside clusters **143** and **144** in good yields. Although these compounds were fairly soluble in water (0.1–1 mM), they showed a tendency to form self-assembled aggregates, as proved by atomic force microscopy (AFM) and dynamic light scattering (DLS) analyses. Moreover, AFM was also employed to study the interaction of **143** with plasmid DNA. In the presence of the glucocluster **143**, the DNA filaments of the supercoiled plasmids aggregated significantly but did not lead to filament condensation or changes in the DNA folding. The lack of the

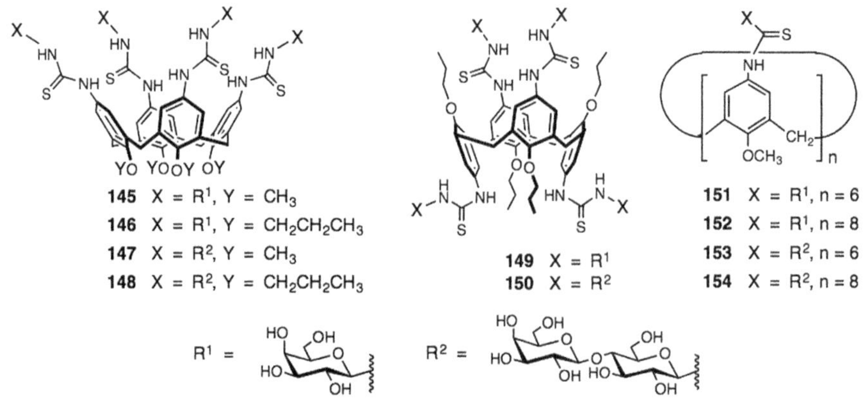

Figure 4.20 Thioureido-tethered glycocalixarene prepared by Ungaro, Gabius and co-workers.[81a]

latter behavior prevented the use of **143** as a synthetic gene delivery system. Turbidimetric experiments were carried out using the hexa- (**143**) and octavalent (**144**) glucoclusters (see Scheme 4.29) and ConA.[79] Although both clusters agglutinated the lectin, the binding of the octamer **144** was considerably stronger than that shown by **143**. On the other hand, the two *N*-glucosylated calixarenes did not bind to the peanut lectin, which is selective for the D-galactoside moieties.

A few years later, the complexation properties of the glucosylthioureidocalix[6]arene **143** and calix[8]arene **144** with several bisphosphonates currently used in the treatment of osteoporosis were studied by Fourier transform ion cyclotron resonance (FT-ICR) MS.[80] It was found that both calixsugars formed complexes with the bisphosphonates, the octavalent derivative **144** showing the highest affinity. On the other hand, in the case of the bisglucosylthioureidocalix[4]arene **120** (see Scheme 4.24) and the tetraglucosylthioureidocalix[4]arene **121** (see Scheme 4.25), the corresponding complexes were not observed, very likely because of the smaller size of the macrocycle.

In 2008, Ungaro's group, in collaboration with Gabius and co-workers, reported on the preparation and biological properties of another series of thioureido-linked calixsugars (Figure 4.20).[81] Exploiting the synthetic approach outlined in Scheme 4.29, they prepared *N*-galactosyl- and *N*-lactosylcalix[4]arenes in cone (**145–148**) and 1,3-alternate conformations (**149, 150**), and also *N*-galactosyl- and *N*-lactosylcalix[6]arenes (**151, 153**) and -calix[8]arenes (**152, 154**).

These galactoside and lactoside clusters were subjected to lectin inhibition solid-phase assays using the plant toxin *Viscum album* agglutinin (VAA). With this aim, the glycoprotein asialofetuin was adsorbed on a microtiter plate while the biotinylated VAA and the glycoclusters, at various concentrations, were in solution. This study revealed that, among the galactoclusters under investigation, only the hexamer **151** and the octamer **152** inhibited the binding of VAA to the immobilized asialofetuin more efficiently than the

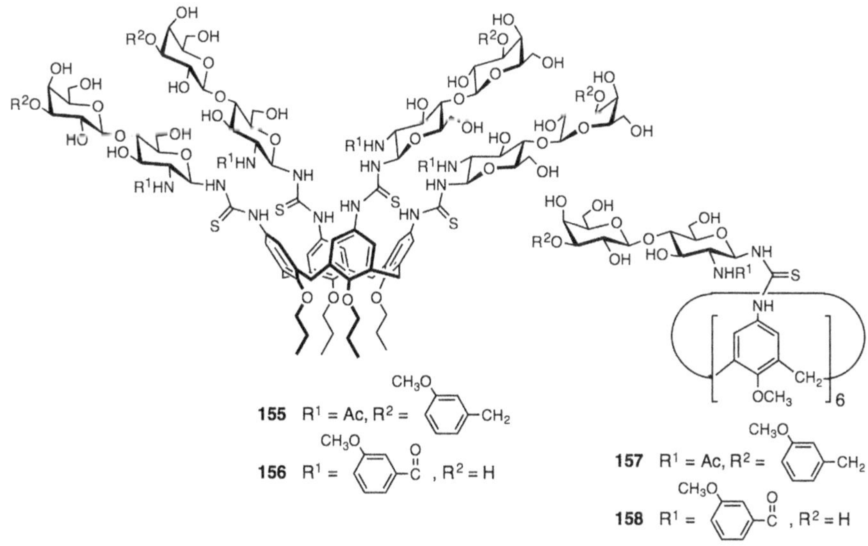

Figure 4.21 Lactosamine-bearing calix[4]arene and calix[6]arene derivatives.

corresponding monomeric ligand, *i.e.* free D-galactose. When similar experiments were performed with the lactoside clusters, it was found that also in this case the hexavalent and octavalent derivatives **153** and **154**, respectively, were stronger inhibitors than free D-lactose. Moreover, the tetravalent lactoside clusters **147**, **148** and **150** were slightly stronger VAA inhibitors than D-lactose.

Solid-phase assays were also employed to evaluate the inhibition activity of the lactoside clusters towards human galectin-1, -3 and -4, important factors in tumor progression. Galectin-1 and -3 were inhibited at a variable extent by the lactosylated calixarenes, whereas galectin-4 was strongly inhibited by these compounds, the hexamer **153** (see Figure 4.20) and the bis-lactoside calix[4]arene derivative **124** (see Scheme 4.26) being the most active glycoclusters (IC_{50} values 300-fold lower than that of free D-lactose).[81a] Aiming to develop potential medical applications, the activity of the above-mentioned glycoclusters was studied by means of cell-binding assays. Under these conditions, the lactosylated calixarenes were found to be active for plant lectin VAA and human galectins, although the IC_{50} values determined in the solid-phase assays did not automatically translate into a direct ranking in the cell assays.

In order to establish the best ligand substitution and ligand valency for the binding to human galectins and VAA, Ungaro and co-workers prepared two calix[4]arenes in a fixed cone conformation (**155**, **156**) and two conformationally mobile calix[6]arenes (**157**, **158**) bearing D-lactosamine moieties linked to the scaffold by a thiourea spacer (Figure 4.21).[82] The sugar ligand displayed bulky aromatic substituents at the N-2 or O-3′ positions in order to improve the affinity toward the lectins.

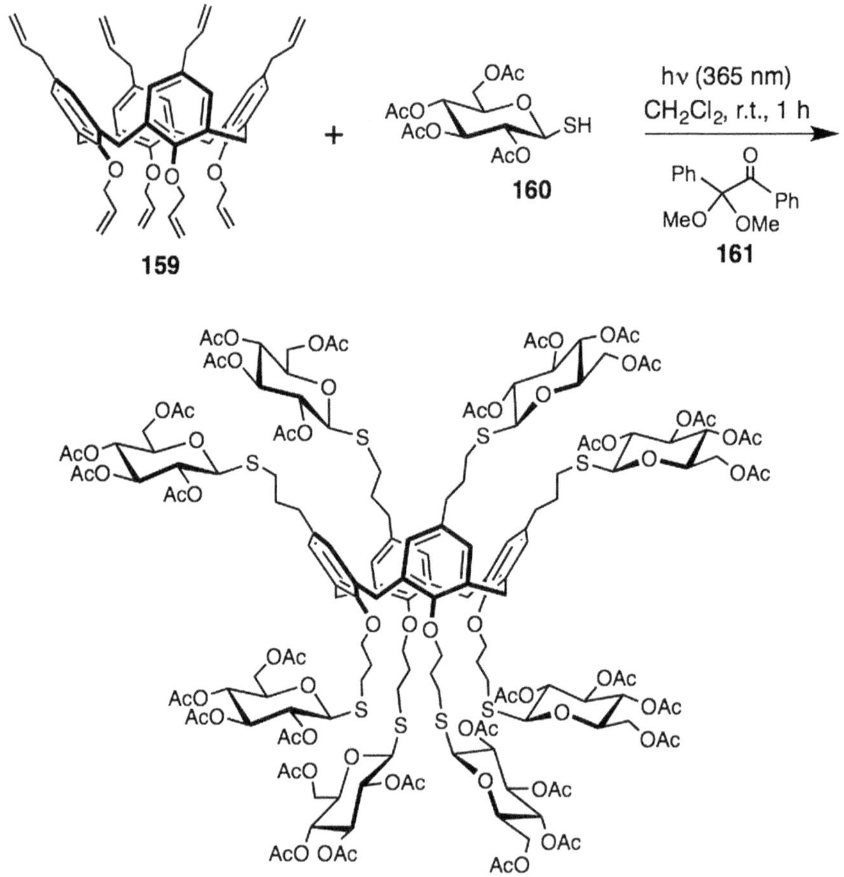

162 (67%)

Scheme 4.30

4.11 Calixsugar Assembly *via* Thiol–Ene Coupling

The free-radical addition of thiols to terminal alkenes, known as thiol–ene coupling, is a century-old reaction that proceeds with anti-Markovnikov regioselectivity and leads to linear thioethers.[83] This ligation tool was exploited by Dondoni and co-workers for the preparation of calix[4]arene *S*-glycosides.[84] Typically, the photoinduced (λ_{max} = 365 nm) addition of the glucosylthiol **160** to the octaallylcalixarene **159** in the presence of 2,2-dimethoxy-2-phenylacetophenone (DPAP) (**161**) as the sensitizer, gave the octavalent glycocluster **162** in 67% yield, corresponding to a 95% average yield for each thiol–ene coupling (Scheme 4.30).

An efficient dual labeling of a calix[4]arene platform with two different carbohydrate fragments at the lower and upper rims was carried as shown in Scheme 4.31. The method involved a sequence of thiol–ene coupling and Cu(I)-catalyzed azide–alkyne cycloaddition. Thus, tetraene- and

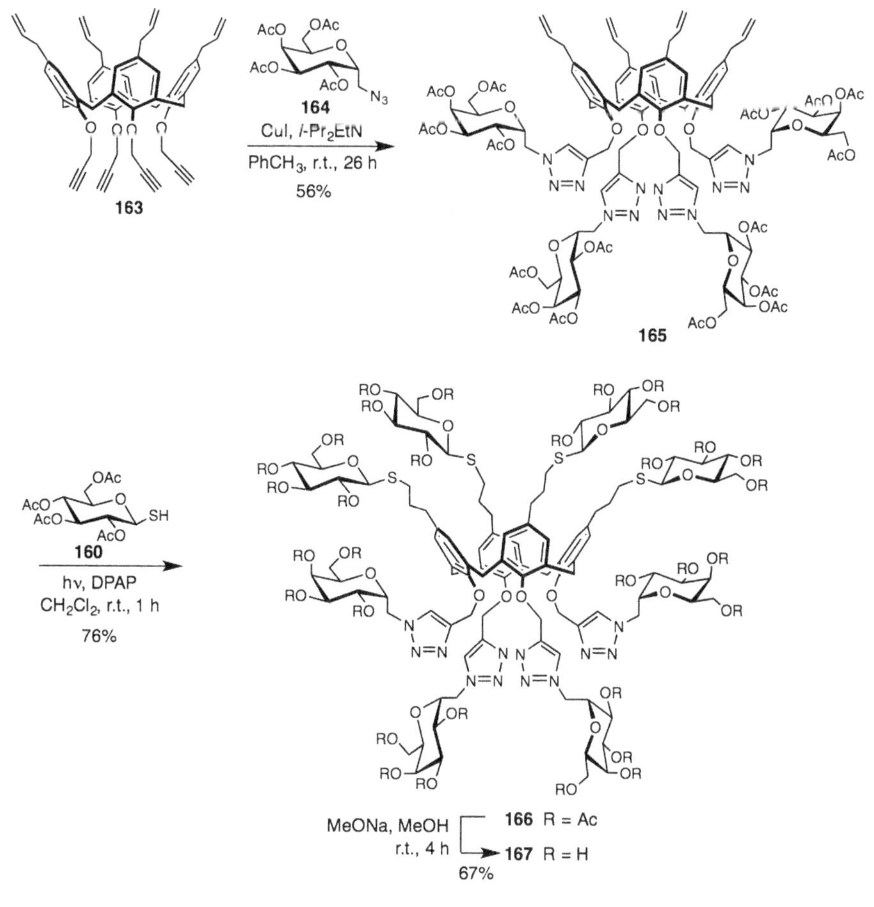

Scheme 4.31

tetrayne-functionalized calix[4]arene **163** was first reacted with galactosylmethyl azide **164** and then with glucosylthiol **160**. Compound **167** thus prepared was the first example of a differentially glycosylated calixarene. Given the specificity of carbohydrate–lectin recognition, it was anticipated that **167** can interact with two different lectins at the same time.

4.12 Miscellanea

The glycocalixarenes described in the previous sections feature the carbohydrate moiety covalently linked to the calixarene scaffold through the anomeric carbon *via* C–O, C–N, C–S and C–C bonds. However, some calixarene–sugar derivatives in which the sugar units are linked to the platform at the non-anomeric position have also been reported.

In order to exploit the amphiphatic properties of calixsugars as hosts for ions and neutral molecules, Dondoni *et al.* synthesized rigid structures in which a

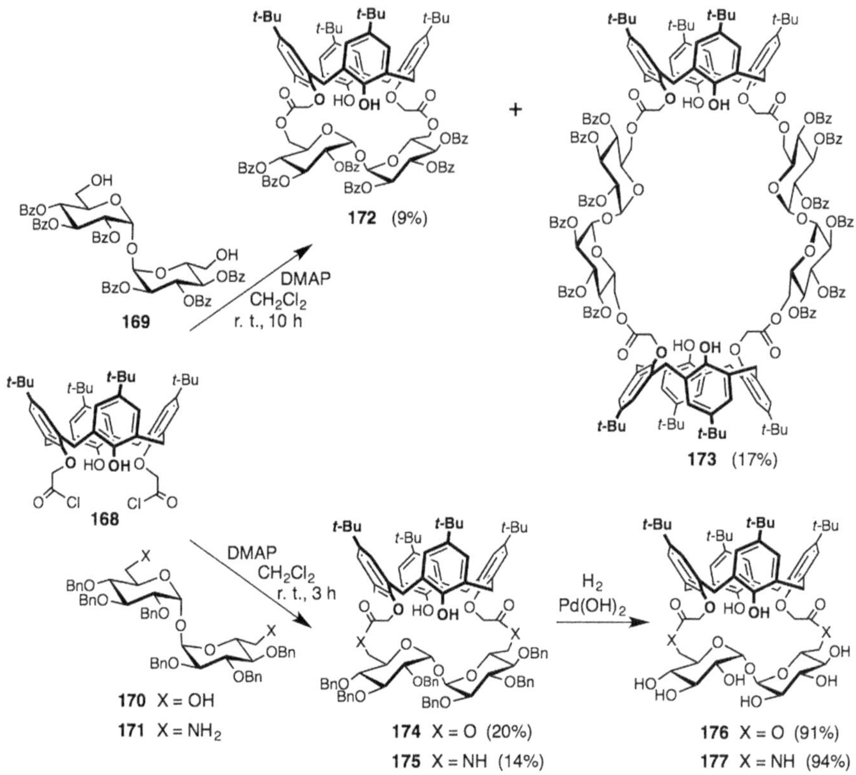

Scheme 4.32

cavity is formed by bridging two 1,3-distal lower rim sites of a calix[4]arene molecule with a disaccharide or by assembling bis-calix[4]arene systems through lower rim–lower rim carbohydrate linkages.[85] The readily available α,α-trehalose diols **169** and **170** and the diamine **171** were selected as disaccharide units for the construction of novel macrocycles *via* the formation of ester and amide linkages with the bis-carboxylic acid chloride **168** (Scheme 4.32). The receptor ability of all the prepared calixsugars towards neutral and charged molecules was investigated by ^{1}H NMR analysis. Only the perbenzoylated double calixsugar **173** in $CDCl_3$ solution showed substantial interaction with imidazole, and it appeared to be inert towards substituted derivatives (1-methylimidazole, benzimidazole) and various nitrogen-containing heterocycles such as pyrrole, pyridine, thiazole, 2-aminopyrimidine and 1,8-diazabicyclo[5.4.0]undec-7-ene (DBU). For the determination of association constant (K_{as}) values, eight different host and guest concentrations were considered using as a probe three proton signals (H-2 and H-6 of the sugar unit and one aromatic proton). The resulting K_{as} value of 11.1 ± 2.5 M^{-1} was calculated by non-linear regression.

Neri and co-workers reported on the so called '*p*-bromodienone route' as a conceptually novel approach for the direct introduction of nucleophiles at the calixarene upper rim using easily accessible calixarene *p*-bromodienone

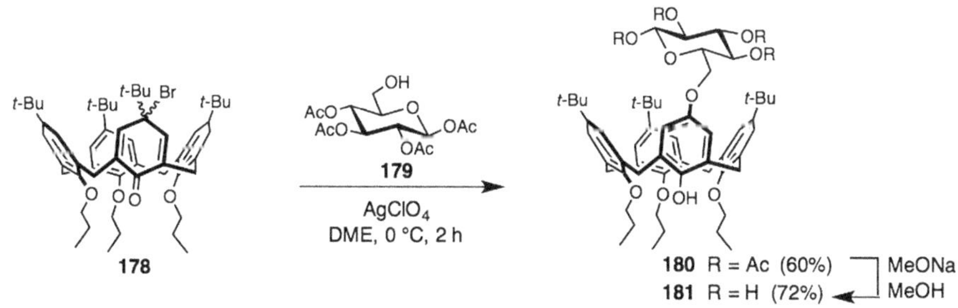

Scheme 4.33

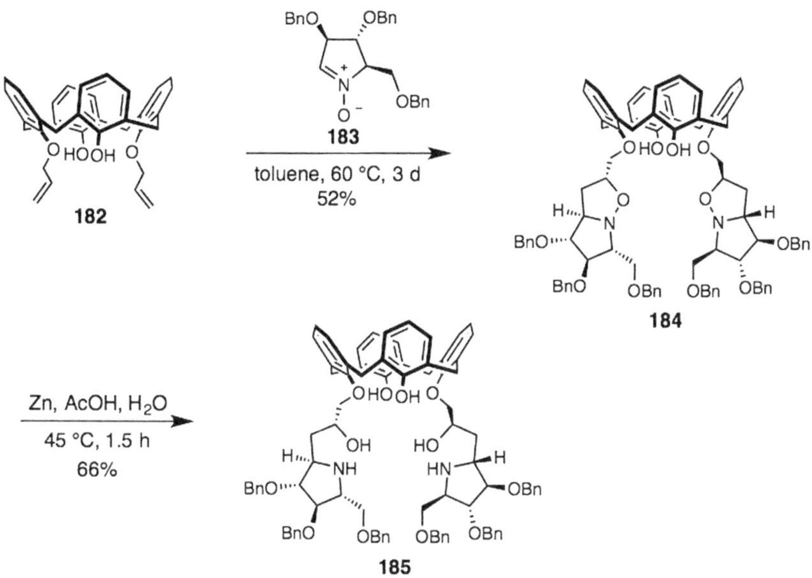

Scheme 4.34

derivatives.[86] In the *p*-bromodienone route, derivatives such as **178** (Scheme 4.33) undergo a silver-mediated nucleophilic substitution and a subsequent rearomatization with a range of different *O*-nucleophiles (alcohols and carboxylates) to give *p*-alkoxy- or *p*-acyloxycalixarenes. In order to expand the potential of the *p*-bromodienone route, the same group[87] investigated the use of appropriate nucleophiles containing glyco, peptido, carboxy and alkyne substituents, which could be useful for biomimetic recognition or for further synthetic elaboration. They described the bromine substitution of **178** with the primary hydroxyl group of 1,2,3,4-tetra-*O*-acetyl-β-D-glucose **179** (10 equiv.) to give the corresponding calixarene–glucose derivative **180** in 60% yield, which was deacetylated to afford the glycocalixarene **181** (Scheme 4.33). Evidently the scope of this method is limited to the preparation of monoglycosylated calix[4]arene derivatives, which, however, can be useful as monovalent ligands in lectin recognition studies.

A special kind of glycocalixarenes in which iminosugar residues are linked to the calixarene scaffold by a non-anomeric linkage was prepared by Casnati and co-workers.[88] The method employed consisted of the 1,3-dipolar cycloaddition reaction of enantiopure cyclic nitrones (*e.g.* **183**) to calix[4]arenes functionalized at the upper or lower rim with two (*e.g.* **182**) or four *O*-allyl groups (Scheme 4.34). N–O bond cleavage of the isoxazolidine ring of the resulting pyrroloisoxazole adduct **184** afforded the iminosugar residue linked to the calixarene scaffold through a hydroxylated alkyl tether (**185**). Unfortunately, attempts to debenzylate the iminosugar fragment by catalytic hydrogenation were unsuccessful, thus preventing the use of these glycocalixarenes in biological assays. This was a serious hurdle to assess the efficiency of these compounds as multivalent glycosidase inhibitors. Nevertheless, the route is now open for preparing suitable iminosugar–calixarene conjugates *via* the nitrone-based methodology.

4.13 Conclusions

Overall, the studies considered in the above sections reveal that the development of synthetic methods for glycocalixarenes by a variety of ligation strategies has been more deeply investigated than applications in glycobiology. The progress in preparative approaches is well documented by impressive results in the synthesis of new glycoconjugates, especially by the fashionable Cu(I)-catalyzed azide–alkyne cycloaddition. Also, the issue of the glycoside cluster effect has been addressed in various papers.

Numerous synthetic and biological studies have emerged from the simple idea conceived in an Italian laboratory (University of Ferrara, 1994) consisting in introducing carbohydrate moieties at each rim of a calix[4]arene scaffold. The studies that followed in many laboratories have considerably broadened the scope of calixarenes from merely ion receptors to efficient auxiliaries in supramolecular chemistry. It is likely that the latter will be the new direction in which the use of calixsugars will be oriented in the coming years.

References

1. (a) A. Varki, *Glycobiology*, 1993, **3**, 97–130; (b) H. Lis and N. Sharon, *Chem. Rev.*, 1998, **98**, 637–674; (c) C. R. Bertozzi and L. L. Kiessling, *Science*, 2001, **291**, 2357–2364; (d) A. Varki, R. Cummings, J. D. Esko, H. H. Freeze, P. Stanley, C. R. Bertozzi, G. W. Hart and M. E. Etzler, *Essentials of Glycobiology*, 2nd edn, Cold Spring Harbor Laboratory Press, Cold Spring Harbor, NY, 2009.
2. N. Jayaraman, *Chem. Soc. Rev.*, 2009, **38**, 3463–3483.
3. (a) M. Mammen, S. K. Choi and G. M. Whitesides, *Angew. Chem., Int. Ed.*, 1998, **37**, 2754–2794; (b) R. T. Lee and Y. C. Lee, *Glycoconjugate J.*, 2000, **17**, 543–551; (c) J. E. Gestwicki, C. W. Cairo, L. E. Strong, K. A. Oetjen and L. L. Kiessling, *J. Am. Chem. Soc.*, 2002, **124**, 14922–14933.

4. (a) Y. C. Lee and R. T. Lee, *Acc. Chem. Res.*, 1995, **28**, 321–327; (b) L. L. Kiessling, J. E. Gestwicki and L. E. Strong, *Curr. Opin. Chem. Biol.*, 2000, **4**, 496–503; (c) J. J. Lundquist and E. J. Toone, *Chem. Rev.*, 2002, **102**, 555–578.
5. (a) Y. M. Chabre and R. Roy, *Adv. Carbohydr. Chem. Biochem.*, 2010, **63**, 165–393; (b) J. L. Jiménez Blanco, C. Ortiz Mellet and J. M. García Fernández, *Chem. Soc. Rev.*, 2013, **42**, 4518–4531.
6. (a) I. J. Krauss, J. G. Joyce, A. C. Finnefrock, H. C. Song, V. Y. Dudkin, X. Geng, J. D. Warren, M. Chastain, J. W. Shiver and S. J. Danishefsky, *J. Am. Chem. Soc.*, 2007, **129**, 11042–11044; (b) O. Renaudet, *Mini-Rev. Org. Chem.*, 2008, **5**, 274–286; (c) M. C. Galan, P. Dumy and O. Renaudet, *Chem. Soc. Rev.*, 2013, **42**, 4599–4612.
7. (a) P. M. Rendle, A. Seger, J. Rodrigues, N. J. Oldham, R. R. Bott, J. B. Jones, M. M. Cowan and B. G. Davis, *J. Am. Chem. Soc.*, 2004, **126**, 4750–4751; (b) T. K. Dam, H. J. Gabius, S. Andre, H. Kaltner, M. Lensch and C. F. Brewer, *Biochemistry*, 2005, **44**, 12567–12571; (c) K. J. Doores, D. P. Gamblin and B. G. Davis, *Chem. Eur. J.*, 2006, **12**, 656–665; (d) H. Nagai, A. Onoda, T. Matsuo and T. Hayashi, *Supramol. Chem.*, 2010, **22**, 57–64.
8. (a) M. J. Cloninger, *Curr. Opin. Chem. Biol.*, 2002, **6**, 742–748; (b) P. Wu, M. Malkoch, J. N. Hunt, R. Vestberg, E. Kaltgrad, M. G. Finn, V. V. Fokin, K. B. Sharpless and C. J. Hawker, *Chem. Commun.*, 2005, 5775–5777; (c) C. D. Heidecke and T. K. Lindhorst, *Chem. Eur. J.*, 2007, **13**, 9056–9067; (d) J.-L. Reymond, M. Bergmann and T. Dabre, *Chem. Soc. Rev.*, 2013, **42**, 4814–4822.
9. (a) R. Narain and S. P. Armes, *Biomacromolecules*, 2003, **4**, 1746–1758; (b) L. Bes, S. Angot, A. Limer and D. M. Haddleton, *Macromolecules*, 2003, **36**, 2493–2499; (c) I. Otsuka, B. Blanchard, R. Borsali, A. Imberty and T. Kakuchi, *ChemBioChem*, 2010, **11**, 2399–2408.
10. (a) J. E. Gestwicki, L. E. Strong, S. L. Borchardt, C. W. Cairo, A. M. Schnoes and L. L. Kiessling, *Bioorg. Med. Chem.*, 2001, **9**, 2387–2393; (b) D. Cade, E. Ramus, M. Rinaudo, R. Auzely-Velty, T. Delair and T. Hamaide, *Biomacromolecules*, 2004, **5**, 922–927; (c) M. Reynolds, M. Marradi, A. Imberty, S. Penadés and S. Pérez, *Chem. Eur. J.*, 2012, **18**, 4264–4273; (d) M. Marradi, F. Chiodo, I. Garcia and S. Penadés, *Chem. Soc. Rev.*, 2013, **42**, 4728–4745.
11. (a) J. F. Nierengarten, J. Iehl, V. Oerthel, M. Holler, B. M. Illescas, A. Munoz, N. Martin, J. Rojo, M. Sanchez-Navarro, S. Cecioni, S. Vidal, K. Buffet, M. Durka and S. P. Vincent, *Chem. Commun.*, 2010, **46**, 3860–3862; (b) S. Cecioni, V. Oerthel, J. Iehl, M. Holler, D. Goyard, J.-P. Praly, A. Imberty, J. F. Nierengarten and S. Vidal, *Chem. Eur. J.*, 2011, **17**, 3252–3261; (c) Y. Chen, A. Star and S. Vidal, *Chem. Soc. Rev.*, 2013, **42**, 4532–4542.
12. (a) D. A. Fulton and J. F. Stoddart, *Bioconjugate Chem.*, 2001, **12**, 655–672; (b) A. Dondoni and A. Marra, *Chem. Rev.*, 2010, **110**, 4949–4977; (c) F. Sansone and A. Casnati, *Chem. Soc. Rev.*, 2013, **42**, 4623–4639.

13. (a) P. R. Ashton, R. Königer, J. F. Stoddart, D. Alker and V. D. Harding, *J. Org. Chem.*, 1996, **61**, 903–908; (b) D. A. Fulton and J. F. Stoddart, *Org. Lett.*, 2000, **2**, 1113–1116; (c) T. Furuike, S. Aiba and S.-I. Nishimura, *Tetrahedron*, 2000, **56**, 9909–9915; (d) M. Gómez-García, J. M. Benito, A. P. Butera, C. Ortiz Mellet, J. M. García Fernández and J. L. Jiménez Blanco, *J. Org. Chem.*, 2012, **77**, 1273–1288; (e) A. Martinez, C. Ortiz Mellet and J. M. García Fernández, *Chem. Soc. Rev.*, 2013, **42**, 4746–4773.
14. (a) F. Morvan, A. Meyer, A. Jochum, C. Sabin, Y. Chevolot, A. Imberty, J.-P. Praly, J.-J. Vasseur, E. Souteyrand and S. Vidal, *Bioconjugate Chem.*, 2007, **18**, 1637–1643; (b) L. Moni, G. Pourceau, J. Zhang, A. Meyer, S. Vidal, E. Souteyrand, A. Dondoni, F. Morvan, Y. Chevolot, J.-J. Vasseur and A. Marra, *ChemBioChem*, 2009, **10**, 1369–1378; (c) N. Spinelli, E. Defrancq and F. Morvan, *Chem. Soc. Rev.*, 2013, **42**, 4557–4573.
15. (a) D. Heyl, E. Rikowski, R. C. Hoffmann, J. J. Schneider and W.-D. Fessner, *Chem. Eur. J.*, 2010, **16**, 5544–5546; (b) B. Trastoy, M. E. Pérez-Ojeda, R. Sastre and J. L. Chiara, *Chem. Eur. J.*, 2010, **16**, 3833–3841; (c) M. Lo Conte, S. Staderini, A. Chambery, N. Berthet, P. Dumy, O. Renaudet, A. Marra and A. Dondoni, *Org. Biomol. Chem.*, 2012, **10**, 3269–3277; (d) A. Marra, S. Staderini, N. Berthet, P. Dumy, O. Renaudet and A. Dondoni, *Eur. J. Org. Chem.*, 2013, 1144–1149.
16. For chemistry, structure and complexing ability of calixarenes, see: (a) C. D. Gutsche, in *Calixarenes, Monographs in Supramolecular Chemistry*, Vol. 1, ed. J. F. Stoddart, Royal Society of Chemistry, Cambridge, 1989; (b) J. Vicens and V. Böhmer, eds, *Calixarenes, a Versatile Class of Macrocyclic Compounds*, Kluwer, Dordrecht, 1991; (c) V. Böhmer, *Angew. Chem., Int. Ed. Engl.*, 1995, **34**, 713–745; (d) M. Takeshita and S. Shinkai, *Bull. Chem. Soc. Jpn.*, 1995, **68**, 1088–1097; (e) C. D. Gutsche, in *Calixarenes Revisited, Monographs in Supramolecular Chemistry*, Vol. 1, ed. J. F. Stoddart, Royal Society of Chemistry, Cambridge, 1998; (f) L. Mandolini and R. Ungaro, eds, *Calixarenes in Action*, Imperial College Press, London, 2000.
17. A. Marra, M. C. Scherrmann, A. Dondoni, A. Casnati, P. Minari and R. Ungaro, *Angew. Chem., Int. Ed.*, 1994, **33**, 2479–2481.
18. Unmodified calix[4]arenes are conformationally mobile macrocycles that adopt a cone conformation because of strong hydrogen bonds among the hydroxy groups.
19. A. Dondoni, A. Marra, M. C. Scherrmann, A. Casnati, F. Sansone and R. Ungaro, *Chem. Eur. J.*, 1997, **3**, 1774–1782.
20. The *n*-propylation of the lower rim hydroxy groups inhibits the oxygen-through-the-annulus-rotation and allows the isolation of the blocked cone conformational isomer: K. Iwamoto, K. Araki and S. Shinkai, *J. Org. Chem.*, 1991, **56**, 4955–4962.
21. For a review outlining the most common lectins and their carbohydrate-binding partners, see: M. Ambrosi, N. R. Cameron and B. G. Davis, *Org. Biomol. Chem.*, 2005, **3**, 1593–1608.
22. A. Dondoni, M. Kleban, X. Hu, A. Marra and H. D. Banks, *J. Org. Chem.*, 2002, **67**, 4722–4733.

23. A. Marra, A. Dondoni and F. Sansone, *J. Org. Chem.*, 1996, **61**, 5155–5158.
24. For a review, see: A. Dondoni and A. Marra, *Chem. Commun.*, 1999, 2133–2145.
25. (a) A. Dondoni, M. C. Scherrmann, A. Marra and J. L. Delépine, *J. Org. Chem.*, 1994, **59**, 7517–7520; (b) A. Dondoni, A. Marra, I. Rojo and M. C. Scherrmann, *Tetrahedron*, 1996, **52**, 3057–3074; (c) A. Dondoni, S. Daninos, A. Marra and P. Formaglio, *Tetrahedron*, 1998, **54**, 9859–9874; (d) A. Dondoni, A. Marra, M. C. Scherrmann and V. Bertolasi, *Chem. Eur. J.*, 2001, **7**, 1371–1382.
26. For reviews, see: (a) A. Dondoni, *Synthesis*, 1998, 1681–1706; (b) A. Dondoni and A. Marra, *Chem. Rev.*, 2004, **104**, 2557–2600.
27. S. J. Meunier and R. Roy, *Tetrahedron Lett.*, 1996, **37**, 5469–5472.
28. R. Roy and J. M. Kim, *Angew. Chem., Int. Ed.*, 1999, **38**, 369–372.
29. U. Schädel, F. Sansone, A. Casnati and R. Ungaro, *Tetrahedron*, 2005, **61**, 1149–1154.
30. W. B. Turnbull, B. L. Precious and S. W. Homans, *J. Am. Chem. Soc.*, 2004, **126**, 1047–1054.
31. A. Bernardi, L. Carrettoni, A. Grosso Ciponte, D. Monti and S. Sonnino, *Bioorg. Med. Chem. Lett.*, 2000, **10**, 2197–2200.
32. D. Arosio, I. Vrasidas, P. Valentini, R. M. J. Liskamp, R. J. Pieters and A. Bernardi, *Org. Biomol. Chem.*, 2004, **2**, 2113–2124.
33. D. Arosio, M. Fontanella, L. Baldini, L. Mauri, A. Bernardi, A. Casnati, F. Sansone and R. Ungaro, *J. Am. Chem. Soc.*, 2005, **127**, 3660–3661.
34. O. Ouerfelli, J. D. Warren, R. M. Wilson and S. J. Danishefsky, *Expert Rev. Vaccines*, 2005, **4**, 677–685.
35. (a) R. Lo-Man, S. Vichier-Guerre, S. Bay, E. Dériaud, V. Huteau, D. Cantacuzène and C. Leclerc, *J. Immunol.*, 2001, **166**, 2849–2854; (b) L. Cipolla, M. Rescigno, A. Leone, F. Peri, B. La Ferla and F. Nicotra, *Bioorg. Med. Chem.*, 2002, **10**, 1639–1646; (c) R. Lo-Man, S. Vichier-Guerre, R. Perraut, E. Dériaud, V. Huteau, L. BenMohamed, O. M. Diop, P. O. Livingston, S. Bay and C. Leclerc, *Cancer Res.*, 2004, **64**, 4987–4994; (d) S. Grigalevicius, S. Chierici, O. Renaudet, R. Lo-Man, E. Dériaud, C. Leclerc and P. Dumy, *Bioconjugate Chem.*, 2005, **16**, 1149–1159.
36. C. Geraci, G. M. L. Consoli, E. Galante, E. Bousquet, M. Pappalardo and A. Spadaro, *Bioconjugate Chem.*, 2008, **19**, 751–758.
37. E. Bousquet, A. Spadaro, M. S. Pappalardo, R. Bernardini, R. Romeo, L. Panza and G. Ronsisvalle, *J. Carbohydr. Chem.*, 2000, **19**, 527–541.
38. G. M. L. Consoli, G. Granata, V. Cafiso, S. Stefani and C. Geraci, *Tetrahedron Lett.*, 2011, **52**, 5831–5834.
39. C. Félix, H. Parrot-Lopez, V. Kalchenko and A. W. Coleman, *Tetrahedron Lett.*, 1998, **39**, 9171–9174.
40. F. Pérez-Balderas and F. Santoyo-González, *Synlett*, 2001, 1699–1702.
41. For reviews, see: (a) M. H. D. Postema, *C-Glycoside Synthesis*, CRC Press, Boca Raton, FL, 1995; (b) J.-M. Beau and T. Gallagher, *Top. Curr. Chem.*, 1997, **187**, 1–4; (c) X. Yuan and R. J. Linhardt, *Curr. Top. Med. Chem.*, 2005, **5**, 1393–1430.

42. A. Dondoni, M. Kleban and A. Marra, *Tetrahedron Lett.*, 1997, **38**, 7801–7804.
43. (a) V. V. Rostovtsev, L. G. Green, V. V. Fokin and K. B. Sharpless, *Angew. Chem., Int. Ed.*, 2002, **41**, 2596–2599; (b) C. W. Tornøe, C. Christensen and M. Meldal, *J. Org. Chem.*, 2002, **67**, 3057–3062.
44. (a) H. C. Kolb, M. G. Finn and K. B. Sharpless, *Angew. Chem., Int. Ed.*, 2001, **40**, 2004–2021; (b) H. C. Kolb and K. B. Sharpless, *Drug Discov. Today*, 2003, **8**, 1128–1137.
45. A. Dondoni and A. Marra, *J. Org. Chem.*, 2006, **71**, 7546–7557.
46. N. A. Rodios, *J. Heterocycl. Chem.*, 1984, **21**, 1169–1173.
47. A. Vecchi, B. Melai, A. Marra, C. Chiappe and A. Dondoni, *J. Org. Chem.*, 2008, **73**, 6437–6440.
48. S. P. Bew, R. A. Brimage, N. L'Hermite and S. V. Sharma, *Org. Lett.*, 2007, **9**, 3713–3716.
49. The structures of these compounds reported in the original paper of Bew, *et al.* are incorrect because the carbohydrate moiety shown is maltose instead of lactose as stated in the text.
50. A. Marra, L. Moni, D. Pazzi, A. Corallini, D. Bridi and A. Dondoni, *Org. Biomol. Chem.*, 2008, **6**, 1396–1409.
51. L. Moni, S. Rossetti, M. Scoponi, A. Marra and A. Dondoni, *Chem. Commun.*, 2010, **46**, 475–477.
52. J. M. de la Fuente and S. Penadés, *Biochim. Biophys. Acta, Gen. Subj.*, 2006, **1760**, 636–651.
53. (a) K. Kalyanasundadaram and M. Grätzel, *Coord. Chem. Rev.*, 1998, **177**, 347–314; (b) L. Ojamäe, C. Aulin, H. Pedersen and P.-O. Käll, *J. Colloid Interface Sci.*, 2006, **296**, 71–78.
54. L. Moni, G. Pourceau, J. Zhang, A. Meyer, S. Vidal, E. Souteyrand, A. Dondoni, F. Morvan, Y. Chevolot, J.-J. Vasseur and A. Marra, *ChemBioChem*, 2009, **10**, 1369–1378.
55. S. Cecioni, R. Lalor, B. Blanchard, J.-P. Praly, A. Imberty, S. E. Matthews and S. Vidal, *Chem. Eur. J*, 2009, **15**, 13232–13240.
56. D. Sicard, S. Cecioni, M. Iazykov, Y. Chevolot, S. E. Matthews, J.-P. Praly, E. Souteyrand, A. Imberty, S. Vidal and M. Phaner-Goutorbe, *Chem. Commun.*, 2011, **47**, 9483–9485.
57. S. Cecioni, S. Faure, U. Darbost, I. Bonnamour, H. Parrot-Lopez, O. Roy, C. Taillefumier, M. Wimmerová, J.-P. Praly, A. Imberty and S. Vidal, *Chem. Eur. J*, 2011, **17**, 2146–2159.
58. S. Cecioni, S. S. Matthews, H. Blanchard, J.-P. Praly, A. Imberty and S. Vidal, *Carbohydr. Res.*, 2012, **356**, 132–141.
59. Lactosylcalix[4]arenes **99** and **101** were reported for the first time by Galante, *et al.* (see ref. 60); however, the relevant optical rotations and HRMS data were missing. Both sets of analytical data were given in the paper by Cecioni *et al.* (ref. 58) published a few months later.
60. E. Galante, C. Geraci, S. Sciuto, V. L. Campo, I. Carvalho, R. Sesti-Costa, P. M. M. Guedes, J. S. Silva, L. Hill, S. A. Nepogodiev and R. A. Field, *Tetrahedron*, 2011, **67**, 5902–5912.

61. M. E. Giorgi, L. Ratier, R. Agusti, A. C. C. Frasch and R. M. de Lederkremer, *Glycoconjugate J.*, 2010, **27**, 549–559.
62. J. P. Chinta and C. P. Rao, *Carbohydr. Res.*, 2013, **369**, 58–62.
63. S. Aleandri, A. Casnati, L. Fantuzzi, G. Mancini, G. Rispoli and F. Sansone, *Org. Biomol. Chem.*, 2013, **11**, 4811–4817.
64. J. Garcia-Hartjes, S. Bernardi, C. A. G. M. Weijers, T. Wennekes, M. Gilbert, F. Sansone, A. Casnati and H. Zuilhof, *Org. Biomol. Chem.*, 2013, **11**, 4340.
65. A. Dondoni and A. Marra, *Tetrahedron*, 2007, **63**, 6339–6345.
66. (a) R. Huisgen, *Angew. Chem., Int. Ed. Engl*, 1963, **2**, 565–632; (b) R. N. Butler, in *Comprehensive Heterocylic Chemistry*, Vol. 4, ed. A. R. Katritzky, C. W. Rees and E. F. V. Scriven, Pergamon Press, Oxford, 1966.
67. Z. P. Demko and K. B. Sharpless, *Angew. Chem., Int. Ed.*, 2002, **41**, 2110–2113.
68. For accounts mainly dealing with the work carried out by this group, see: (a) A. Casnati, F. Sansone and R. Ungaro, *Acc. Chem. Res.*, 2003, **36**, 246–254; (b) L. Baldini, A. Casnati, F. Sansone and R. Ungaro, *Chem. Soc. Rev.*, 2007, **36**, 254–266.
69. C. Saitz-Barria, A. Torres-Pinedo and F. Santoyo-González, *Synlett*, 1999, 1891–1894.
70. F. Sansone, E. Chierici, A. Casnati and R. Ungaro, *Org. Biomol. Chem.*, 2003, **1**, 1802–1809.
71. M. Torvinen, E. Kalenius, F. Sansone, A. Casnati and Janne Jänis, *J. Mass Spectrom.*, 2011, **46**, 787–793.
72. G. M. L. Consoli, F. Cunsolo, C. Geraci, T. Mecca and P. Neri, *Tetrahedron Lett.*, 2003, **44**, 7467–7440.
73. S. Viola, G. M. L. Consoli, S. Merlo, F. Drago, M. A. Sortino and C. Geraci, *J. Neurochem.*, 2008, **107**, 1047–1055.
74. G. M. L. Consoli, F. Cunsolo, C. Geraci and V. Sgarlata, *Org. Lett.*, 2004, **6**, 4163–4166.
75. M. G. J. ten Cate, M. Omerovic, G. V. Oshovsky, M. Crego-Calama and D. N. Reinhoudt, *Org. Biomol. Chem.*, 2005, **3**, 3727–3733.
76. K. Krenek, M. Kuldová, K. Hulíková, I. Stibor, P. Lhoták, M. Dudic, J. Budka, H. Pelantová, K. Bezouska, A. Fiserová and V. Kren, *Carbohydr. Res.*, 2007, **342**, 1781–1792.
77. T. K. Lindhorst and C. Kieburg, *Angew. Chem., Int. Ed. Engl*, 1996, **35**, 1953–1956.
78. K. Bezouska, V. Kren, C. Kieburg and T. K. Lindhorst, *FEBS Lett.*, 1998, **426**, 243–247.
79. F. Sansone, L. Baldini, A. Casnati and R. Ungaro, *Supramol. Chem.*, 2008, **20**, 161–168.
80. M. Torvinen, E. Kalenius, F. Sansone, A. Casnati, P. Turhanen and J. Jänis, *Supramol. Chem.*, 2012, **24**, 228–233.
81. (a) S. André, F. Sansone, H. Kaltner, A. Casnati, J. Kopitz, H.-J. Gabius and R. Ungaro, *ChemBioChem*, 2008, **9**, 1649–1661. The D-*gluco* stereoisomer of **145** was prepared using the same synthetic approach; see:

(b) M. Torvinen, R. Neitola, F. Sansone, L. Baldini, R. Ungaro, A. Casnati, P. Vainiotalo and E. Kalenius, *Org. Biomol. Chem.*, 2010, **8**, 906–915.
82. S. André, C. Grandjean, F.-M. Gautier, S. Bernardi, F. Sansone, H.-J. Gabius and R. Ungaro, *Chem. Commun.*, 2011, **47**, 6126–6128.
83. (a) K. Griesbaum, *Angew. Chem., Int. Ed. Engl*, 1970, **9**, 273–287; (b) A. Dondoni, *Angew. Chem., Int. Ed*, 2008, **47**, 8995–8997.
84. M. Fiore, A. Chambery, A. Marra and A. Dondoni, *Org. Biomol. Chem.*, 2009, **7**, 3910–3913.
85. A. Dondoni, X. Hu, A. Marra and H. D. Banks, *Tetrahedron Lett.*, 2001, **42**, 3295–3298.
86. F. Troisi, T. Pierro, C. Gaeta and P. Neri, *Org. Lett.*, 2009, **11**, 697–700.
87. C. Gaeta, F. Troisi, C. Talotta, T. Pierro and P. Neri, *J. Org. Chem.*, 2012, **77**, 3634–3639.
88. F. Cardona, G. Isoldi, F. Sansone, A. Casnati and A. Goti, *J. Org. Chem.*, 2012, **77**, 6980–6988.

CHAPTER 5

Dendritic Glyco Architectures – From H-Bond-Driven Molecular Interactions to Their Potential Use in Brain Disease Therapy

DIETMAR APPELHANS*[a], JOSEP CLADERA[b], MARK ROGERS[c], AND BRIGITTE VOIT*[a,d]

[a]Leibniz-Institut für Polymerforschung Dresden eV, Hohe Strasse 6, 01069 Dresden, Germany; [b]Biophysics Unit and Centre of Studies in Biophysics, Department of Biochemistry and Molecular Biology, Universitat Autònoma de Barcelona, 08193 Bellaterra, Catalonia, Spain; [c]School of Biology and Environmental Science, University College Dublin, Belfield, Dublin 4, Ireland; [d]Organic Chemistry of Polymers, Technische Universität Dresden, 01062 Dresden, Germany
*E-mail: applhans@ipfdd.de, voit@ipfdd.de

5.1 Introduction

The terms linear[1] and dendritic[2] glycopolymers automatically imply their use in studying protein–carbohydrate interactions.[3,4] A hallmark of many protein–carbohydrate interactions is multivalency.[1] Multivalency exists in

RSC Polymer Chemistry Series No. 15
Glycopolymer Code: Synthesis of Glycopolymers and Their Applications
Edited by C. Remzi Becer and Laura Hartmann

Published by the Royal Society of Chemistry, www.rsc.org

different forms[5–7] where the molecular recognition of carbohydrate units on various oligosaccharide architectures by, for example, surfaces of proteins and cells is undoubtedly a crucial step to stop or to trigger biological processes *in vitro* and *in vivo*.[1] This can lead to a multiple high-affinity interaction with cell surfaces which will then be irreversible. Moreover, low-affinity multivalent interactions are kinetically labile and allow the cleavage of cell interactions to find the right interacting counterpart.[1] Such dynamic interaction characteristics with varying interaction strengths are of invaluable merit when searching for new application fields of glycodendrimers. Here, the H-bond active sugar part of the glycodendrimers will guide interactions but is not considered to be therapeutically active as is the case in other applications where multivalent specific molecular recognition of sugar units, *e.g.* by cell surfaces, plays an important role. Recent developments in treating and analysing brain diseases[8–16] revealed that various kinds of glycodendrimers are more promising materials for administration in this field (Alzheimer and prion disease) than their non-oligosaccharide-modified dendritic counterparts.[17–22] This is mainly because glycodendrimers show much lower toxicity than their parent counterparts in concentration-dependent *in vitro* and *in vivo* toxicity studies.[8–13] In addition to having high biocompatibility, therapeutics used in brain diseases must interact effectively with brain disease proteins and peptides that agglomerate in the brain to form amyloid plaques, such as exist in Alzheimer and prion disease, and thus lead to the adverse effects of the illness. Strategies discussed in this regard include a competitive interaction of the therapeutic agent with the peptides, protecting them from aggregation by solubilizing and stabilizing small oligomeric peptides, or even deaggregating already formed protein agglomerates. In addition, the therapeutic agent has to be able to cross the blood–brain barrier easily.

This chapter summarizes the use of glycodendrimers as therapeutic materials in brain diseases and outlines the reasons for their success in this new biomedical application field where complex biological interactions play a crucial role. In this respect, the chameleon-like or adaptable interaction characteristics of glycodendrimers against low and high molecular weight biomolecules, liposomes and cells are presented. The participating glycodendrimers can be considered as weakly interacting materials, which may allow for dynamic and reversible interactions with biologically active materials, an essential prerequisite for potential application as an anti-Alzheimer agent. Overall, H-bonding in combination with or without ionic interactions of the involved glycodendrimers can be responsible for initiating, stopping and/or manipulating the aggregation process of β-sheet Alzheimer and prion peptides, but also may induce the proteolytic degradation of protease-resistant prion proteins on different levels. Finally, this chapter highlights that other H-bond interactions are characteristic of glycodendrimers, in contrast to those known from the understanding of the ‘glycocodes’ through multivalency.[23]

5.2 Molecular Architectures and Characteristics for Molecular Interactions

For the successful use of dendritic glycopolymers as potential agents/drugs in the treatment of brain diseases (*e.g.* prion disease and Alzheimer disease), it is desirable to understand some of their fundamental interaction characteristics in solution, but also in general their *in vitro* interactions in the presence of biologically active molecules (*e.g.* proteins and nucleic acids). This is an indispensable prerequisite when administering these materials against various cells and *in vivo*, but also against peptides, which tend to aggregate into fibrillar structures. Therefore, the structural composition/shape of dendritic glycopolymers and their physicochemical properties will be considered in the following.

Generally, dendritic glycopolymers based on dendritic polyamine core molecules are characterized by a dense or open oligosaccharide shell (Figure 5.1).[8,24,25] Simplifying the terms dense and open-shell dendritic glycopolymers for both classes (Figure 5.1), the term 'dense shell' is preferentially attributed in the case of polypropylenimine (PPI) and polyethylenimine (PEI) scaffolds to the presence of disubstituted primary amino groups by oligosaccharide units, while the secondary amino groups in the PEI scaffold are also substituted by one oligosaccharide unit. For the open-shell architecture, primary amino groups are only substituted by one oligosaccharide unit for both dendritic polyamine scaffolds. Thus, the surface charge or charge density of PPI glycodendrimers ranges from positive for open-shell glycodendrimers to neutral for dense-shell glycodendrimers (Figure 5.1).[8,9,25] Dense-shell glycodendrimers can also be negatively charged after an additional sulfation.[9] In contrast, oligosaccharide-modified hyperbranched PEIs always possess pH-dependent cationic surface charge and charge density over a broad pH range [from 2 to the isoelectric point (8–9)]. The intensity of these cationic charge characteristics is mainly tailored by the different oligosaccharide architectures A, B and C (Figure 5.1).[24,26] Architecture A, a dense-shell architecture, has the lowest cationic charge in comparison with architecture C, an open-shell architecture, possessing the largest cationic charge within this series. Furthermore, these dendritic glycopolymers can be described as isolated macromolecules in aqueous solution and under physiological conditions.[8,24,25,27,28] Finally, the diameters of the PPI glycodendrimers range from about 2 nm for the first generation up to 8 nm for the highest generation,[8,10] whereas the hyperbranched counterparts (having a PEI core of 25 000 g mol^{-1}) have diameters up to 10–12 nm.[24] In this context, diameters of oligosaccharide-modified hyperbranched PEIs are not really changeable by varying the pH owing to the molecular architecture and shape.[27] A similar solution behaviour of the dense- and open-shell glycodendrimers to that found for the oligosaccharide-modified hyperbranched PEIs is assumed.

Moreover, small-angle X-ray scattering (SAXS) and dynamic light scattering (DLS) studies of oligosaccharide-modified hyperbranched PEIs revealed

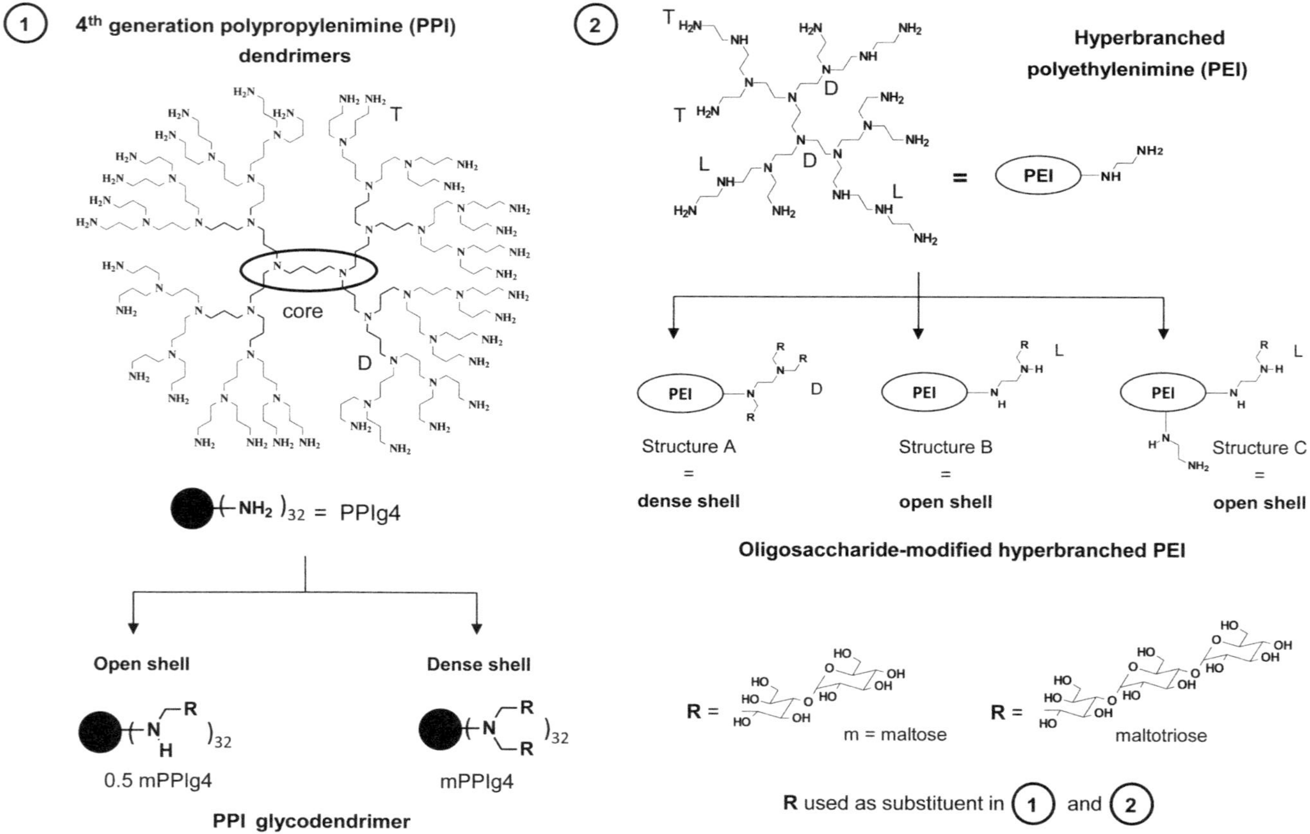

Figure 5.1 Structures of polypropylenimine (PPI) dendrimer (1) and hyperbranched polyethylenimine (PEI) (2). Simplified structure of open- and dense-shell PPI glycodendrimer (1) and oligosaccharide-modified hyperbranched PEI (2). Adapted and reproduced with permission from ref. 24. Copyright 2009 American Chemical Society.

that this class of dendritic glyco-macromolecules can be considered to have a core–shell architecture.[27] This means that the oligosaccharide units of the open and dense shell are really located in the outer sphere of this dendritic glycopolymers. Hence this class of dendritic glycopolymers can theoretically undergo both ionic and H-bond-driven interactions, depending on the shell density.

Results from several analytical methods [polyelectrolyte titration experiments, streaming potential pH titration, zeta potential, small-angle neutron scattering (SANS), but also molecular modelling and theoretical calculations on the backfolding properties of oligosaccharide units on the dendritic scaffold] support the existence of a core–shell architecture for the dense-shell glycodendrimers (Figure 5.2).[8–10] In polyelectrolyte titration experiments, the

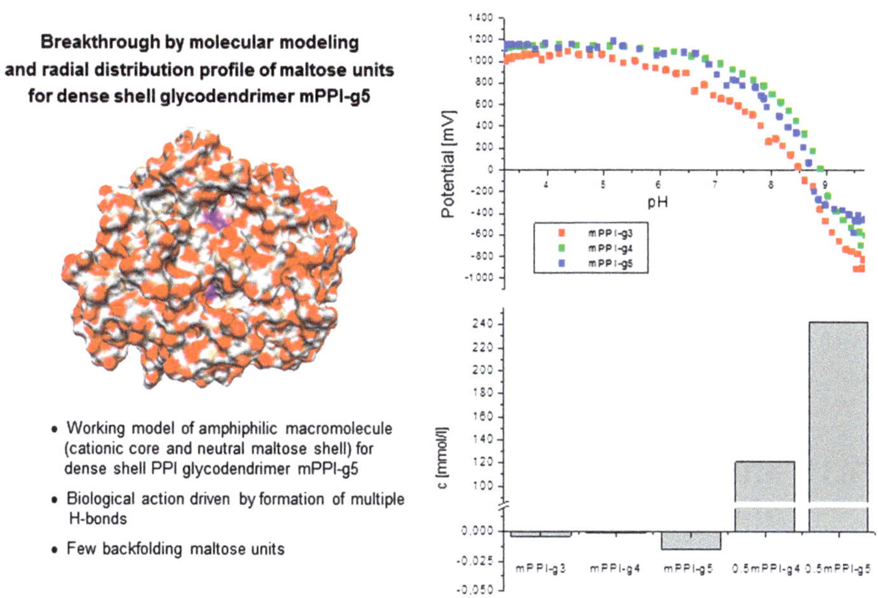

Figure 5.2 H-bond-active dense-shell glycodendrimer for biological interactions and applications considered as an amphiphilic macromolecule with a cationic core and neutral oligosaccharide shell. Left: image of dense-shell glycodendrimer mPPI-g5 after molecular modelling and radial distribution profile of maltose units on the PPI dendrimer surface. The purple colour in the mPPI-g5 image indicates the presence of back-folded maltose units even more embedded/wrapped by the dendritic PPI scaffold. Top right: streaming potential pH profiles of dense-shell glycodendrimers mPPI-g3–mPPI-g5. Bottom right: total charge of dendrimers as charge density (C at theoretical 1 mM glycodendrimer solution) of mPPI-g3–mPPI-g5 and 0.5 mPPI-g5 was determined by polyelectrolyte titration experiments [anionic polyelectrolyte, sodium polyethanesulfonate, and cationic polyelectrolyte, poly(allyldimethylammonium chloride), at pH 7 in 50 mM Tris–HCl buffer]. Adapted and reproduced with permission from ref. 11 and 9. Copyright 2010 and 2013 American Chemical Society.

most impressive result was that an anionic polyelectrolyte is not able to compensate for the cationic charge of the PPI core molecule in the dense-shell glycodendrimers (Figure 5.2).[8] This also means that the anionic polyelectrolyte macromolecule cannot penetrate the dense oligosaccharide shell and only very small molecules such as water or other nutrients are able to diffuse through the dense oligosaccharide shell of glycodendrimers. In this context, the cationic charge of the PPI core molecule in dense-shell glycodendrimers was only determinable by pH streaming potential measurements (Figure 5.2)[11] and not by zeta potential measurements under physiological conditions.[8] From these results one can consider a dense-shell glycodendrimer to be an amphiphilic macromolecule with a cationic core and a neutral shell. From the theoretical point of view, preferred H-bond-driven surface interactions of the oligosaccharide shell should be responsible for the biological interactions of dense-shell glycodendrimers. In line with this, molecular modelling and theoretical calculations confirmed our visual image of a dense oligosaccharide shell for this class of glycodendrimers where only few oligosaccharide units are backfolded and wrapped by the dendritic PPI scaffold (indicated by the purple colour in the mPPIg5 image in Figure 5.2) when dense-shell glycodendrimers are modelled in a water droplet.[9] On the other hand, open-shell glycodendrimers are characterized by a cationic surface charge and an H-bond-active surface (Figure 5.2).[8,11]

Summarizing the molecular properties of dendritic glycopolymers, all of them possess spherical shapes and (most of them) core–shell architectures with diameters in the lower nanometre range (up to 12 nm) and are theoretically suited to undergo ionic and/or H-bond-driven interactions.

5.3 Molecular and Biological Interactions of Dendritic Glycopolymers

Here we present some characteristic molecular and biological interactions of dendritic glycopolymers pointing to common features but also differences between hyperbranched and perfectly branched core–shell glyco architectures.

In a very recent study[28] of the complexation capability of core–shell architectures based on PEI cores in aqueous solution, very surprising results were obtained, indicating that the molecular composition and charge of the analyte molecules preferentially determine the amount of complexed analyte molecules (Figure 5.1, structures A–C) when it is ensured that non-complexed analyte molecules are fully separated under shear forces during an ultrafiltration step. Small and weakly anionic analyte molecules allow moderate or strong ionic interactions with the cationic PEI core of the core–shell glyco architecture and thus for a large amount of complexed analyte, whereas cationic or neutral analyte molecules show a low uptake owing to too weak interactions. Therefore, no H-bond-driven interactions of the glyco shell between analyte molecules and core–shell architectures could be observed under these experimental conditions and, clearly, ionic interactions induced by the cationic charge of the PEI core dominate, independent of the shell coverage in structure A, B or C (Figure 5.1) and even under a high molar excess of the analytes.

In contrast to the observations on glyco structures with a hyperbranched PEI core, the interaction of the PPI core containing open and dense glyco shell with ATP molecules, considered as strong anions with a bulky 3D structure, depends strongly on the density and hence the shielding capacity of the glyco shell (Figure 5.3).[29] Only the cationic open-shell glycodendrimer is able to undergo ionic interactions with this low molecular weight drug, while the neutral charge of the dense oligosaccharide shell of the glycodendrimer prevents any kind of interactions with the strongly anionic ATP molecules in aqueous solution, but also under physiological conditions (Figure 5.3). One can state that this 'bulky-3D' low molecular weight drug cannot penetrate the dense oligosaccharide shell but also cannot form any kind of H-bonds with the dense oligosaccharide shell.[29] In contrast, the anionic more planar dye molecule ANS is suited to undergo ionic interactions with the dense-shell glycodendrimers (Figure 5.3).[24] In line with this, a similar complexation behaviour of glycodendrimers with the anionic drug Mant-ATP, which also can be considered 'bulky-3D,' is observable.[30] The observed complexation capabilities of dense- and open-shell glycodendrimers confirm our assumption above that for core–shell glyco architectures the uptake of small analytes is dominated by the possibility of ionic interactions with the core, and H-bonding of the glyco shell can be neglected.

Furthermore, core–shell glyco architectures are suited to guide the *in situ* fabrication of small nanoparticles of size 1–5 nm, *e.g.* Au and CdSe. Whereas neutral dense-shell glycodendrimers, having a small size of a few nanometres, stabilize the surface of such nanoparticles through decoration with several

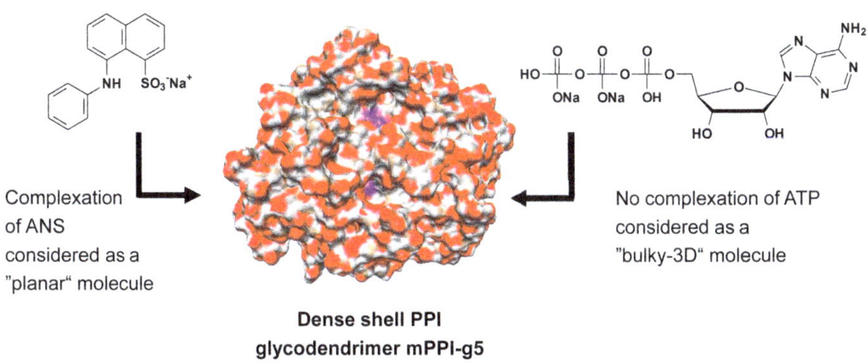

Figure 5.3 Limited complexation behaviour of fifth-generation dense-shell glycodendrimers mPPIg5 in aqueous solution and PBS buffer at pH 7.4. Visualization of dense-shell glycodendrimer mPPI-g5 by molecular modelling. Adapted and reproduced with permission from ref. 9. Copyright 2013 American Chemical Society.

dendritic macromolecules,[31,32] Au nanoparticles are fabricated within the hyperbranched PEI scaffold with a molar mass of 25 000.[33,34] In this context, a prominent result was the stabilization of anionic green fluorescent nanodiamonds conjugated with antibodies by cationic open-shell glycodendrimers for the selective targeting of mitochondria and actin surface.[35] Cell-internalized hybrid material, composed of cationic open-shell glycodendrimer and anionic green fluorescent nanodiamond, does not result in any uncontrolled aggregation processes after cell uptake, in contrast to protamine–nanodiamond or liposome–nanodiamond hybrid materials.[35]

The stabilization ability towards various 'hard' metallic and inorganic nanoparticles gave first indications that especially open- and dense-shell glycodendrimers might also be used for studying the interactions with globular 'soft' biological materials such as proteins. Of special interest here were first, the determination of the nature and strength of these interactions, and second, to use specific strong interactions of glycodendrimers for inhibiting protein aggregation or even for destabilizing already formed protein aggregates. This knowledge is of the greatest importance for the potential application of glycodendrimers in the treatment of brain diseases such as prion disease, where specific protein interactions and aggregations play a decisive role. To be able to apply our glycodendrimers successfully as therapeutic agents in brain diseases, one has to consider and ultimately understand multiple interactions of dendritic glycopolymers that may suppress protein aggregation processes such as destabilizing the secondary/tertiary structures of mutated brain proteins, which will require strong interactions or keeping oligomeric structures soluble by dynamic aggregation/disaggregation processes caused by weak interactions.

With this in mind, the parental unmodified PPI dendrimers with primary amine surface groups and open- and dense-shell glycodendrimers have been used to explore their interactions with various proteins.[8,36,37] In a first study, parental PPI dendrimers showed a stronger interaction against human serum albumin (HSA) molecules than dense-shell glycodendrimers.[24] This was explained by the fact that dense-shell glycodendrimers can only interact with HSA through weak multiple H-bonds. In line with this, a gradually decreasing interaction strength from parental PPI dendrimer to open-shell glycodendrimer and to dense-shell glycodendrimer was observed towards the cationic lysozyme and nearly neutral liver alcohol dehydrogenase. Hence in these studies it was observed that all dendrimers studied show to some extent destabilizing effects on the secondary structure of various charged or non-charged proteins. However, most prominently, the interaction characteristics of PPI dendrimers and open-shell glycodendrimers are enhanced, probably due to the non-shielded cationic charge of the dendritic core.[36,37]

From these studies of the molecular interaction properties of glycodendrimers towards proteins,[8,36,37] one can draw the following conclusions. First, glycodendrimer interactions with cationic, neutral or anionic proteins are mainly tailored by their oligosaccharide shell density. Thus, dense-shell glycodendrimers can only undergo H-bond-driven interactions of the oligosaccharide shell with the protein surface. This results in a significantly lower interaction strength with proteins compared with PPI and their open-shell counterparts and usually leads to low destabilization of the secondary structure of proteins, allowing for a dynamic, meaning reversible, interaction profile. As discussed in the literature, especially the latter point may be of great importance if drugs are to be successfully applied in the treatment of prion disease.[9,11] Second, interactions of open-shell glycodendrimers are governed by both their remaining cationic surface charge from the dendritic core and their H-bond-active oligosaccharide shell. The dominance of either kind of interaction depends on several parameters such as pH and charge of the protein, but also surface patchiness of the protein, providing areas with differences in charge, hydrophobicity and H-bond-active groups.

For the potential administration of dendritic glycopolymers as anti-Alzheimer and anti-prion agents, one other essential piece of knowledge is their interaction characteristics against phospholipid membrane structures. Various anionic vesicles, composed of neutral and anionic phospholipids, have been prepared to study their interactions with neutral and cationic dendritic glycopolymers.[38,39] Applying a weakly cationic PEI core–shell glyco architecture (diameter 5 nm, Figure 5.1, structure B) in aqueous solution, anionic vesicle morphogenesis from vesicular (diameter 60 nm) into tubular network structures is observable (Figure 5.4), while the cationic unmodified hyperbranched PEI generally induces nearly no morphological transformation of anionic vesicles.[38] This example indicates that the open oligosaccharide shell as an H-bond-active surface in the cationic PEI core–shell glyco architecture is the driving force for the initiation of morphological transformation processes, whereas the cationic charge of the corresponding core macromolecules plays a minor role in the complex transformation process.[38] Additionally, these spherical and compact core–shell glyco architectures[38] are strongly involved in keeping together the formed tubular network structures. Recent results from our laboratories revealed that the PEI core–shell glyco architectures are also able to induce the formation of elongated interconnected tubular structures from giant vesicles. The most surprising result is that such vesicular morphogenesis is observable with an optical light microscope in real time. This allows us to consider specific dendritic glycopolymers as artificial tubulating proteins[40,41] that are able to imitate specific biological processes where Nature needs a large armada of several proteins.[40,41]

It is important to note that no vesicle morphogenesis of neutral and anionic phospholipid vesicles (diameter 120–140 nm) is observable when open- or dense-shell glycodendrimers are used instead of the PEI core–shell glyco architectures.[39] Differential scanning calorimetry (DSC) and fluorescence studies revealed, on the other hand, that neutral dense-shell glycodendrimers show stronger interactions against vesicles than the cationic open-shell glycodendrimers. A possible explanation is that dense-shell

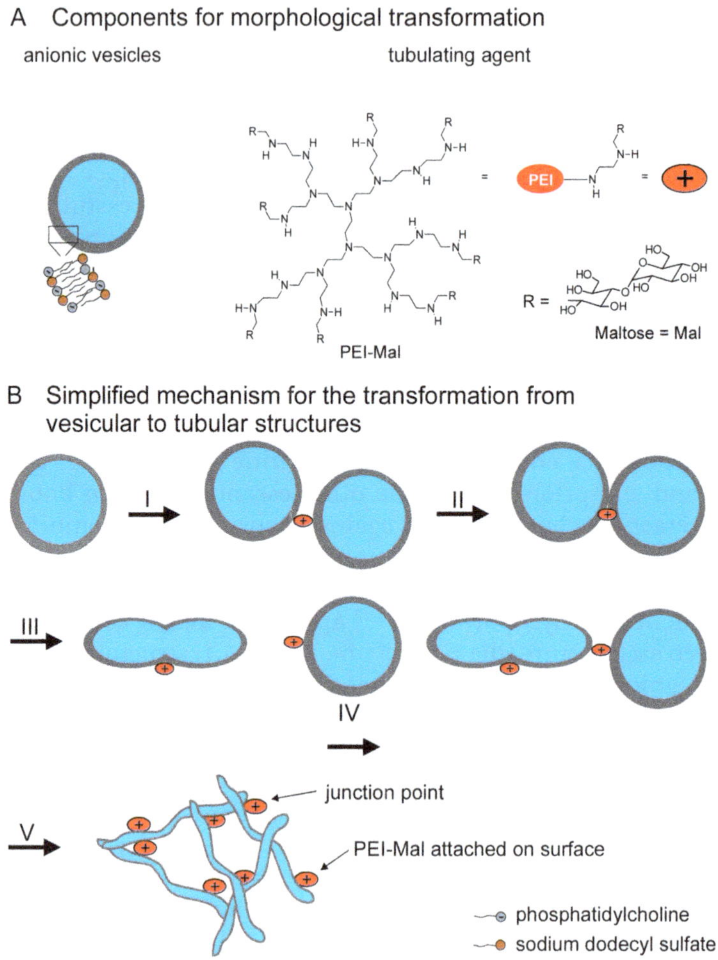

Figure 5.4 (A) Anionic vesicles composed of phosphatidylcholine and sodium dodecyl sulfate and maltose-modified polyethylenimine (PEI-Mal) as components for morphological transformation and (B) simplified mechanism for the transformation of globular vesicles into a tube-like network in the presence of PEI-Mal. Description of I–V for Figure 5.1, structure B: I = fixation of two anionic vesicles by cationic PEI-Mal; II = starting fusion process of two vesicles by cationic PEI-Mal; III = elongated micelles formed by two vesicles which is governed by PEI-Mal; IV = formation of larger elongated micelles for establishing tubular structures; V = formation of tubular network obtained from I – IV. Reproduced from ref. 38 with permission from The Royal Society of Chemistry.

glycodendrimers offer a more stable and/or larger H-bond-active surface against vesicle surfaces than open-shell glycodendrimers.[39]

From the above results, one can conclude that especially dense-shell glycodendrimers may allow for adaptive interactions with different biological materials and simplified biological systems, where the H-bond-active surface dominates.

5.4 Biocompatibility, Cellular Uptake Processes and Crossing the Blood–Brain-Barrier

This part of the chapter is focused on the biocompatibility, cellular uptake processes and ability to cross the blood–brain barrier of dendritic glycopolymers. Especially the latter is also essential for the potential administration of these macromolecules in brain disease therapy.

Various architectures of dendritic glycopolymers were tested to elucidate whether they exhibit alteration effects on the cell viability of normal and cancerous cell lines.[9–11,26,42–47] Looking at the concentration/dose- and architecture-dependent cell viability of all dendritic glycopolymers (Figure 5.1), it was found as a clear trend that preferentially dense-shell architectures are characterized by a markedly reduced toxicity, including no or very low genotoxicity,[43] but also allowing good proliferation of red blood cells and lymphocytes.[44] But also a weak cationic charge of dendritic glycopolymers can be compensated by various cell lines without affecting their cell viability. In a recent study, chameleon-like properties of open-shell glycodendrimers were evaluated *in vitro*.[47] This weakly cationic core–shell glyco architecture is highly biocompatible with normal cell lines, whereas their presence in cancerous cells preferentially induces apoptosis. Moreover, good biocompatibility of both classes of core–shell glyco architectures was proven also *in vivo*.[48,49] Again, dense oligosaccharide shell glycodendrimers provide the highest biocompatibility after administration in rats,[48] whereas open oligosaccharide structures based on PEI core–shell architectures are better addressable in mice than their dense-shell architectures.[49] The latest results may indicate a slight degradation of the dense oligosaccharide shells when larger doses are administrated to mice. Overall for both material classes, a dose- and sugar decoration-dependent behaviour *in vivo* was observed, where the growth and body weight of the animals were not affected until the concentration administered was too high. Moreover, in the case of PEI core–shell glyco architectures, high efficacy for *in vivo* gene delivery was also observable.[49]

5.5 Applications in Brain Disease

After the elucidation of their structural and physical characteristics and also their interactions with biological components and cells, recent biomedical applications of glycodendrimers in the field of brain disease are described. Explicitly, it is shown that glycodendrimers can act as promising polymeric therapeutic agents and in diagnostics for Alzheimer disease (AD) and prion disease.

Both brain diseases, AD and prion disease, are 'conformational diseases.' In line with this, amyloid plaques are characteristic features of the central nervous system (CNS) affected by neurodegenerative diseases.[50,51] These amyloid plaques are composed of peptides or proteins, mostly assembled into fibrillar filaments. One of the main characteristics of these peptides and proteins is that they are rich in β-sheet structure and that they form characteristic aggregates, known as amyloid aggregates. A major interest in the research area dedicated to neurodegeneration is the search for compounds able to interfere with and/or block the formation of the amyloid aggregated species related to cell toxicity.[52] This was also an early motivation to check the potential of glycodendrimers as disaggregation agents.

5.5.1 Glycodendrimers as Anti-Alzheimer Agents

The pathology of AD is a progressive and irreversible one with only minimal repair and regenerative capacity of the brain tissue. Hence the neurons of some brain areas gradually deteriorate and die.[53]

According to the amyloid cascade hypothesis, in an unhealthy brain[54] (Figure 5.5) amyloid peptides (Aβ peptides), the main component of the characteristic amyloid plaques, are directly implicated in the generation of the toxicity that leads to cell death. In particular, cell toxicity is considered to be associated with globular oligomeric species that are intermediates in the polymerization process that ends with the formation of the amyloid fibrils found in senile plaques. Alzheimer's Aβ peptides are a cleavage product of the amyloid protein precursor (APP). APP belongs to a group of transmembrane protein isoforms in the CNS. It can be processed by three different secretases. When cleaved by α- and β-secretases, a non-amyloidogenic peptide is produced. However, when cleaved by β- and γ-secretases, a mixture of 40 and 42 residues-long amyloid peptides [Aβ(1–40) and Aβ(1–42)] is produced, showing a high tendency to aggregate into the extracellular space of the CNS.

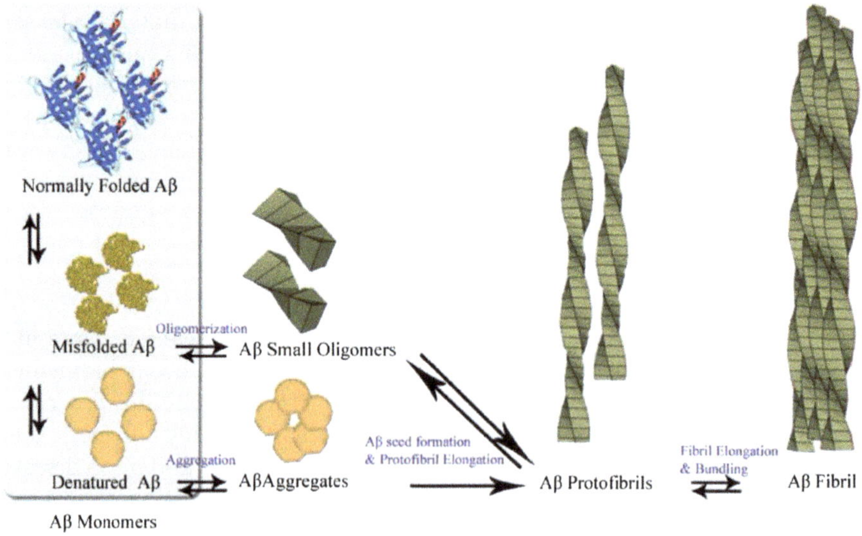

Figure 5.5 Diagram of amyloidogenesis process. Sequential formation of amyloid aggregates, protofibrils and fibrils. Reprinted with permission from ref. 53. Copyright 2008 IOS Press.

Amyloid fibril formation has been intensively studied *in vitro*.[54,55] Fibril formation follows a nucleation-dependent polymerization process (Figure 5.6) that can be monitored using the dye thioflavin (Figure 5.7). During the lag phase of the typically sigmoid-shaped kinetics, different forms of non-fibrillar, low and high molecular weight intermediates are formed (Figure 5.6). Results indicate that these non-fibrillar intermediates can be cytotoxic whereas the fibrils have no or very low toxicity.[56]

One of the challenges in finding effective drugs in AD is to find molecular agents capable of interfering with the amyloid polymerization process and blocking the toxic elements or impeding their formation. Theoretically, interfering with fibril formation by an anti-Alzheimer agent may imply the formation of non-fibrillar toxic species, including possible toxic dendrimer–amyloid complexes.[12] Hence, it is desirable that any anti-Alzheimer agent does not generate toxic species with Aβ in the presence of neuronal cells. The characteristics of the complex between the dendrimer and the amyloid peptide depend on the molar ratio and pH, among other parameters. Two types of aggregates have been identified: (i) clumped fibrils and (ii) amorphous non-fibrillar aggregates (Figure 5.8).[12] Based on these results, the observed amorphous non-fibrillar aggregates are similar to the amorphous amyloid peptide aggregates described as so-called granular non-fibrillar aggregates (GNAs),[57] with regard to their morphology and toxicity against neuronal cells.[12,13]

Polyamidoamine (PAMAM) and phosphorus dendrimers have been proven to exhibit anti-amyloidogenic properties by interfering with amyloid fibril

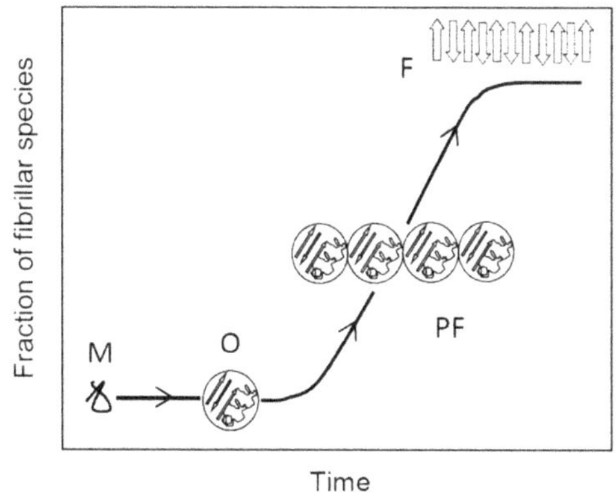

Figure 5.6 Schematic illustration of the amyloid fibril formation kinetics (nucleation-dependent polymerization process). Monomeric peptides or proteins (M) combine during a lag phase into oligomeric nuclei (O). The transition between monomers and nuclei during this so-called nucleation phase may take place through the formation of different oligomeric intermediates (dimers, trimers, *etc.*). The nucleation phase is followed by an elongation phase during which protofibrillar structures (PF) form by combination of nuclei. Finally, protofibrils convert into amyloid fibrils (F). Reproduced from ref. 12 with permission from The Royal Society of Chemistry.

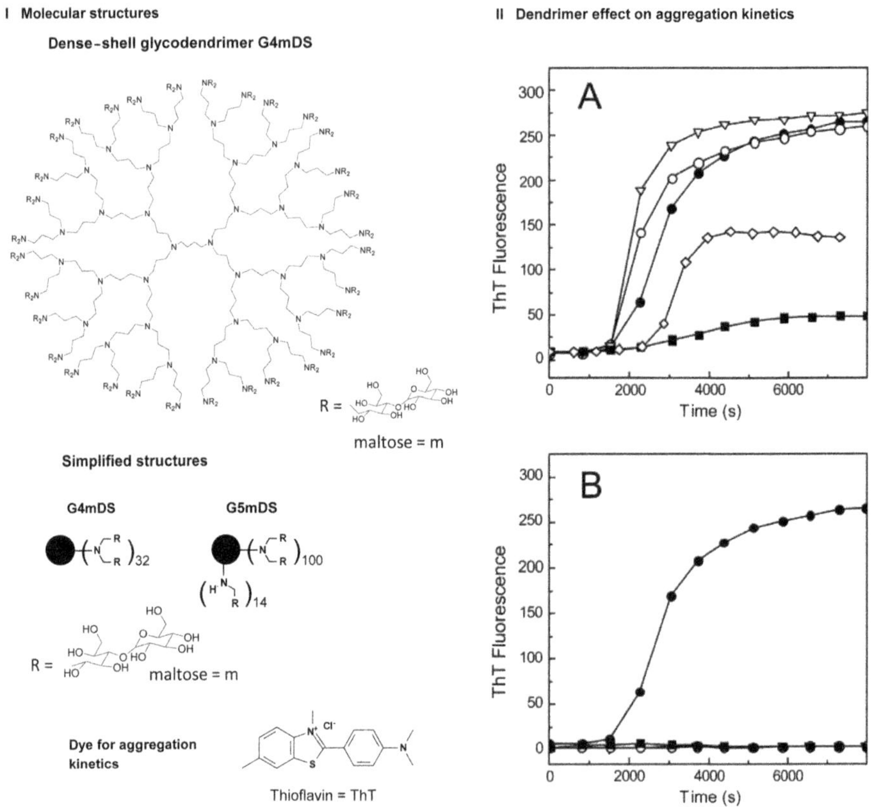

Figure 5.7 Anti-amyloidogenic properties of dense-shell glycodendrimers using the dye thioflavin as an optical sensor for following nucleation-dependent polymerization process of Aβ(1–40) into Aβ fibrils. I, Molecular structures used in this study. II, Dendrimer effect on kinetics: effect of PPI-maltose dendrimers on the aggregation kinetics of Aβ(1–40). Fibril formation was measured by monitoring ThT fluorescence variation. (A) Effect of PPI-G4-Mal on Aβ(1–40) fibril formation; (B) effect of PPI-G5-Mal on Aβ(1–40) fibril formation. (•) Aβ(1–40) 25 μM (control); (Δ) Aβ(1–40) + dendrimer at dendrimer-peptide ratio = 0.1; (○) Aβ(1–40) + dendrimer at dendrimer-peptide ratio = 1: (◊) Aβ(1–40) + dendrimer at dendrimer-peptide ratio = 5; (■) Aβ(1–40) + dendrimer at dendrimer-peptide ratio = 10. Conditions: temperature 37 °C and pH 7.4. Adapted and reproduced with permission from ref. 13. Copyright 2011 American Chemical Society.

formation by either accelerating fibril formation or by slowing and/or inhibiting fibril formation.[12] However, the dendrimers described so far in relation to AD turned out to be cytotoxic. In contrast, fourth- and fifth-generation dense-shell glycodendrimers are non-toxic to PC12 and SH-SY5Y cells up to 50 μM dendrimer concentration and have a low toxic effect between 50 and 100 μM.[13] Both cell lines are normally used as neuronal models to test cell toxicity.[58,59] With the important advantage of very low cytotoxicity, the

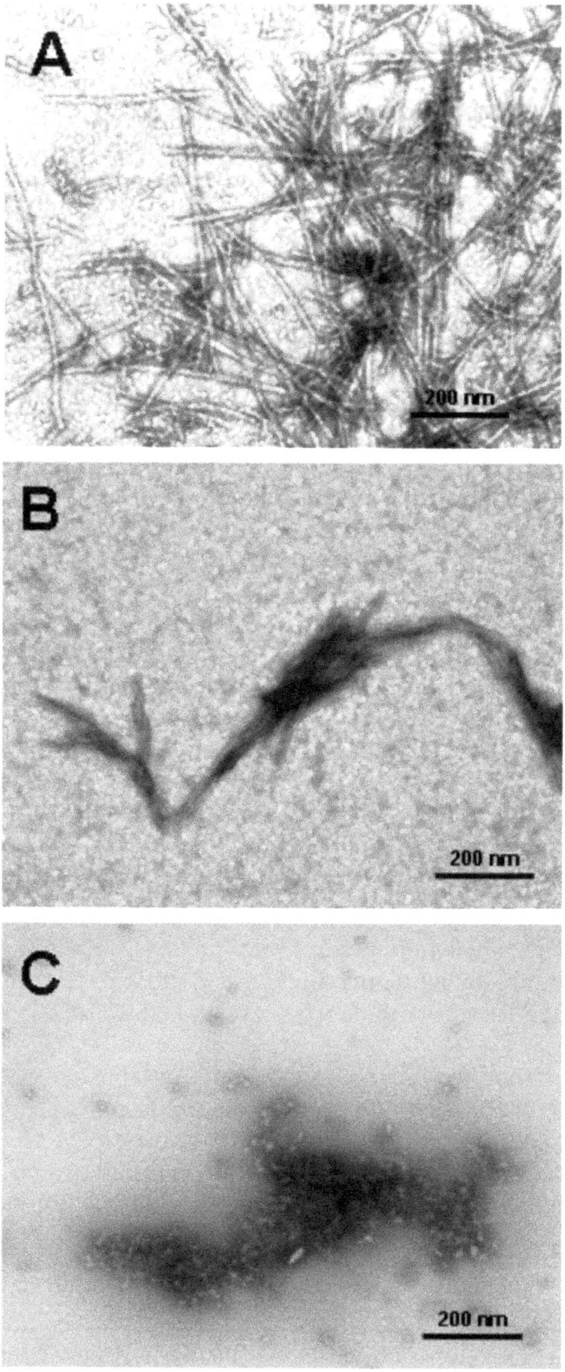

Figure 5.8 Transmission electron micrographs of Aβ(1–40) incubated at pH 7.4 (coexistence of fibrils and globular oligomers) (A), Aβ(1–40) incubated at pH 7.4 in the presence of PPI-G4-Mal (detection of clumped fibrils) (B)

capacity of dense-shell glycodendrimers to interfere with the nucleation-dependent aggregation of Aβ(1–40) (Figure 5.7) is very similar to the capacity reported for PAMAM dendrimers. At low dendrimer:peptide ratios there is a slight acceleration effect of fourth-generation glycodendrimer on the polymerization process, whereas with increasing dendrimer:peptide ratio the amount of fibrils at the end of the process is reduced (Figure 5.9). Surprisingly, fifth-generation glycodendrimers have a complete inhibitory effect on Aβ(1–40) fibril formation at any dendrimer/peptide ratio (Figure 5.9).[13]

Characterization of glycodendrimer/Aβ(1–40) aggregates by transmission electron microscopy revealed the presence of different morphologies (Figure 5.8). Compared with normal fibrils (Figure 5.8(A)), clumped fibrils can be observed when using low fourth-generation glycodendrimer:peptide ratios (Figure 5.8(B)). GNAs are generally generated by any high dendrimer:-peptide ratios for both glycodendrimers. Furthermore, the cell viability of GNAs and clumped fibrils against PC12 cells was tested and it was shown that GNAs based on fourth- or fifth-generation glycodendrimer are toxic and clumped fibrils based on fourth-generation glycodendrimers (Figure 5.8(B)) are non-toxic.[13]

In summary, dense-shell glycodendrimer interference with the Aβ(1–40)'s fibril formation process has to yield dendrimer/peptide non-toxic morphologies for its application *in vivo*. Moreover, in order to make any application of glycodendrimers *in vivo*, their interaction *in vitro* with the Aβ(1–42) peptide should also be studied. The amount of Aβ(1–42) in relation to Aβ(1–40) seems to be an important parameter in the affected brains, so the influence of dendrimers on this ratio is an important parameter.[10,60]

In a recent study,[10] the effect of glycodendrimers on the toxicity of Aβ(1–42) and human AD brain extract towards SH-SY5Y neuroblastoma cells was addressed. Furthermore, their anti-amyloidogenic properties *in vivo* were also tested. In agreement with previous results,[13] fourth-generation glycodendrimers [open-shell architectures with maltose (G4mOS) and maltotriose (G4m-IIIOS); dense-shell architectures with maltose (G4mDS)] and fifth-generation glycodendrimers with dense maltose shell (G5mDS) were chosen for this study.

It was shown that first, all glycodendrimers, except G4m-IIIOS, are able to screen out Aβ(1–42) toxicity on SH-SY5Y neuroblastoma cells. On the other hand, dense-shell glycodendrimers G4mDS and G5mDS even compensate human AD brain extract toxicity on SH-SY5Y neuroblastoma cells, whereas open-shell glycodendrimers G4mOS and G4m-IIIOS do not reduce the toxic effect of human AD brain extract on neuronal cell lines.[10]

---

and Aβ(1–40) incubated at pH 7.4 in the presence of PPI-G5-Mal (detection of granular non-fibrillar, amorphous aggregates) (C). Aβ(1–40) and Aβ(1–40)–dendrimer complexes were incubated for 12 h at pH 7.4 and 37 °C before the preparation of the microscopy grids. Reproduced with permission from ref. 13. Copyright 2011 American Chemical Society.

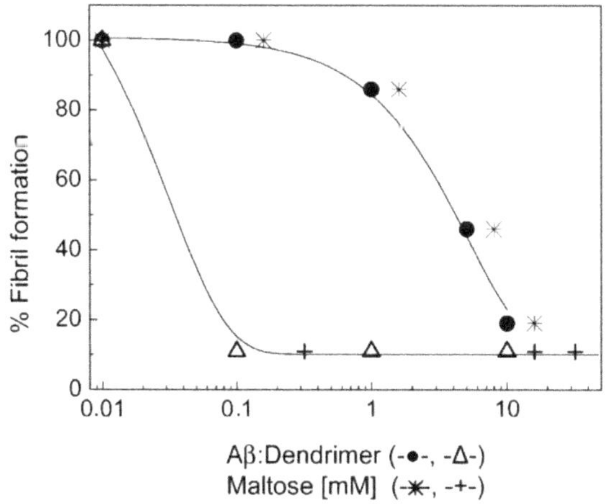

Figure 5.9 Differential effects of PPI-G4-Mal and PPI-G5-Mal dendrimers on inhibition of Aβ(1–40) fibril formation. The percentage fibril formation is represented as a function of dendrimer:peptide ratio and total maltose concentration. (●) Effect of PPI-G4-Mal as a function of the dendrimer-peptide ratio; (*) effect of PPI-G4-Mal as a function of maltose concentration; (Δ) effect of PPI-G5-Mal as a function of the dendrimer:peptide ratio; (+) effect of PPI-G5-Mal as a function of maltose concentration. The percentage fibril formation was determined from the ThT fluorescence values at the plateaus of the curves in Figure 5.6. Reproduced with permission from ref. 13. Copyright 2011 American Chemical Society.

As mentioned above, intranasally administered glycodendrimers (G4mDS, G4mOS and G5mDS) can successfully cross the blood–brain barrier of mice.[10] Hence G4mDS and G4mOS was used further to check their effect on soluble Aβ(1–42)–Aβ(1–40) mixtures with regard to minimizing amyloid plaques and effecting cognitive ability of APP/PS1 mice. However, Aβ plaque quantification in the neocortex of APP/PS1 mice revealed significantly higher levels after chronic treatment with both glycodendrimers, especially for G4mOS, compared with control APP/PS1 mice treated with phosphate-buffered saline (PBS) alone. Moreover, there was also no improvement in, but preservation of, the cognitive ability of APP/PS1 mice, and the level of soluble Aβ(1–42) and Aβ(1–40) in APP/PS1 mice was not be increased and was similar to that found in the control APP/PS1 mice.[10]

Based on these first results on studying the anti-amyloidogenic properties of glycodendrimers *in vitro* and *in vivo*, one can emphasize the very high cytocompatibility of the open- and dense-shell glycodendrimers and their efficient disaggregation properties against human AD brain extracts.[10] However, the structures studied so far have not been effective in *in vivo* studies of APP/PS1 mice, hence there is a need for improvement, *e.g.* by introducing other functional units in the shell of glycodendrimers.

5.5.2 Glycodendrimers as Anti-Prion Agents

Prion protein functions in the CNS are directed at protecting neurons from extracellular attacks and they are mostly/partly integrated in the cell membrane of neuronal cells. Mainly, prion proteins act by hampering neuronal cell death by catching/regulating Cu(II) ion metabolisms, decreasing oxidative stress and defending against any apoptotic attacks from extracellular matrix.[61]

Prion diseases are a group of invariably fatal neurodegenerative diseases that include Creutzfeldt–Jacob disease (CJD) in humans, bovine spongiform encephalopathy (BSE) in cattle, scrapie in sheep and goats and chronic wasting disease in deer and elk.[62] Generally, prion diseases are characterized by either misfolding or the mutation of an endogenous glycoprotein called the prion protein (PrP). In healthy hosts, PrP molecules are folded into a normally cellular conformation denoted cellular PrP^{C}, whereas in unhealthy hosts PrP molecules adopt an aggregation-prone, protease-resistant scrapie conformation termed PrP^{SC}. Moreover, PrP^{SC} molecules are essential components of infectious prions.[63–66] In this context, prion infectivity sometimes arises from a mere protein aggregate that can penetrate mucosal barriers, circumvent innate and adoptive immunity and travel across the blood–brain barrier eventually to provoke brain disease.[67]

In an interaction study of dense-shell glycodendrimers with prion peptide 185–208, derived from the human prion protein, it was shown for the first time that dense-shell glycodendrimers can generally suppress the formation of fibrillar structures of β-sheet peptides in a concentration-dependent manner and can act as a potential anti-amyloid agent.[8] At very low concentration of dense-shell glycodendrimers, an enhanced nucleation-dependent polymerization mechanism of prion peptide 185–208 is still observable, as known from fibrillar formation processes of pure Alzheimer and prion peptides.[50,51] The anti-prion activity of dense-shell glycodendrimers increases significantly with increasing generation and concentration, leading finally to a total suppression of fibrillar formation for the highest generation applied. The molecular interaction between dense-shell glycodendrimers and β-sheet peptides is mainly governed by H-bond-driven interactions of the sugar shell.

Accumulation of PrP^{SC}, often in the form of amyloid plaques,[68–71] is accompanied by neuropathologic changes primarily in the CNS, spongiform vacuolation in the (cerebral) grey matter of the brain, neuronal loss and astrocytic proliferation (astrocytic gliosis).[11,72,73] Moreover, PrP^{SC} is often the major and sole component of infectious prion agents.[62,64] Currently, no effective therapy is available for these brain diseases.[69,73,74]

In this context, polyamine dendrimers were successfully investigated as anti-prion agents. They reduced the levels of protease-resistant PrP^{SC} (PrP^{res}) in scrapie-infected cells and pre-existing aggregates of prions in mice bioassays. Thus, the level of infectivity could be abolished by the dendritic polyamine scaffolds.[74,75] On the other hand, these cationic dendritic polyamine scaffolds demonstrate high toxicity, limiting their usefulness as an effective anti-prion therapeutic agents *in vitro* and *in vivo*. Stimulated by the enhanced biocompatibility of open- and dense-shell glycodendrimers[9–11,26,42–49] and their promising tuneable interaction properties against biological entities (*e.g.* proteins and cells), several studies have examined whether glycodendrimers can help to destabilize the protease-resistant form of PrP^{res},[76,77] differentiate prion strains using various dendrimers and act as anti-prion drugs for inhibiting PrP^{C} conversion into PrP^{SC}.

First, the toxicity of PPI dendrimers (third- and fifth-generation) were examined in a mouse neuroblastoma cell line (N2a) using various PPI glycodendrimers and parental PPI dendrimers.[11] Following treatment for 4 days, N2a cells retained high cell viability against glycodendrimers, at concentrations up to 150 mg mL^{-1}, compared with parental PPI dendrimers (<2.5 μg mL^{-1}).

After the evaluation of the effect of dendrimer concentration on cell viability, a comparative study of PPI dendrimer anti-prion activity against PrP^{SC}-infected N2a (ScN2a) cells was carried out using third-, fourth- and fifth-generation PPI dendrimers decorated with amino, maltose or maltotriose surface groups, respectively. The dendrimers were exposed for 3 days to ScN2a in the presence of proteinase K followed up by a preparative procedure for final analysis by immunoplot.[11] The PrP^{res} used originated from an RML prion strain based on the genetic background of scrapie. The relative level of PrP^{res} after incubation with various dendrimers was determined by digestion with proteinase K. From these results, the following anti-prion activities of PPI dendrimers were determined: First, third-generation parent PPI dendrimers and glycodendrimers with dense maltose shell are not able to participate in the reduction of the PrP^{res} level despite using increasing dendrimer concentrations against ScN2a cells. On the other hand, fifth-generation dendrimers with peripheral amino or dense maltose shell showed a dose-dependent reduction of PrP^{res} level, both showing the same anti-prion activity. Finally, anti-prion activity within the series of third-, fourth- and fifth-generation dense-shell glycodendrimers decorated with maltose units showed that only the fifth-generation glycodendrimers provide the desirable significant anti-prion activity. This anti-prion activity of fifth-generation glycodendrimers was further supported and confirmed by simultaneously carried out dose- and time-dependent experiments. Finally, on comparing the anti-prion activity of fifth-generation dense-shell glycodendrimers with maltose and maltotriose shell glycodendrimers, no difference in effectiveness was observed. Both glycodendrimers provide the same desired activity against PrP^{res} level. Therefore, it is not necessary to use dense-shell glycodendrimers with maltotriose units for initiating anti-prion activity.[11]

Summarizing these results, the most efficient anti-prion activity is only available with fifth-generation dendrimers and not by lower dendrimer generations but is independent of the presence or absence of any glyco shell.[11] This indicates impressively that cationic surface interactions coming from amine units of the PPI core are not required for this activity and maltose-decorated PPI dendrimers can completely adopt the function of parent PPI dendrimers for reducing the PrP^{res} level in SCN2a cells or in scrapie-infected brain homogenates from mice. Overall, it also means that fifth-generation dense-shell glycodendrimer can destabilize the secondary/tertiary structure of the protease-resistant PrP^{SC}, thus facilitating proteolytic digestion by proteinase K.

Different prion strains are characterized by different incubation times and lesion profiles, induced in the hosts.[78–80] Different stains have different interspecies transmission properties and different pathogenicity, especially in humans.[81,82] Hence it is desirable to distinguish between prion strains in a rapid and simply carried out diagnosis approach. Generally, the identification of prion strains is based on several different diagnostic approaches.[14]

Different glycodendrimers may interact differently with distinct prion strains, making it possible to characterize these stains based on the different properties. With this in mind, a study to explore the anti-prion activity of dendrimers, differing in size, charge and surface composition, compared with one known prion strain was undertaken. This was subsequently extended to demonstrate that different strains could be distinguished based on their profile of interaction with different glycodendrimers.[14] This may offer the unique chance to identify and investigate various infectious prion strains in the same diagnostic approach.

In order to establish a dendrimer-based prion strain differentiation approach,[14] mouse brain homogenates with well-characterized prion strains (ME7, 22A, RML and 79A with genetic background from scrapie and 301C and 301V with genetic background from BSE) were treated with increasing concentrations of various dendrimers: neutral and H-bond-active fifth-generation dense-shell glycodendrimer (mPPIg5), cationic and H-bond-active fifth-generation open-shell glycodendrimer (0.5 mPPIg5), cationic fourth-generation PPI (PPIg4) dendrimer and cationic fifth-generation polyamidoamine (PAMAMg5) dendrimer. After treatment with dendrimers, the level of PrP^{res} was determined by proteinase K digestion, immunobloting and densitometry.

Figure 5.10 summarizes the results of the differentiation of prion strains by dendrimers. It can be concluded that first, all cationic dendrimers are more efficient than the dense-shell glycodendrimer mPPIg5 at inducing proteinase K sensitivity in PrP^{res}. mPPIg5 generates proteinase sensitivity of prion strain RML only at high dendrimer concentration ($>50\ \mu M$) Second, a common proteolytic digestion of protease-resistant PrP^{res} for all cationic dendrimers is observable with decreasing degree: 79A, RML > ME7, 22A > 301C, 301V. It seems likely that 301C and 301V with a BSE genetic background are the most protease-resistant PrP^{res}. This is mainly governed by their compact and stable protein conformation. On the other hand, there are no strong dendrimer interactions with 301C and 301V proteins to initiate any destabilization effects on the conformation of PrP^{res} from 301V and 301C strains.

In conclusion, at least two or three dendrimers with different interaction profiles against known prion strains are needed to identify unknown prion (strains) in the future. Until now, the success of this bioassay has been

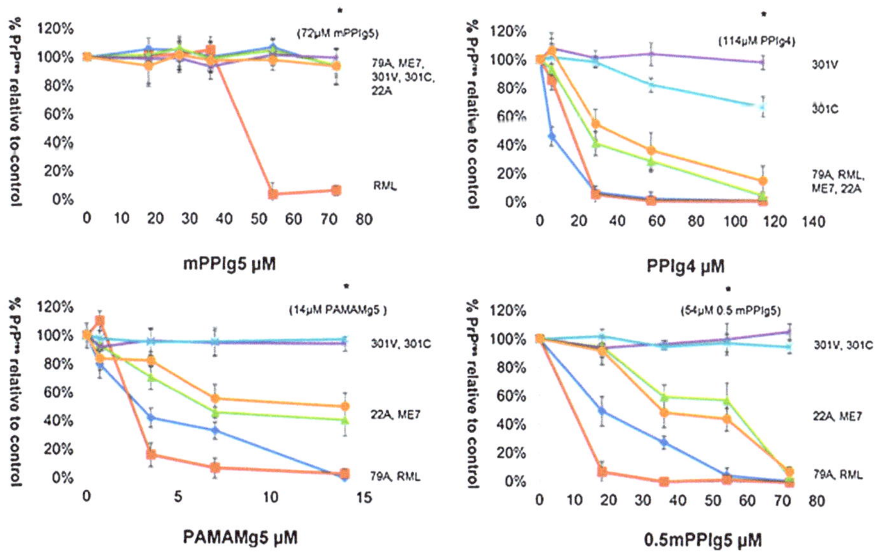

Figure 5.10 Differentiation of prion strains. Immunoblot images of dendrimer-treated brain homogenates were analysed by densitometry for percentage PrP^{res} levels that remained after dendrimer treatment and protease digestion, calculated relative to a non-dendrimer-treated control. —◆—, 79A; —■—, RML; —▲—, ME7; —×—, 301V; —✳—, 301C; —●—, 22A. Error bars represent SD; $n = 2$ biological repeats. *, Statistically significant difference between prion strains at the concentrations indicated ($p < 0.01$). The mean levels of PrP^{res} remaining for each prion strain after treatment with 72 μM mPPIg5, 114 μM PPIg4, 14 μM PAMAMg5 or 54 μM 0.5 mPPIg5 were compared using a one-way ANOVA. Prion strains 79A, RML, 301V and 301C can be statistically differentiated; 22A and ME7 can be differentiated from the other prion strains but not from one another. Reproduced from ref. 14 with permission from Wiley-VCH.

limited to prion strains adapted from scrapie, and for prion strains with other genetic backgrounds new dendrimers with stronger interaction characteristics against PrP^{res} are needed to destabilize protease-resistant PrP^{res}. Overall, a fast and simple diagnostic assay for the differentiation of prion strains was established with glycodendrimers in comparison with other known diagnostic assays in this field of identifying prion strains.[83–85]

Motivated by these results, further efforts were made to fabricate the next generation of glycodendrimers characterized by stronger destabilization effects on prion strains that would facilitate the proteolytic protein kinase K digestion of protease-resistant PrP^{res}.[9] Figure 5.11 illustrates the new generation of so-called 'hairy' glycodendrimers used in a previous study: cationic and H-bond-active fourth-generation dense-shell glycodendrimer decorated with alkyl-spacered tryptophan units (mPPIg4-Tryp), cationic and H-bond-active fifth-generation dense-shell glycodendrimer with PEG-spacered amino units (mPPIg5-PEG-NH_2), cationic and H-bond-active fifth-generation

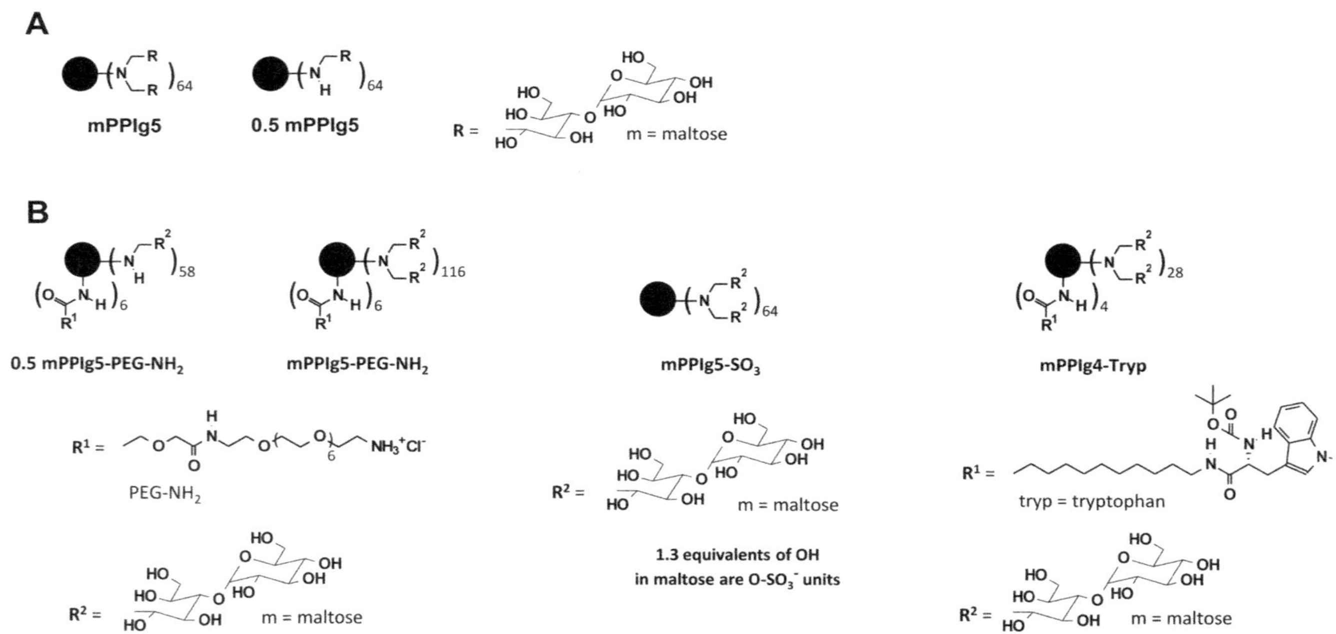

Figure 5.11 (A) Fifth-generation dense maltose shell glycodendrimer (mPPIg5) and fifth-generation open maltose shell glycodendrimer (0.5 mPPIg5). (B) Novel dendrimers utilized for checking their anti-prion activity: mPPIg5-PEG-NH_2 is a dense-shell glycodendrimer PPIg5 modified through the addition of six $NH_3^+Cl^-$ surface groups on PEG spacers. 0.5 mPPIg5-PEG-NH_2 is an open-shell glycodendrimer 0.5 mPPIg5 modified through the addition of $NH_3^+Cl^-$ surface groups on PEG spacers. mPPIg4-Tryp is a dense-shell glycodendrimer mPPIg4 modified through the addition of four tryptophan groups on aliphatic spacers. mPPIg5-SO_3 is a dense-shell glycodendrimer mPPIg5 modified through the addition of 1–2 sulfate units per maltose unit in the outer shell of mPPIg5. Reproduced with permission from ref. 9. Copyright 2013 American Chemical Society.

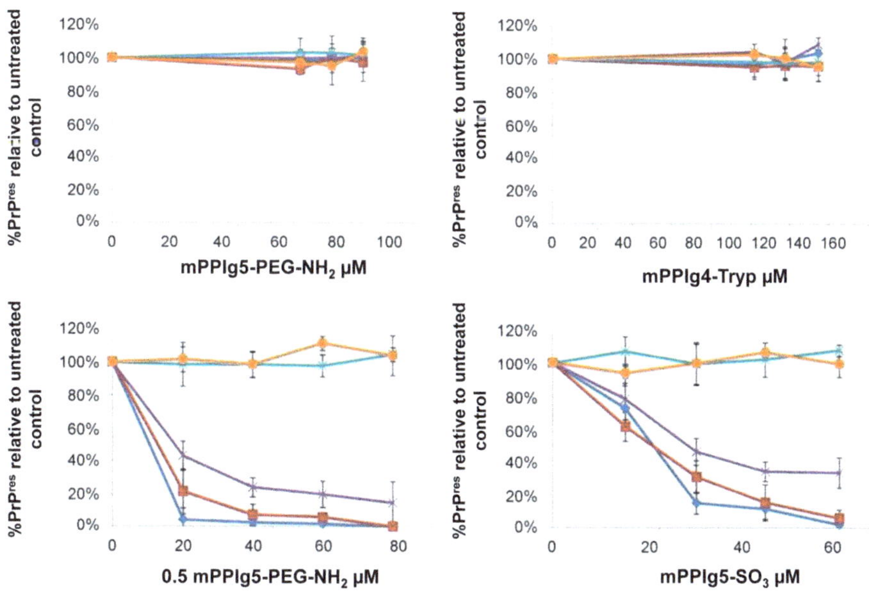

Figure 5.12 Anti-prion activity of novel dendrimer shown in Figure 5.11. The percentage PrP^{res} levels remaining after dendrimer treatment and protease digestion were calculated relative to a non-dendrimer-treated control by densitometry. —◆— 79A; —■— RML; —✕— ME7; —✳— 301C; —●— 301V. Error bars represent SD; $n = 3$. Reproduced with permission from ref. 9. Copyright 2013 American Chemical Society.

open-shell glycodendrimer with PEG-spacered amino units (0.5 mPPIg5-PEG-NH$_2$), and, in addition, anionic dense-shell glycodendrimer decorated with additional sulfate units (mPPIg5-SO$_3$).[9] For comparison purposes, PPIg4 and mPPIg5 (Figures 5.1 and 5.2) from a previous study[11,14] were also used. The introduction of cationic PEG-spacered amino units in mPPIg5-PEG-NH$_2$ and 0.5 mPPIg5-PEG-NH$_2$, and also cationic alkyl-spacer tryptophan units in mPPIg4-Tryp, was predicted to increase the strength of molecular interactions of new glycodendrimers with prion strains for destabilizing the conformation of PrP^{res}, making it susceptible to proteolytic proteinase K digestion. Similarly, the introduction of strong anionic sulfate units in the sugar shell of mPPIg5-SO$_3$ was also predicted to enhance the molecular interactions with prion strains as a counterpart to the cationic parent polyamine dendrimers successfully applied in the differentiation of prion strains.[14]

Figure 5.12 shows the anti-prion activity of the new dendrimers mPPIg5-PEG-NH$_2$, 0.5 mPPIg5-PEG-NH$_2$, mPPIg4-Tryp and mPPIg5-SO$_3$ against the prion strains RML, 79A and ME7 with a common scrapie genetic background, and also 301C and 301V with a common BSE genetic background. In contrast with the previous studies,[11,14] dense-shell glycodendrimer mPPIg5-PEG-NH$_2$ (Figure 5.11) shows absolutely no anti-prion activity against RML strain as found for the neutral fifth-generation dense-shell glycodendrimer mPPIg5

(Figure 5.11), and also no anti-prion activity against all other strains. A similar interaction characteristic was also found for the cationic fourth-generation dense-shell glycodendrimer mPPIg4-Tryp. The introduction of PEG-spacered amino units in the open-shell glycodendrimer does not result in any increased anti-prion activity, but in a slightly lower level of 0.5 mPPIg5-PEG-NH_2 (Figure 5.11) in comparison with 0.5 mPPIg5 (Figure 5.11). Surprisingly, the anionic glycodendrimer mPPIg5-SO_3 provides a similar anti-prion activity against all prion strains as known from the cationic PPI dendrimer PPIg4 (Figure 5.1),[14] but only at a slightly higher concentration.

Overall, the idea behind the PEG-spacered amino units was that the amino groups may better penetrate the molecular structure of PrP^{res} to induce attractive interaction forces between glycodendrimers 0.5 mPPIg5-PEG-NH_2 or mPPIg5-PEG-NH_2 and PrP^{res}. However, the anti-prion activity study at pH 7.4 revealed that hairy-like glycodendrimers did not result in any enhanced anti-prion activity. Hence the few attached PEG-spacered amino units in open- and dense-shell glycodendrimers seem to act only as slightly repellent PEG chains. On the other hand, negatively charged dendrimers are able to act as a substitute for cationic polyamine dendrimers with nearly the same anti-prion activity. Moreover, to confirm the high potential of anionic mPPIg5-SO_3 anti-prion activity, comparison experiments with mPPIg5, PPIg4 and mPPIg5-SO_3 at pH 7.4 and 3 were carried out. Generally, the anti-prion activities of all dendrimers on the same level were also confirmed by pH 3.[9]

Finally, one other interesting point throws further light on the chameleon-like interaction behaviour of PPI glycodendrimer against biological entities. Whereas neutral dense-shell glycodendrimer mPPIg5 did not show anti-prion activity in the diagnostic assay against prion strain 22A or 22L *in vitro*, mPPIg5 (Figure 5.11) cured cells infected with prion strain 22L.[15] This surprising result was illustrated by an improved bioassay, referred to as nSCA, which evaluated mPPIg5 as an intracellular anti-prion drug. This anti-prion drug activity was also compared with that of well-known drugs such as suramin and STI571, and mPPIg5 showed similar activity in comparison with the low-molecular weight drugs.[15] On the other hand, the anionic glycodendrimer mPPIg5-SO_3 did not show any anti-prion drug activity in an intracellular setting, when cells were infected with prion strain RML.[9] This further highlights the promise of sugar-modified dendrimers as anti-prion therapeutic agents.

5.6 Conclusions

This chapter has demonstrated the manifold and partly conflicting complexation, interaction and biological characteristics of PPI glycodendrimers with different oligosaccharide shell densities. To understand the characteristics of glycodendrimers in terms of complexation with low molecular weight drugs and interactions with nanoparticles and liposomes, examples of oligosaccharide-modified hyperbranched PEI were also shown to exhibit the unique and specific characteristics of open and dense-shell glycodendrimers. One major advantage of all of these oligosaccharide-decorated dendritic polyamine scaffolds is their very high *in vitro* and *in vivo* biocompatibility compared with the parent dendritic polyamine macromolecules.

Overall, the surface composition, charge density, generation and shell density of glycodendrimers dictate their final complexation and interaction characteristics with drug and analyte molecules and larger biological molecules and entities. Open- and dense-shell glycodendrimers often show conflicting complexation and interaction characteristics, especially under *in vitro* conditions. This obvious difference is related to the molecular composition, where dense-shell glycodendrimers can be considered as amphiphilic macromolecules with a cationic core and neutral oligosaccharide shell allowing no resulting surface charge, whereas open-shell glycodendrimers can exhibit weak cationic interactions. Therefore, the interactions and biological properties of dense-shell glycodendrimers with respect to proteins and liposomes are mainly governed by their multiple H-bond-active shell. The contribution of the H-bond-active shell becomes more dominant when the generation number increases. Moreover, PPI glycodendrimers are also able to cross the blood–brain barrier by using different modes of administration (intraperitoneal and intranasal). Their high biocompatibility, the ability to cross the blood–brain barrier and the weak H-bond-driven interactions towards proteins and enzymes make dense-shell glycodendrimers as logical choice for consideration as anti-amyloidogenic agents in neurodegenerative diseases.

The studies described in this chapter confirmed that various glycodendrimers are impressively characterized by interesting anti-Alzheimer amyloid peptide and anti-prion properties. However, remarkable differences in the interaction profiles of glycodendrimers with respect to both brain diseases, Alzheimer and prion disease, exist, depending on the precise design of the molecular structure. Neutral dense-shell glycodendrimers are superior materials with regard to anti-amyloidogenic properties in the treatment of Alzheimer disease, whereas they play a very minor role in the degradation of the protease-resistant prion proteins (PrP^{res}) in the presence of proteases. In this context, glycodendrimers with cationic or anionic surface charge, however, interact more strongly against PrP^{res} and induce the desirable degradation of PrP^{res} in the presence of proteases.

Nevertheless, further efforts are still needed to optimize the surface composition of dense- and open-shell glycodendrimers for their potential use as anti-Alzheimer and anti-prion agents under various *in vitro* and *in vivo* conditions. However, it is likely that in the future tailor-made glycodendrimers may be involved in the successful treatment of brain diseases by fabricating non-toxic and soluble aggregates as oligomeric/polymeric entities in Alzheimer disease and by enhancing the destabilization on PrP^{res} in order to favour proteolytic digestion in prion disease.

References

1. L. L. Kiessling and J. C. Grim, *Chem. Soc. Rev.*, 2013, **42**, 4476–4491.
2. Y. M. Chabre and R. Roy, in *Advances in Carbohydrate Chemistry and Biochemistry*, vol. 63, ed. D. Horton, Elsevier Academic Press, Oxford, 2010, pp. 168–393.
3. P. Albersheim and B. S. Valent, *J. Cell Biol.*, 1978, **78**, 627–643.
4. Y. C. Lee, *FASEB J.*, 1992, **6**, 3193–3200.
5. W. I. Weiss and K. Drickamer, *Annu. Rev. Biochem.*, 1996, **65**, 441–473.
6. J. M. Rini, *Annu. Rev. Biophys. Biomol. Struct.*, 1995, **24**, 551–577.
7. H.-J. Gabius, S. André, J. Jiménez-Barbero, A. Romero and D. Solis, *Trends Biochem. Sci.*, 2011, **36**, 298–313.
8. B. Klajnert, D. Appelhans, H. Komber, N. Morgner, S. Schwarz, S. Richter, B. Brutschy, M. Ionov, A. K. Tonkikh, M. Bryszewska and B. Voit, *Chem.-Eur. J.*, 2008, **14**, 7030–7041.
9. J. M. McCarthy, B. Rasines Moreno, D. Filippini, H. Komber, M. Marek, M. Cernescu, B. Brutschy, D. Appelhans and M. S. Rogers, *Biomacromolecules*, 2013, **14**, 27–37.
10. O. Klementieva, E. Aso, D. Filippini, N. Benseny-Cases, M. Carmona, S. Juves, D. Appelhans, J. Cladera and I. Ferrer, *Biomacromolecules*, 2013, **14**, 3570–3580.
11. M. Fischer, D. Appelhans, S. Schwarz, B. Klajnert, M. Bryszewska, B. Voit and M. Rogers, *Biomacromolecules*, 2010, **11**, 1314–1325.
12. D. Appelhans, N. Benseny-Cases, O. Klementieva, M. Bryszewska, B. Klajnert, J. Cladera, in *Dendrimers in Biomedical Applications*, ed. B. Klajnert, L. Peng, V. Cena, RSC Press, Cambridge, 2013, pp. 1–13.
13. O. Klementieva, N. Benseny-Cases, A. Gella, D. Appelhans, B. Voit and J. Cladera, *Biomacromolecules*, 2011, **12**, 3903–3909.
14. J. McCarthy, B. Rasines Moreno, D. Appelhans and M. Rogers, *Adv. Healthcare Mater.*, 2012, **1**, 768–772.
15. J. M. McCarthy, M. Franke, U. K. Resenberger, S. Waldron, J. C. Simpson, J. Tatzelt, D. Appelhans and M. S. Rogers, *PloS One*, 2013, **8**, e55282.
16. M. F. Ottaviani, M. Cangiotti, L. Fiorani, S. Lucchi, T. Wasiak, D. Appelhans and B. Klajnert, *Curr. Med. Chem.*, 2012, **19**, 5907–5921.
17. B. Klajnert, M. Cortijo-Arellano, M. Bryszewska and J. Cladera, *Biochem. Biophys. Res. Commun.*, 2006, **339**, 577–582.
18. B. Klajnert, J. Cladera and M. Bryszewska, *Biomacromolecules*, 2006, **7**, 2186–2191.
19. B. Klajnert, M. Cortijo-Arellano, J. Cladera and M. Bryszewska, *Biochem. Biophys. Res. Commun.*, 2006, **345**, 21–28.
20. B. Klajnert, M. Cortijo-Arellano, J. Cladera, J. P. Majoral, A. M. Caminade and M. Bryszewska, *Biochem. Biophys. Res. Commun.*, 2007, **364**, 20–25.
21. B. Klajnert, M. Cangiotti, S. Calici, J. P. Majoral, A. M. Caminade, J. Cladera, M. Bryszewska and M. F. Ottaviani, *Macromol. Biosci.*, 2007, **7**, 1065–1074.
22. B. Klajnert, M. Cangiotti, S. Calici, M. Ionov, J. P. Majoral, A. M. Caminade, J. Cladera, M. Bryszewska and M. F. Ottaviani, *New J. Chem.*, 2009, **33**, 1087–1093.

23. O. Renaudet and R. Roy, *Chem. Soc. Rev.*, 2013, **42**, 4515–4517.
24. D. Appelhans, H. Komber, M. Abdul Quadir, A. Loos, S. Schwarz, S. Richter, A. Aigner, M. Müller, R. Haag and B. Voit, *Biomacromolecules*, 2009, **10**, 1114–1124.
25. D. Appelhans, U. Oertel, R. Mazzeo, H. Komber, J. Hoffmann, S. Weidner, B. Brutschy, B. Voit and M. F. Ottaviani, *Proc. R. Soc. London, Ser. A*, 2010, **466**, 1489–1513.
26. S. Höbel, A. Loos, D. Appelhans, S. Schwarz, J. Seidel, B. Voit and A. Aigner, *J. Control. Release*, 2011, **149**, 146–158.
27. A. F. Thünemann, R. Bienert, D. Appelhans and B. Voit, *Macromol. Chem. Phys.*, 2012, **213**, 2362–2369.
28. S. Tripp, D. Appelhans, C. Striegler and B. Voit, *Chem. Eur. J.*, 2014, **20**, 8314–8319.
29. A. Szulc, D. Appelhans, B. Voit, M. Bryszewska and B. Klajnert, *New J. Chem.*, 2012, **36**, 1610–1615.
30. A. Szulc, D. Appelhans, B. Voit, M. Bryszewska and B. Klajnert, *J. Fluoresc.*, 2013, **23**, 349–356.
31. T. Pietsch, D. Appelhans, N. Gindy, B. Voit and A. Fahmi, *Colloids Surf., A*, 2009, **341**, 93–102.
32. A. Fahmi, T. Pietsch, D. Appelhans, N. Gindy and B. Voit, *New J. Chem.*, 2009, **33**, 703–706.
33. A. Köth, J. Koetz, D. Appelhans and B. Voit, *Colloid Polym. Sci.*, 2008, **286**, 1317–1327.
34. A. Köth, B. Tiersch, D. Appelhans, M. Gradzielski, H. Cölfen and J. Koetz, *J. Disper. Sci. Technol.*, 2012, **33**, 52–60.
35. M. Mkandawire, A. Pohl, T. Gubarevich, V. Lapina, D. Appelhans, G. Rödel, W. Pompe, J. Schreiber and J. Opitz, *J. Biophoton.*, 2009, **2**, 596–606.
36. M. Ciolkowski, B. Pałecz, D. Appelhans, B. Voit, B. Klajnert and M. Bryszewska, *Colloids Surf., B*, 2012, **95**, 103–108.
37. M. Ciolkowski, I. Halets, D. Shcharbin, D. Appelhans, B. Voit, B. Klajnert and M. Bryszewska, *New J. Chem.*, 2012, **36**, 1992–1999.
38. A. Köth, D. Appelhans, B. Tiersch and J. Koetz, *Soft Matter*, 2011, **7**, 10581–10584.
39. D. Wrobel, D. Appelhans, M. Signorelli, D. Fessas, B. Voit and J. Maly, submitted.
40. Y. Tanaka-Takiguchi, M. Kinoshita and K. Takiguchi, *Curr. Biol.*, 2009, **19**, 140–145.
41. P. D. Blood and G. A. Voth, *Proc. Natl. Acad. Sci. U. S. A.*, 2006, **103**, 15068–15072.
42. A. Janaszewska, K. Mączyńska, G. Matuszko, D. Appelhans, B. Voit, B. Klajnert and M. Bryszewska, *New J. Chem.*, 2012, **36**, 428–437.
43. B. Ziemba, G. Matuszko, D. Appelhans, B. Voit, M. Bryszewska and B. Klajnert, *Biopolymers*, 2012, **97**, 642–648.
44. B. Ziemba, I. Halets, D. Shcharbin, D. Appelhans, B. Voit, M. Bryszewska and B. Klajnert, *J. Biomed. Mater. Res., Part A*, 2012, **100**, 2870–2880.
45. A. Filimon, L. E. Sima, D. Appelhans, B. Voit and G. Negroiu, *Curr. Med. Chem.*, 2012, **19**, 4955–4968.

46. I. Franiak-Pietryga, E. Ziółkowska, B. Ziemba, D. Appelhans, B. Voit, M. Szewczyk, J. Góra-Tybor, T. Robak, B. Klajnert and M. Bryszewska, *Mol. Pharm.*, 2013, **10**, 2490–2501.
47. B. Ziemba, I. Franiak-Pietryga, M. Pion, D. Appelhans, M.-Á. Muñoz-Fernández, B. Voit, M. Bryszewska and B. Klajnert-Maculewicz, *Int. J. Pharm.*, 2014, **461**, 391–402.
48. B. Ziemba, A. Janaszewska, K. Ciepluch, M. Krotewicz, W. A. Fogel, D. Appelhans, B. Voit, B. Klajnert and M. Bryszewska, *J. Biomed. Mater. Res., Part A*, 2011, **99**, 261–268.
49. D. Gutsch, D. Appelhans, S. Höbel, B. Voit and A. Aigner, *Mol. Pharm.*, 2013, **10**, 4666–4675.
50. C. Haass and D. J. Selkoe, *Nat. Rev. Mol. Cell Biol.*, 2007, **8**, 101–112.
51. S. B. Prusiner, *Dev. Biol. Stand.*, 1991, **75**, 55–74.
52. C. W. Cairo, A. Strzelec, R. M. Murphy and L. L. Kiessling, *Biochemistry*, 2002, **41**, 8620–8629.
53. A. Nazem and G. A. Mansoori, *J. Alzheimer's Dis.*, 2008, **13**, 199–223.
54. J. Hardy and D. J. Selkoe, *Science*, 2002, **297**, 353–356.
55. N. Benseny, M. Cocera and J. Cladera, *Biochem. Biophys. Res. Commun.*, 2007, **361**, 916–921.
56. K. N. Dahlgreen, A. M. Manelli, W. B. Stine, Jr, L. K. Baker, G. A. Krafft and M. J. LaDu, *J. Biol. Chem.*, 2002, **277**, 32047–32053.
57. N. Benseny-Cases, O. Klementieva, J. Maly and J. Cladera, *Curr. Alzheimer Res.*, 2012, **9**, 962–971.
58. J. Bieschke, S. J. Siegel, Y. Fu and J. W. Kelly, *Biochemistry*, 2008, **47**, 50–59.
59. O. Simakova and N. J. Arispe, *J. Neurosci.*, 2007, **27**, 13719–13729.
60. S. W. Synder, U. S. Ladror, W. S. Wade, G. T. Wang, L. W. Barrett, E. D. Matayoshi, H. J. Huffaker, G. A. Krafft and T. F. Holzman, *Biophys. J.*, 1994, **67**, 1216–1228.
61. A. Sakudo, in *Prions: Current Progress in Advanced Research*, ed. A. Sakudo and T. Onodera, Caister Academic Press, Norfolk, 2013, pp. 5–10.
62. S. B. Prusiner, *Proc. Natl. Acad. Sci. U. S. A.*, 1998, **95**, 13363–13383.
63. S. Suppattapone, J. R. Piro and J. R. Rees, *CNS Neurol. Disord.: Drug Targets*, 2009, **8**, 323–328.
64. S. B. Prusiner, *Science*, 1982, **216**, 136–144.
65. J. Castilla, P. Saa, C. Hetz and C. Soto, *Cell*, 2005, **121**, 195–206.
66. N. R. Deleault, B. T. Harris, J. R. Rees and S. Supattapone, *Proc. Natl. Acad. Sci. U. S. A.*, 2007, **104**, 9741–9746.
67. T. Onodera and A. Sakudo, in *Prions: Current Progress in Advanced Research*, ed. A. Sakudo and T. Onodera, Caister Academic Press, Norfolk, UK, 2013, pp. 1–3.
68. A. Aguzzi, F. Baumann and J. Bremer, *Annu. Rev. Neurosci.*, 2008, **31**, 439–477.
69. M. Nunziante, S. Gilch and H. M. Schatzl, *ChemBioChem*, 2003, **4**, 1268–1284.
70. S. B. Prusiner, R. Gabizon and M. P. McKinely, *Acta Neuropathol.*, 1987, **72**, 299–314.

71. E. Beck and P. M. Daniel, in *Prions: Novel Infectious Pathogens Causing Scrapie and Creutzfeldt–Jakob Disease*, ed. S. B. Prusier and M. P. McKinely, Academic Press, New York, 1987, pp. 331–385.
72. J. Collinge, *Annu. Rev. Neurosci.*, 2001, **24**, 519–550.
73. J. Collinge, *J. Neurol., Neurosurg. Psychiatry*, 2005, **76**, 906–917.
74. H. A. Kretzschmar, *Eur. Arch. Psychiatry Clin. Neurosci.*, 1999, **249**, Suppl. 3 56–63.
75. S. Supattapone, H. O. Nguyen, F. E. Cohen, S. B. Prusiner and M. R. Scott, *Proc. Natl. Acad. Sci. U. S. A.*, 1999, **96**, 14529–14534.
76. K. K. Hsiao, D. Groth, M. Scott, S. L. Yang, H. Serban, D. Rapp, D. Foster, M. Torchia, S. J. Dearmond and S. B. Prusiner, *Proc. Natl. Acad. Sci. U. S. A.*, 1994, **91**, 9126–9130.
77. G. C. Telling, P. Parchi, S. J. DeArmond, P. Cortelli, P. Montagna, R. Gabizon, J. Mastrianni, E. Lugaresi, P. Gambetti and S. B. Prusiner, *Science*, 1996, **274**, 2079–2082.
78. E. Bruce and H. Fraser, *Curr. Top. Microbiol. Immunol.*, 1991, **172**, 125–138.
79. M. E. Bruce, *Br. Med. Bull.*, 1993, **49**, 822–838.
80. H. Fraser and A. G. Dickson, *J. Comp. Pathol.*, 1973, **83**, 29–40.
81. R. G. Will, J. W. Ironside, M. Zeidler, K. Estibeiro, S. N. Cousens, P. G. Smith, A. Alperovitch, S. Poser, M. Pocchiari and A. Hofman, *Lancet*, 1996, **347**, 921–925.
82. B. I. Lasmezas, J. G. Fournier, V. Nouvel, H. Boe, D. Marce, F. Lamoury, N. Kopp, J. J. Hauw, J. Ironside, M. Bruce, D. Dormont and J. P. Deslys, *Proc. Natl. Acad. Sci. U. S. A.*, 2001, **98**, 4142–4147.
83. C. Weissmann, *Folia Neuropathol.*, 2009, **47**, 104–113.
84. S. Mahal, C. Barker, C. Demczyk, E. Smith, E. Julius and C. Weissmann, *Proc. Natl. Acad. Sci. U. S. A.*, 2007, **104**, 20908–20913.
85. D. Peretz, M. R. Scott, D. Groth, R. A. Williamson, D. R. Williamson, D. R. Burton, F. E. Cohen and S. B. Prusiner, *Protein Sci.*, 2001, **10**, 854–863.

CHAPTER 6

Self-Assembly of Glycopolymers: From Nano-Objects to Hydrogels

YU ZHAO[a], GAOJIAN CHEN[b], AND GUOSONG CHEN*[a]

[a]State Key Laboratory of Molecular Engineering of Polymers, Department of Macromolecular Science, Fudan University, Shanghai 200433, China;
[b]Center for Soft Condensed Matter Physics and Interdisciplinary Research, Soochow University, Suzhou 215006, China
*E-mail: guosong@fudan.edu.cn

6.1 Introduction

Self-assembly plays a basic and irreplaceable role in Nature, of which amazing examples exist everywhere, *e.g.* DNA forms a double helix, then complexes with histone octamers forming chromatin, which will compact into fiber *via* a two-step self-assembly process.[1] Protein can even self-assemble into many kinds of structures with various functionalities, for instance, the large heterocomplex ribosome, myosin filaments, crystalline bacterial S layers and membranes containing two-dimensional arrays of bacteriorhodopsin or porins.[2] One can imagine that being a relatively huge family of crucial macromolecules, sugars will also contribute to Nature through varieties of self-assembled structures. However, probably because the chemical structures of sugars are much more complicated than those of proteins and the interactions between different sugar molecules currently are difficult to

RSC Polymer Chemistry Series No. 15
Glycopolymer Code: Synthesis of Glycopolymers and Their Applications
Edited by C. Remzi Becer and Laura Hartmann

Published by the Royal Society of Chemistry, www.rsc.org

identify, our understanding of this huge family, particularly their self-assembly behavior, is so far limited.

From the viewpoint of polymer science, we suggest that there are two key aspects worth exploring for finally deciphering the code behind the self-assembly of sugars in Nature. (1) Glycocalyx, which is a dense layer of complex carbohydrates, exists universally on both prokaryotic and eukaryotic cells.[3] The success of block copolymer self-assembly provides a great opportunity to mimic this type of structure *via* various synthetic 'sweet' polymers. (2) Polysaccharides provide the structural backbone of many living creatures, from bacteria to fungi and from plants to marine organisms. In these systems, sugars are compactly packed, with strong intermolecular interactions. To understand or mimic this type of structure, more attention needs to be paid to self-assembled nano-objects with glyco-inside structures.

For constructing 'glycocalyx mimics' and 'glyco-inside structures,' glycopolymers are a simplified but powerful building block. Polymer scientists are able to design new polymeric sugars without complicated chemical synthesis; the materials obtained will exhibit the characteristics of polymers and their self-assembly process can be understood by means of classical theory of macromolecular self-assembly. Moreover, the nano-objects obtained can be utilized as drug- or protein-delivery vehicles and glucose sensors, which now seem promising and attractive. In this chapter, we first summarize the general behavior of self-assembly of glycopolymers into nano-objects followed by a section focusing on their drug-delivery applications. Then the self-assembled glyco-inside structure obtained *via* a new protection–deprotection self-assembly procedure is described. Finally, some hydrogels made of glycopolymer or lectin are briefly discussed. It should be mentioned that the focus in this chapter is on glycopolymers with a synthetic polymeric backbone.

6.2 Self-Assembly of Glycopolymers in Solution

Macromolecular self-assembly (MSA) has been an active and fruitful research field since the 1980s, and especially in the 21st century, promoted by the remarkable developments in controlled radical polymerization in polymer chemistry, *e.g.* ATRP (atom-transfer radical polymerization) and RAFT (reversible addition–fragmentation chain transfer polymerization) and driven by the demands in bio-related investigations and applications.[4] As a part of MSA, research on the self-assembly of glycopolymers in solution started with several early reports around 2000.[5] However, after more than 10 years of development, although there is some accumulation of related research papers, work on the self-assembly of glycopolymers is still limited, whereas that of the traditional amphiphilic block copolymers with a variety of architectures has developed very rapidly.

Self-assembly of glycopolymers first started as a small branch of macromolecular self-assembly, in which the glyco-units play the role of biocompatible hydrophilic components such as poly(ethylene glycol) (PEG) and poly(acrylic

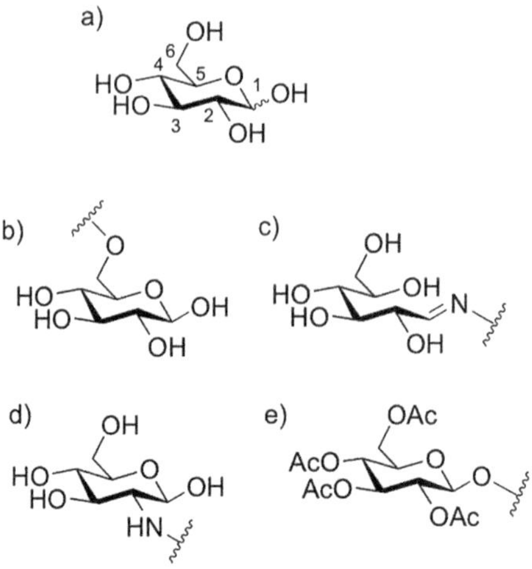

Figure 6.1 (a) A typical structure of glucopyranose with the carbons on pyranose ring numbered clockwise. (b–e) Possible modifications on monomer or polymer main chain of sugars.

acid) (PAA) do. For this purpose, convenient synthetic procedures for introducing glyco-units into the polymer chain were required. In this respect, sugars have more chemical handles than nucleic acids and amino acids. For the most common form of monosaccharides, *i.e.* hexopyranose, one primary hydroxyl group, three secondary hydroxyl groups and one hemiacetal are available for modification (Figure 6.1(a)). The only primary hydroxyl group (6-OH) has less steric hindrance, resulting in easier modification without protective groups *via* a one-step synthesis (Figure 6.1(b)). More popularly, amine-functionalized monomers react with an anomeric center (1-OH) and give open-chain derivatives (Figure 6.1(c)). Bearing one amino group among several hydroxyl groups, commercially available amino sugars, *e.g.* glucosamine (2-amino-2-deoxyglucose), are also widely used to obtain glycopolymers for self-assembly owing to the facile modification *via* the primary amine (Figure 6.1(d)). Several years ago, acetates began to be employed in the preparation of sugar monomers, because acyl sugars are less polar and easier to handle than free sugars and the stereochemistry of the product can be controlled in some sugars after protection. The fully acyl-protected sugar can be directly coupled to commercially available 2-hydroxyethyl methacrylate-type monomers under weak acid catalysis, resulting in a mixed α-/β-stereochemistry or the pure α- or β-anomer of sugars (Figure 6.1(e)). The stereochemistry is different on different sugar backbones. In some cases, the acetates were removed after polymerization of acrylate monomers. The base used for deacylation may attack the ester bond close to the polymer backbone, resulting in free acid pendant groups instead of the desired glyco-units. In these cases, the glycopolymers are excellent hydrophilic components from the viewpoint of drug delivery, for insulin release and glucose sensors, according to their biocompatibility, degradability and possible targeting properties.

On the other hand, the biomimetic aspect of the self-assembly of glycopolymers, although seemingly very attractive and demanding, emerged fairly late compared with the above-mentioned topics. The idea was proposed mainly by two different research groups.[6,7] For this purpose, the sugars need to be connected to the polymer backbone *via* its natural linkage, *i.e.* an anomeric center with well-controlled stereochemistry, to ensure its biological function, at least the binding ability to lectins. Here, the discussion on glycopolymer self-assembly in solution is divided into three sub-sections: general behavior of self-assembly, glyconanoparticles for drug/protein-delivery applications and self-assembled nano-objects mimicking glycocalyx. So far, almost all of the related nanostructures reported have sugars on their surface, hence the above three categories in fact mainly concern the glyco-outside type of structure; its counterpart, the glyco-inside structure, is discussed in the final section.

6.2.1 General Behavior of Self-Assembly of Glycopolymer-Containing Block Copolymers

Studies of the self-assembly of glycopolymers are based on the achievements of the self-assembly of amphiphilic block copolymers. In the mean time, progress in controlled radical polymerization has provided satisfactory glycopolymers with a well-controlled molecular structure and narrow polydispersity index. Among the many common monosaccharides, glucose is used as a representative in most studies. Modification of glucose is simpler than that of other monosaccharides, owing to the dominant equatorial hydroxyl groups on the pyranose ring. Moreover, the stereochemistry of glucose on the anomeric center is straightforward, *i.e.* β-selectivity is greatly preferred because of steric effects and neighboring group participation with acetate on 2-OH. In one representative early study by Li *et al.* with the block copolymer polystyrene-*b*-poly[(2-β-D-glucopyranosyloxy)ethyl acrylate] (PS-*b*-PGEA), different morphologies of aggregates, such as micelle-like spheres, vesicles, tubules and even large compound micelles, with diameters from 100 nm to 2 μm, were observed on varying the copolymer composition.[5] In that study, not only copolymer composition but also other effects controlling morphology were noted, including copolymer concentration and common solvent. Similarly, as reported by You and Schlaad, a block copolymer containing PS as the hydrophobic block and 1,2-polybutadiene randomly modified with β-D-1-thioglucopyranose as the hydrophilic block was transformed into vesicles in water with a diameter around 120 nm.[8]

More recently, 1,2-polybutadiene-*b*-poly(ethylene oxide) randomly modified with thioglucopyranose using the same method (PB_x-*b*-PEO_y; $x = 65$, $y = 212$ and $x = 68$, $y = 34$, with only 6–9% of the double bonds unmodified with sugars) gave large unilamellar vesicles, which were characterized by dynamic/static light scattering (DLS/SLS), transmission electron microscopy (TEM) and small-angle X-ray scattering (SAXS).[9] Owing to the high M_w of sugars, the hydrophilic weight ratio of the block copolymers was found to be exceptionally high ($w_{hydrophilic} > 0.5$), which made the observed vesicle morphology surprising, because compact micelles are expected at this high $w_{\text{hydrophilic}}$. Moreover, the diameter of the vesicles formed by the more hydrophilic copolymer PB_{65}-*b*-PEO_{212} was about two times larger than that of PB_{68}-*b*-PEO_{34}, *i.e.* the smaller the packing parameter, the smaller the curvature, which seemed contrary to the conventional trend of the opposite relation between packing parameter of the amphiphile and curvature of the aggregates. The authors explained this peculiarity with an asymmetric model of the self-assembled vesicles with β-glucopyranoside on the outer layer and PEO on the inner layer.

In addition to the conventional synthetic hydrophobic blocks, polypeptides are also employed as hydrophobic block connecting to glycopolymers, with the advantages of better biodegradability, biocompatibility, tunable functionality and defined secondary structures than synthetic hydrophilic polymers.[10] As the helical conformation of the hydrophobic peptide segment poly(γ-benzyl-L-glutamate) (PBLG) provides an alternative way to form and stabilize vesicles, PBLG-*b*-poly(galactosylated propargylglycine) (PBLG-*b*-PGG) copolymers were prepared with different hydrophilic weight ratios. It was found that irrespective of the hydrophilic weight ratio, coexistence of spherical and worm-like structures was observed when water, *i.e.* a non-solvent for the hydrophobic segment, was added to a solution of PBLG-*b*-PGG in the common solvent dimethyl sulfoxide (DMSO), as shown in Figure 6.2(a–c) (water in DMSO). The coexistence of morphologies was explained by kinetically trapped metastable morphologies, *i.e.* when the solvent diffusion was faster than the self-assembly, no matter how the molecular composition varied. In contrast, the opposite approach. *i.e.*, adding the DMSO solution to water (DMSO in water) promoted the formation of spherical structures without worm-like assemblies (Figure 6.2(d) and (e)).[11] The mixture of spherical micelles and vesicles observed in Figure 6.2(f), formed by the block copolymers with the longest PGG block, indicated the upper limit of the hydrophilic weight ratio for polymersome formation with these diblock copolymers. In addition to block copolymers, *via* efficient post-polymerization modification, polylysine was modified with sugars, affording random amphiphilic block polymers that also self-assembled in water.[12]

6.2.2 Glyconanoparticles for Drug/Protein Delivery Applications

So far, studies of this category have been the most prolific among the three branches of glycopolymer self-assembly. The water solubility, biocompatibility and biodegradability of sugars make the 'sweet' polymer a popular protective shell for various drug-delivery vehicles. Moreover, some of the acceptors on a tumor cell surface recognize a pair of adjacent hydroxyl groups of the sugar backbone; for example, the well-known ASGPR (asialoglycoprotein receptor) of Hep G2 cells recognizes different constitutional isomers of sugars with the expected role of targeting. On the other hand, with the prevalence of diabetes, detection of blood sugar levels and the controlled delivery of insulin are important research topics. For example, as a representative

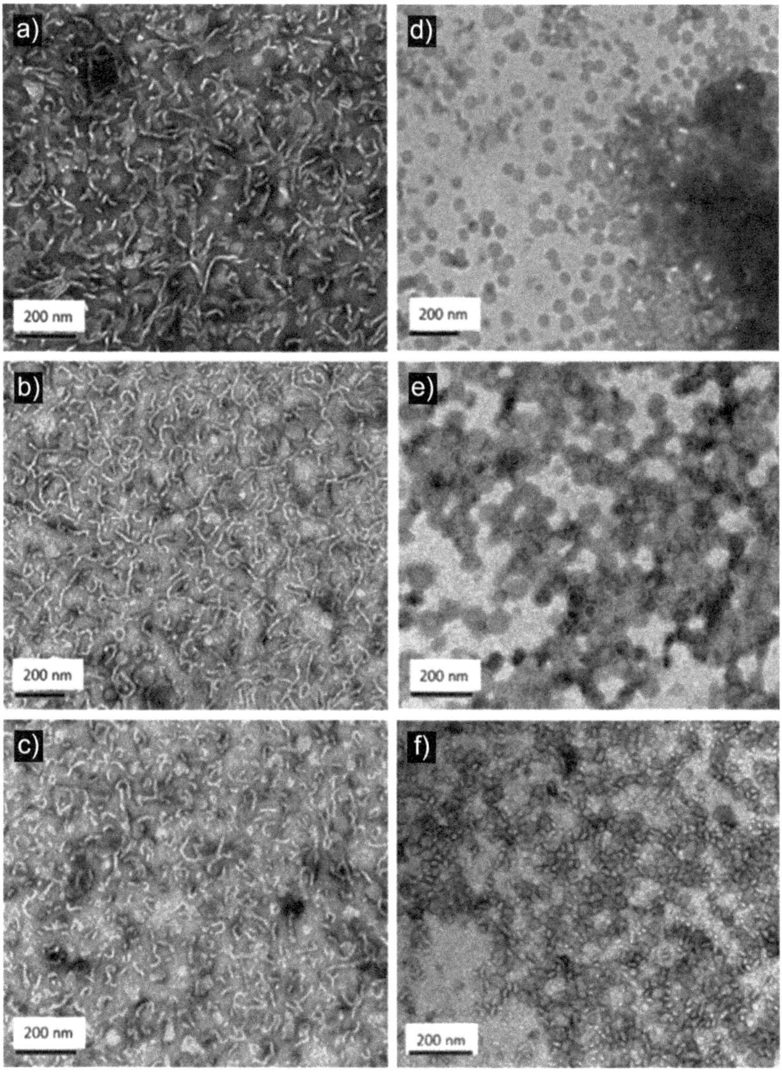

Figure 6.2 TEM images of samples obtained by instantaneously adding (a–c) water in DMSO and (d–f) DMSO in water. (a, d) $PBLG_{20}$-*b*-PGG_{18}; (b, e) $PBLG_{20}$-*b*-PGG_{25}; (c, f) $PBLG_{20}$-*b*-PGG_{32}. Source: ref. 11.

way to prepare sugar-responsive nanoparticles, a glycopolymer may complex with phenylboronic acid *via* two *cis*-hydroxyl groups or α,β-diol of the sugar open-chain structure, forming sugar-responsive nanoparticles.[13,14] During this self-assembly process, insulin can be immobilized inside the nanoparticle, followed by successful glucose-responsive release.[13]

In this respect, sugars linked *via* 6-hydroxyl groups and 2-amino groups are widely used owing to their facile synthetic procedures, required for a possible delivery vehicle. It is worth mentioning that Narain and Armes[15] developed

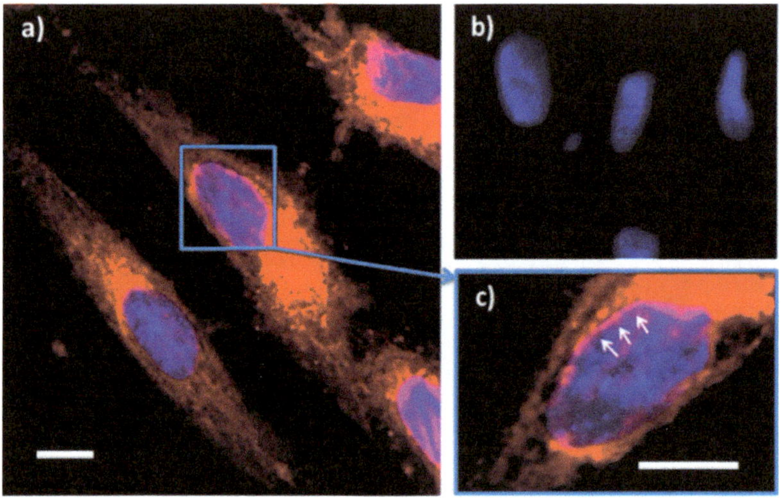

Figure 6.3 Effective intracellular delivery of RhB in HDF cells mediated by galactose-coated vesicles. (a) Confocal microscopy image of live HDF cells. Note the intracellular staining of membranes (red) after exposure to the rhodamine-loaded vesicles; cell nuclei were counter-stained blue using Hoechst 33 342. (b) HDF cells treated with the same vesicles containing no RhB dye. (c) Higher magnification image obtained for (a): effective intracellular delivery of rhodamine dye allows selective staining of the nuclear membrane (white arrows). Scale bar: 50 μm. Source: ref. 19.

an efficient strategy of reacting 2-aminoethyl methacrylate with D-gluconolactone and D-lactobionolactone, generating 2-gluconamidoethyl methacrylate (GAMA) and 2-lactobionamidoethyl methacrylate (LAMA), respectively. The sugar derivative monomers obtained in fact are linear chiral glycerol-like structures, which were further utilized by different research groups.[16]

With possible applications in drug delivery, a spherical nanoparticle is the most common morphology obtained. These sugar-appended nanoparticles show little cytotoxicity, which is very important for drug-delivery applications. Studies have shown that these 'sweet' nanoparticles entered HeLa cells mainly *via* macropinocytosis, without specific binding with receptors on the cell surface.[17] More efficient internalization of these 'sweet' nanoparticles was observed by the cells containing a specific acceptor to sugars on their surface. For example, much more efficient delivery of doxorubicin to Hep G2 cells than HEK293 cells was achieved with galactose-coated nanoparticles. Here the HEK293 cell, which does not contain any sugar-specific protein on its surface, was used as a control.[18] Furthermore, effective intracellular delivery of rhodamine B octadecyl ester (RhB) to HDF cells was mediated by galactose-coated vesicles. HDF cells express galectin, *i.e.* a family of lectin-recognizing galactosides, on their surface. As shown in Figure 6.3, the red color inside the cell (Figure 6.3(a)), especially the selective staining of the nuclear membrane (Figure 6.3(c)), proves the effective intracellular delivery of RhB to HDF cells.[19]

In addition to these sugar-coated nanoparticles or vesicles, more nano-objects with various sizes aimed at drug-delivery vehicles *via* self-assembly of different glycopolymers have been prepared. Kataoka and co-workers reported that block copolymers PEG-*b*-PLA [PLA = poly(lactic acid)] with monosaccharides linked to the end of the PEG *via* 6-OH formed micelles with a diameter of 40 nm.[20] Using poly(ε-caprolactone) (PCL) as the hydrophobic block, amphiphilic block glycopolymers were obtained, which formed micelles with diameters of 20–125 nm,[17] or even much larger aggregates (diameter 400 nm).[21] Other examples of hydrophobic blocks included polycarbonate.[18] poly(*n*-butyl acrylate),[22] poly(propylene oxide) and thermo-responsive poly(ethylene glycol) methacrylate.[23] In addition to diblock linear copolymers, asymmetric modified homopolymers,[24,25] rod-coil block copolymers,[26,27] triblock copolymers,[28] brush copolymers[29] and star-shaped copolymers[30] with the glyco part as the hydrophilic unit have also been employed to form delivery nanoparticles. Besides conventional self-assembly strategies, polymerization-induced self-assembly was found to be a powerful approach to prepare various self-assembled morphologies in concentrated solution for drug-delivery vehicles. Recently, Armes and co-workers[19] reported the synthesis of a series of nano-objects containing sugars, including nanospheres, worm-like micelles and vesicles, by polymerization of monomer 2-hydroxypropyl methacrylate, starting from a 1 : 9 molar mixture of sugar-containing RAFT macromolecular chain transfer agent (MacroCTA) and non-sugar-containing RAFT macroCTA.

By incorporating fluorophores into glycopolymers, functional nanoparticles with potential applications in cell imaging have been reported. For example, random copolymers of 4,4-difluoro-4-bora-3a,4a-diaza-*s*-indacene (BODIPY)-containing monomer and 2-*O*-methacryloyloxyethyl-(2,3,4,6-tetra-*O*-acetyl-β-D-galactopyranoside) (AcGEMA) were synthesized by ATRP for tumor cell imaging. The polymers form micelles with a spherical shape of diameter 210–250 nm.[31] Chen and co-workers reported that a random copolymer of 2-(methacrylamido)glucopyranose (MAG) and methacrylic acid (MAA) could be used as templates to prepare glycopolymer-functionalized silver nanoclusters through microwave irradiation, and the nanoclusters showed efficient binding ability towards K562 cells and inhibited the cell viability in a dose-dependent manner[32] (Figure 6.4(a)). In another example, an efficient one-pot reaction generated a novel glycopolymer–porphyrin conjugate, where the hydrophobic porphyrin was in the middle of the glycopolymer chain and the conjugate then self-assembled into nanoparticles with potential applications in targeted photodynamic therapy[33] (Figure 6.4(b)).

6.2.3 Self-Assembled Vesicles Mimicking *Glycocalyx*

Decades of research in glycobiology have demonstrated that the carbohydrates of glycocalyx play an irreplaceable role in a variety of cellular events, including proliferation, recognition, intercellular communication and immunoregulation (Figure 6.5).[34] It is suggested that glyco-code is the most complicated coding system in organisms, containing large amounts

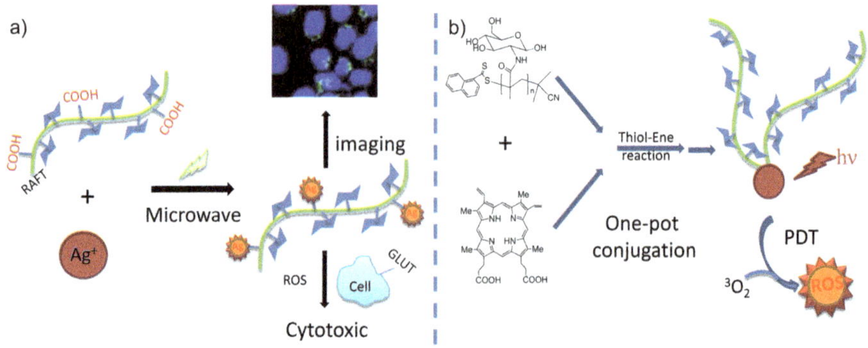

Figure 6.4 (a) Microwave-assisted synthesis of glycopolymer-functionalized silver nanoclusters; (b) one-pot synthesis of glycopolymer–porphyrin conjugate as a photosensitizer for targeted cancer imaging and photodynamic therapy.

of information owing to the structural diversity. Information translation and transcription in glycobiology generally rely on intermolecular binding between oligosaccharide ligands and their specific receptors.[4] It is also known that the binding constant for a 1:1 carbohydrate–lectin complex is rather low; the apparent high activity and specificity are a result of multivalent interactions. The glycocalyx structure on the cell surface and multivalent binding events are the two obvious research targets for current studies on glycocalyx mimicking of glycopolymer self-assembly.

Self-assembled glycopolymers have their own peculiarities, with the appropriate size, sugar density, backbone flexibility, *etc.* Pasparakis and Alexander prepared block copolymers with highly hydrophilic poly(2-glucosyloxyethyl methacrylate) as one block and poly(diethylene glycol methacrylate) as the second block; by using controlled free-radical techniques, the block copolymers assembled into vesicles with a surface displaying glucose functionality, as a mimic of natural cells 'talking' with *Escherichia coli* bacteria.[6] They proposed that the interfacial interaction of glucose-coated vesicles with bacteria might trigger disruption of the vesicular membrane and therefore vesicles containing molecular 'information' could communicate with the bacteria during interaction. The dye ethidium bromide was further used as a model compound and efficient release of this dye was observed after 30 min of incubation. However, currently it cannot be concluded that the specific binding of glucose with *fim*H on the *E. Coli* surface plays a crucial role because this binding was found to be very weak in a previous study.[35]

There is a wide gap in this research area: sugars with a natural linkage on an anomeric center with clear stereochemistry are definitely required, and also more types of monosaccharides, disaccharides and oligosaccharides are highly demanded. It should be emphasized that the binding affinity and specificity in lectin recognition will be unsatisfactory if the carbohydrate moieties are ill-linked, with a resulting loss of the potential in studies of glycobiology. Considering such factors, Chen and co-workers reported

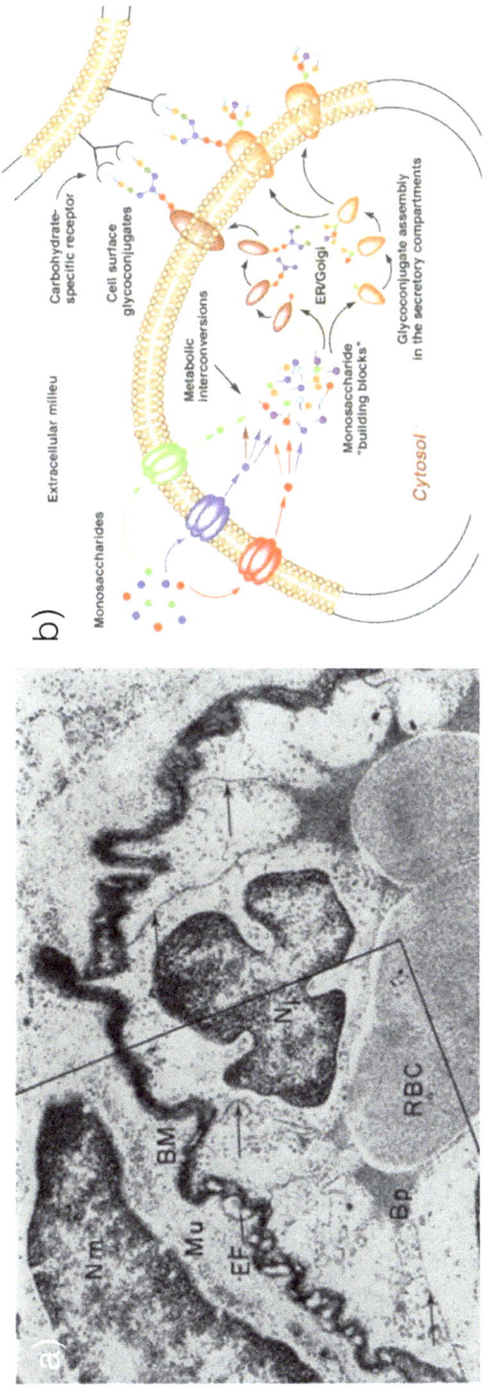

Figure 6.5 (a) TEM image of glycocalyx on a cell surface. (b) Cartoon of the function of the glycocalyx. Source: ref. 50.

glycopolymers with definite configurations, *i.e.* galactopyranoside and glucopyranoside linked *via* a β-*N* linkage on the anomeric center, which was then employed to produce glycocalyx-mimicking vesicles.[7] The vesicles are further prepared *via* the complexation of the glycopolymers with thermally responsive poly(*N*-isopropylacrylamide) (PNIPAm) with a phenylboronic acid end and increasing temperature. Here, in the vesicles the sugar-containing part and the hydrophobic part are connected by a dynamic covalent bond, differing from all the assemblies formed from block copolymers discussed above and with the advantage of easy replacement of different glycopolymers on the vesicle surface. DLS was employed to monitor the binding process between the sugars on the different vesicle surfaces and three lectins, *i.e. Arachis hypogaea* (PNA), *Erythrina cristagalli* (ECA) and concanavalin A (ConA). The vesicles show clear specificities in the sugar–protein interactions (Figure 6.6(a)): V-PGal (vesicles with β-*N*-galactopyranoside) interacts well with PNA and ECA but not with ConA, whereas V-PGlc (vesicles with β-*N*-glucopyranoside) interacts with none of the lectins.

Very recently, a constitutional effect of sugars of the assembled nanoparticles of glycopolymers on lectin binding and cell uptake pathway was demonstrated by Chen and co-workers.[36] Nanoparticles with regioisomers of galactopyranoside, *i.e.* 1-Gal-NP, nanoparticles with galactoside linked from 1-OH and 6-Gal-NP, nanoparticles with galactoside linked from 6-OH were prepared. By using DLS, it was found that only 1-Gal-NP bound to PNA and ECA while 6-Gal-NP did not, showing the dramatic difference between the two. More importantly, both of them could enter Hep G2 cell lines *via* ASGPR-mediated endocytosis, which can be explained by the binding of the acceptor with the *cis* 3-OH and 4-OH on galactopyranoside. By using different fluorescent trackers for early endosome, later endosome and lysosome, it was found that the two kinds of nanoparticles exhibited different pathways inside the cell, *i.e.* 1-Gal-NP passed early endosome, later endosome and finally reached lysosome, whereas 6-Gal-NP stopped at the stage of early endosome (Figure 6.6(b)).

6.2.4 Nano-Objects and Deprotection-Induced Self-Assembly

A glyco-inside nanostructure is as crucial as the glyco-outside form but has been overlooked for a long time. Until now, as far as we know, the structure mentioned by Schlaad *et al.*[9] is the only example, in which the PS-*b*-glucose-grafted PI (17 wt% glucose) self-assembled into vesicles in organic media with diameters of around 250 nm. Preparation of this type of nanostructure is more complex than that of the glyco-outside form. As the glyco-inside nanostructures are formed in organic solvents, it could be difficult to obtain the 'real' morphology because rapid evaporation of the solvent might cause some problems and even give false data under TEM.

Lately, a new strategy for self-assembly of sugar-containing block copolymers leading to the glyco-inside nanostructure was suggested by Chen and co-workers[37] based on protection–deprotection chemistry. In many cases,

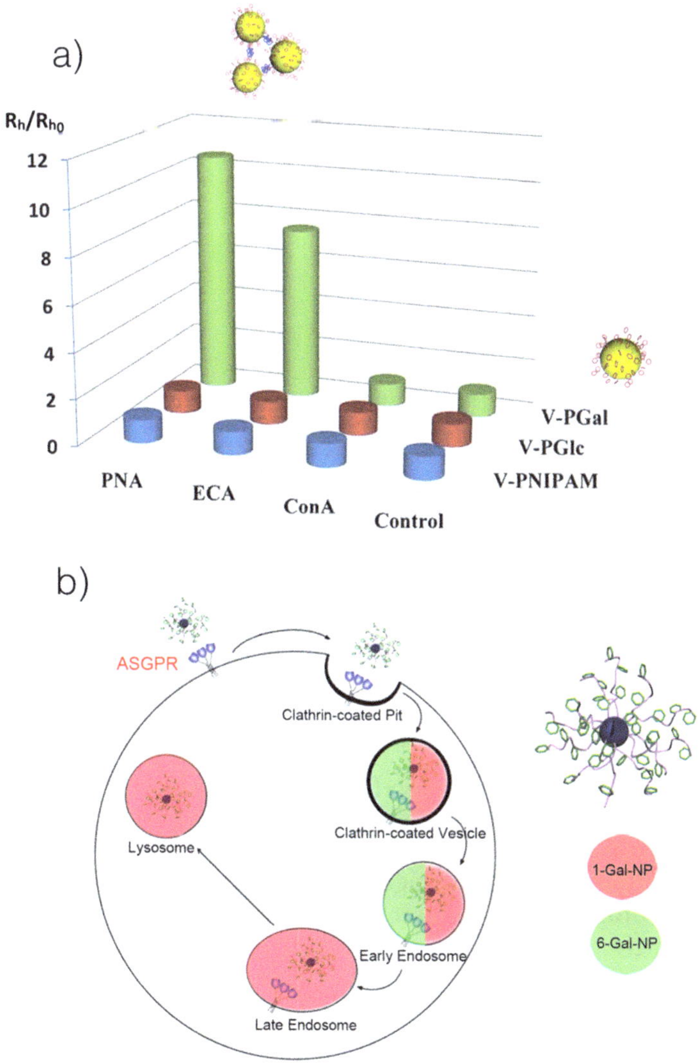

Figure 6.6 (a) Comparison of scattered light intensity after titration of different lectins to different vesicles. (b) Regioisomerism effect shown by different pathways of 1-Gal-NP and 6-Gal-NP after ASGPR-mediated endocytosis.

protection–deprotection of carbohydrate is indispensable in the preparation of glycopolymers, which causes an obvious polarity change of the polymers, but it has been neglected in studies of the self-assembly of glycopolymers. With acyl-protected sugar units as the glyco block, block copolymer PS-*b*-PManAc dissolves in THF as single chains. After deacylation, which is initiated using a carefully chosen catalyst, tetrabutylammonium hydroxide (TBAOH), the unprotected block copolymer PS-*b*-PMan self-assembled into glyco-inside structures with PS as the soluble shell and glycopolymer as the

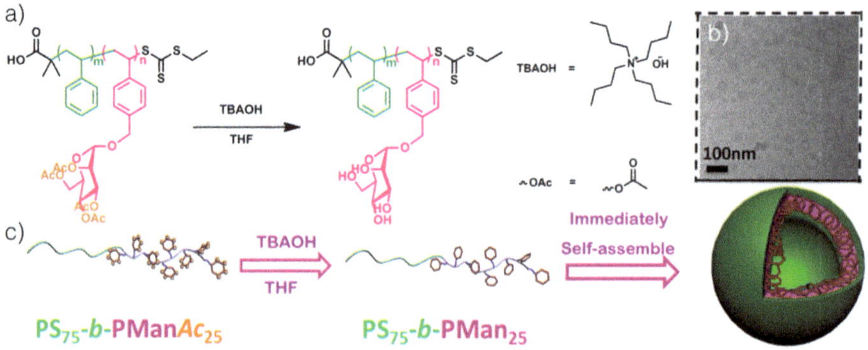

Figure 6.7 (a) Scheme of deprotection on PS-*b*-PManAc to form PS-*b*-PMan and the structure of the catalyst. (b) Cryo-TEM image of self-assembled PS_{75}-*b*-$PMan_{25}$ in THF. (c) Scheme of the deprotection-induced micellization (vesicles) process.

solidified state, *i.e.* the wall of vesicles or core of micelles depending on the weight ratio of the glyco part (Figure 6.7). Although the structure has been successfully prepared with different morphologies, further investigation of the effects of sugar structures on such assembly behavior is essential.

6.3 Hydrogels Formed by Glycopolymers

Hydrogels are important materials for many biomedical applications, including drug delivery, biosensors, cell culture and tissue regeneration.[38,39] Considering macromolecular hydrogels made of sugars, generally there are four types according to the components: (1) hydrogels made from polysaccharides; (2) hydrogels made from saccharide–peptide copolymers; (3) hydrogels made from glycopolymers with simple sugars existing as pendant groups; (4) hydrogels made from carbohydrate–protein interactions, in which sugars are present for binding to proteins. Among these four types, the first is the most popular. Many different types of natural polysaccharides, including agarose, alginate and hyaluronic acid, have been used[40] to build a variety of hydrogels, and this subject has been reviewed in several articles recently.[41] For the second type, elegant work was performed by Guan and co-workers, in which functional saccharide–peptide copolymers were produced from galactaric acid and lysine. As a synthetic polymer composed of natural building blocks, the copolymer led to valuable properties of the resultant hydrogels, *e.g.* as promising synthetic extracellular matrices for regenerative medicine applications.[42] However, strictly galactaric acid is a derivative of a carbohydrate, *i.e.* it is not a real carbohydrate, especially after its copolymerization with lysine, although its backbone is very similar to the carbohydrate open-chain structure with multiple hydroxyl groups.

Regarding hydrogels obtained starting from glycopolymers, few reports could be found. Compared with traditional monomers for controlled

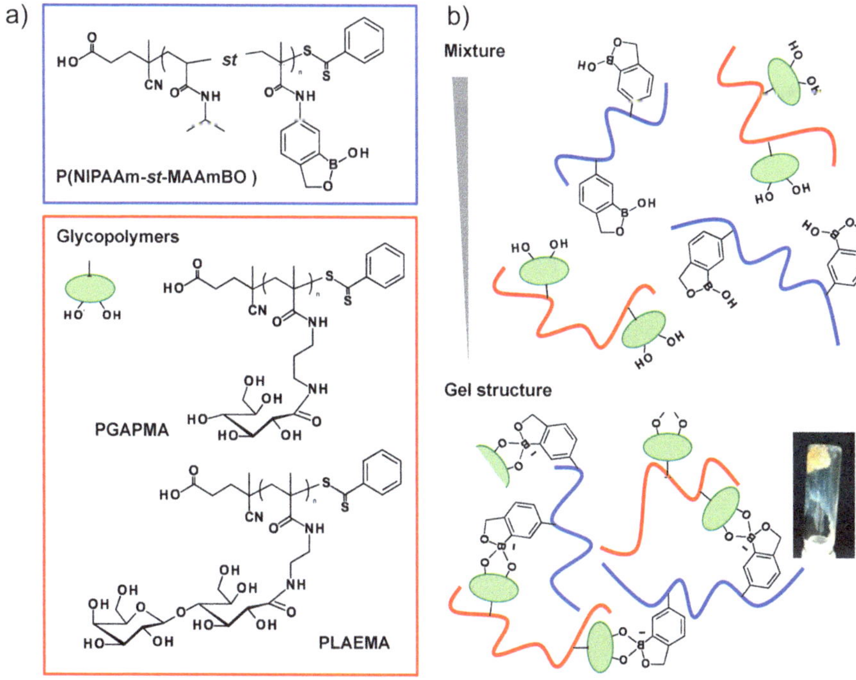

Figure 6.8 (a) Structures of polymers P(NIPAAm-*st*-MAAmBO), PGAPMA and PLAEMA. (b) Gel formation *via* boronic–diol interaction between P(NIPAAm-*st*-MAAmBO) and glycopolymers.

polymerization, which are usually commercially available, glycomonomers require multiple synthetic steps. Further, the anomeric mixture is always a problem for 'sweet' compounds. To prepare a hydrogel, which is a bulky material, a steady supply of monomer is necessary. Lack of a material source could be the most likely reason to explain the current status. Although glycomonomers can be directly copolymerized with a crosslinker and other monomers, the network obtained did not contain a well-characterized glycopolymer as a building block, hence we do not cover here the limited hydrogels obtained by this method. As far as we know, just a few studies that used glycopolymers as building blocks to prepare hydrogels have appeared in the literature.[43]

Recently, a hydrogel prepared by glycopolymers with low polydispersity was reported by Narain and co-workers.[44] Mixing of a copolymer, P(NIPAAm-*st*-MAAmBO), where MAAmBO is a boron-containing moiety, and a glycopolymer, *i.e.* either PGAPMA or PLAEMA with respective monosaccharide and disaccharide pendant groups, resulted in hydrogels driven by the boronic–diol interaction (Figure 6.8). The diol moiety could be either the *cis*-diol on the sugar backbone or the α,β-diol in the open-chain form. In the work reported,[44] a linear glycopolymer prepared by RAFT polymerization was first utilized as a building block to construct hydrogels. Interestingly, owing to the reversible nature of the boronic–diol bond, the hydrogels respond to

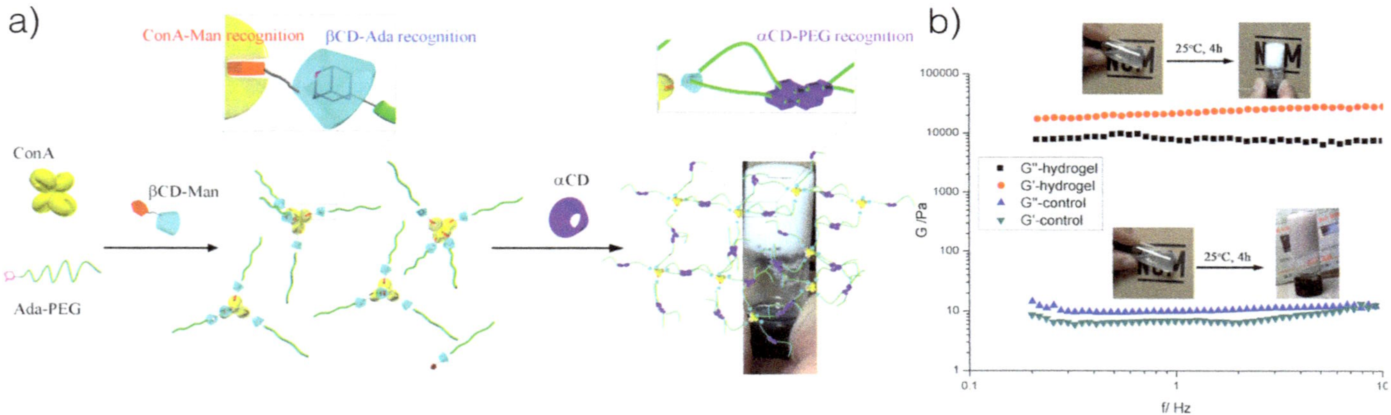

Figure 6.9 (a) Scheme of the formation of supra-conjugate and hydrogel. (b) Storage moduli (G') and viscous moduli (G'') of hydrogel containing 0.5 mM ConA, 2 mM βCD-Man, 2 mM Ada-PEG and 0.1 M αCD, compared to a control sample containing 2 mM βCD-Man, 2 mM Ada-PEG and 0.1 M αCD without ConA. Source: ref. 49.

free glucose. Further, the gel can be dissociated *via* dialysis, as the bonding is reversible and sensitive to water. In addition, since the phase transition of PNIPAm also contributed to hydrogelation, the hydrogel is also responsive to temperature.

In the final type of hydrogel, free proteins act as crosslinking points to transform linear polymers into a 3D network.[45] Various natural or recombinant proteins have been employed for hydrogelations[46] by taking advantage of the molecular recognition between protein and ligand.[47] However, because proteins normally contain 2–4 binding sites on their large surface ($R_h \approx$ 5–10 nm), the crosslinking efficiency is fairly low compared with traditional chemical crosslinking reagents. For the purpose of making sugar-containing hydrogels *via* crosslinking of proteins, lectin–sugar interactions need to be considered. However, the molecular weights of lectins are always high (≥100 kDa), especially higher than those of the popular natural or recombinant proteins used for hydrogelation. This feature brings more obstacles to building lectin-based hydrogels than other protein-based structures. Further, recombination of lectins has not been reported yet. The first hydrogel containing lectin was reported in 2004, in which the glycomonomer glucosyloxyethyl methacrylate was copolymerized with acrylic acid-modified ConA. Since the binding between glucopyranoside and ConA could crosslink the polymers, hydrogel was formed immediately after polymerization.[48] The gel showed good mechanical properties with glucose responses.

The only report demonstrating the crosslinking role of free lectin is from a well-designed system with the same lectin, ConA, as the crosslinker (Figure 6.9).[49] In the hydrogel, ConA was first linked to a dual molecular recognition molecule (βCD-Man), which contains an α-mannopyranoside (Man) and a β-cyclodextrin (CD). The Man moiety binds to ConA, and the β-CD binds to the end adamantane group of modified PEG (Ada-PEG). So *via* this dual linker, a supra-conjugate forms with ConA as the core and PEG as the four arms. Then α-CD is added, which threads onto the PEG chain to form a necklace-like polypseudorotaxane structure (PPR). The PPR chain will aggregate at a certain concentration, which finally drives the supra-conjugate to a hydrogel. The most striking feature of this hydrogel is that the moduli of the supra-conjugate gel are three orders of magnitude higher than those of the control sample (with all components except ConA). Further, the supra-conjugate gel shows clear solid properties with $G' > G''$, whereas the control sample without ConA exhibits liquid properties with $G'' > G'$, although the vial of the control sample can be reversed.

References

1. (a) A. L. Loins, *Science*, 1974, **183**, 330; (b) B. Dorigo, T. Schalch, A. Kulangara, S. Duda, R. R. Schroeder and T. J. Richmond, *Science*, 2004, **306**, 1571.
2. P. Ringler and G. E. Schulz, *Science*, 2003, **302**, 106.
3. A. Rambourg and C. P. Leblond, *J. Cell Biol.*, 1967, **32**, 27.

4. Y. Zhao, F. Sakai, L. Su, Y. Liu, K. Wei, G. S. Chen and M. Jiang, *Adv. Mater*, 2013, **25**, 5215.
5. (a) Z.-C. Li, Y.-Z. Liang, G.-Q. Chen and F.-M. Li, *Macromol. Rapid Commun.*, 2000, **21**, 375; (b) Z.-C. Li, Y.-Z. Liang and F.-M. Li, *Chem. Commun.*, 1999, 1557.
6. G. Pasparakis and C. Alexander, *Angew. Chem. Int. Ed.*, 2008, **47**, 4847.
7. L. Su, Y. Zhao, G. S. Chen and M. Jiang, *Polym. Chem.*, 2012, **3**, 1560.
8. L. You and H. Schlaad, *J. Am. Chem. Soc.*, 2006, **128**, 13336.
9. H. Schlaad, L. You, R. Sigel, B. Smarsly, M. Heydenreich, A. Mantion and A. Masic, *Chem. Commun.*, 2009, 1478.
10. C.-M. Dong and E. L. Chaikof, *Colloid Polym. Sci.*, 2005, **283**, 1366.
11. J. Huang, C. Bonduelle, J. Thévenot, S. Lecommandoux and A. Heise, *J. Am. Chem. Soc.*, 2012, **134**, 119.
12. R. Wang, N. Xu, F-S. Du and Z.-C. Li, *Chem. Commun.*, 2010, **46**, 3902.
13. H. Yang, X. Sun, G. Liu, R. Ma, Z. Li, Y. An and L. Shi, *Soft Matter*, 2013, **9**, 8589.
14. X. Jin, X. Zhang, Z. Wu, D. Teng, X. Zhang, Y. Wang, Z. Wang and C. Li, *Biomacromolecules*, 2009, **10**, 1337.
15. (a) R. Narain and S. P. Armes, *Macromolecules*, 2003, **36**, 4675; (b) R. Narain and S. P. Armes, *Biomacromolecules*, 2003, **4**, 1746; (c) R. Narain, S. P. Armes, *Chem. Commun.*, 2002, 2776.
16. C. Zheng, Q. Guo, Z. Wu, L. Sun, Z. Zhang, C. Li and X. Zhang, *Eur. J. Pharm. Sci.*, 2013, **49**, 474.
17. K.-Y. Peng, M.-Y. Hua and R.-S. Lee, *Carbohydr. Polym.*, 2014, **99**, 710.
18. F. Suriano, R. Pratt, J. P. K. Tan, N. Wiradharma, A. Nelson, Y.-Y. Yang, P. Dubois and J. L. Hedrick, *Biomaterials*, 2010, **31**, 2637.
19. V. Ladmiral, M. Semsarilar, I. Canton and S. P. Armes, *J. Am. Chem. Soc.*, 2013, **135**, 13574.
20. K. Yasugi, T. Nakamura, Y. Nagasaki, M. Kato and K. Kataoka, *Macromolecules*, 1999, **32**, 8024.
21. N. Xu, R. Wang, F.-S. Du and Z.-C. Li, *J. Polym. Sci., Part A: Polym. Chem.*, 2009, **47**, 3583.
22. O. León, V. Bordegé, A. Muñoz-Bonilla, M. Sánchez-Chaves and M. Fernández-García, *J. Polym. Sci., Part A: Polym. Chem.*, 2010, **48**, 3623.
23. (a) G. J. Chen, S. Amajjahe and M. H. Stenzel, *Chem. Commun.*, 2009, 1198; (b) J. Kumar, L. McDowall, G. J. Chen and M. H. Stenzel, *Polym. Chem.*, 2011, **2**, 1879.
24. A. G. Dal Bó, V. Soldi, F. C. Giacomelli, C. Traelet, B. Jean, I. Pignot-Paintrand, R. Borsali and S. Fort, *Langmuir*, 2012, **28**, 1418.
25. B.-S. Kim, D.-J. Hong, J. Bae and M. Lee, *J. Am. Chem. Soc.*, 2005, **127**, 16333.
26. (a) I. Otsuka, C. Travelet, S. Halila, S. Fort, I. Pignot-Paintrand, A. Narumi and R. Borsali, *Biomacromolecules*, 2012, **13**, 1458; (b) S. de Medeiros Modolon, I. Otsuka, S. Fort, E. Minatti, R. Borsali and S. Halila, *Biomacromolecules*, 2012, **13**, 1129.
27. K. Aissou, A. Pfaff, C. Giacomelli, C. Travelet, A. H. E. Müller and R. Borsali, *Macromol. Rapid Commun.*, 2011, **32**, 912.

28. Y. Luo, L. Liu, X. Wang, H. Shi, W. Lv and J. Li, *Soft Matter*, 2012, **8**, 1634.
29. V. Dhaware, A. Y. Shaikh, M. Kar, S. Hotha and S. S. Gupta, *Langmuir*, 2013, **29**, 5659.
30. S. Liu, H. Huang, X.-H. Dai, W. Zhou and C.-M. Dong, *J. Polym. Sci., Part A: Polym. Chem.*, 2009, **47**, 2009.
31. Z. Lu, L. Mei, X. Zhang, Y. Wang, Y. Zhao and C. Li, *Polym. Chem.*, 2013, **4**, 5743.
32. J. Lu, W. Zhang, L. Yuan, W. Ma, X. Li, W. Lu, Y. Zhao and G. J. Chen, *Macromol. Biosci.*, 2014, **14**, 340–346.
33. W. Lu, W. Ma, J. Lu, X. Li, Y. Zhao and G. J. Chen, *Macromol. Rapid Commun.*, 2014, **35**, 827–833.
34. T. W. Rademacher, R. B. Parekh and R. A. Dwek, *Annu. Rev. Biochem.*, 1998, **57**, 785.
35. J. Bouckaert, J. Berglund and M. Schembri, *et al., Mol. Microbiol.*, 2005, **55**, 441.
36. P. Sun, Y. He, M. Lin, Y. Zhao, Y. Ding, G. S. Chen and M. Jiang, *ACS Macro Lett.*, 2014, **3**, 96.
37. L. Su, C. Wang, F. Polzer, Y. Lu, G. Chen and M. Jiang, *ACS Macro Lett.*, 2014, **3**, 534.
38. J. Kopecek and J. Yang, *Angew. Chem. Int. Ed.*, 2012, **51**, 7396.
39. I. Tomatsu, K. Peng and A. Kros, *Adv. Drug Deliv. Rev.*, 2011, **63**, 1257.
40. X. Xu, A. K. Jha, D. A. Harrington, M. C. Farach-Carson and X. Jia, *Soft Matter*, 2012, **8**, 3280.
41. (a) B. Balakrishnan and R. Banerjee, *Chem. Rev.*, 2011, **111**, 4453; (b) S. V. Vlierberghe, P. Dubruel and E. Schacht, *Adv. Mater.*, 2014, **26**, 125.
42. (a) S. W. Liao, T.-B. Yu and Z. Guan, *J. Am. Chem. Soc.*, 2009, **131**, 17638; (b) K. Chawla, T.-B. Yu, L. Stutts, M. Yen and Z. Guan, *Biomaterials*, 2012, **33**, 6052.
43. N. Sakamoto, K. Suzuki, A. Kishida and M. Akashi, *J. Appl. Polym. Sci.*, 1998, **70**, 965.
44. Y. Kotsuchibashi, R. V. C. Agustin, J.-Y. Lu, D. G. Hall and R. Narain, *ACS Macro Lett.*, 2013, **2**, 260.
45. J. D. Ehrick, M. R. Luckett, S. Khatwani, Y. Wei, S. K. Deo, L. G. Bachas and S. Daunert, *Macromol. Chem. Phys.*, 2009, **9**, 864.
46. H. Wang, Y. Shi, L. Wang and Z. Yang, *Chem. Soc. Rev.*, 2013, **42**, 891.
47. X. Liao, G. S. Chen and M. Jiang, *Polym. Chem.*, 2013, **4**, 1733.
48. T. Miyata, A. Jikihara, K. Nakamae and A. Hoffman, *J. Biomater. Sci. Polym. Ed.*, 2004, **15**, 1085.
49. K. Wei, J. Li, G. S. Chen and M. Jiang, *ACS Macro Lett.*, 2013, **2**, 278.
50. C. R. Bertozzi and L. L. Kiessling, *Science*, 2001, **291**, 2357.

CHAPTER 7

Glyconanoparticles: Synthesis and Biomedical Applications

YINAN WANG[a], STEPHEN QUAN[a], PIYUSH KUMAR[b], AND RAVIN NARAIN*[a]

[a]Department of Chemical and Materials Engineering, University of Alberta, Edmonton, Alberta T6G 2V4, Canada; [b]Oncologic Imaging, Department of Oncology, University of Alberta, Cross Cancer Institute, Edmonton, Alberta T6G 1Z2, Canada
*E-mail: narain@ualberta.ca

7.1 Introduction

Carbohydrates play a pivotal role in molecular recognition events and biological metabolic processes, but they are the least understood and exploited in biomedical research. Carbohydrates can be classified into four different groups, monosaccharides, disaccharides, oligosaccharides and polysaccharides, based on the number of constituent sugar molecules linked together. Within living organisms, polysaccharides can serve as dense energy storage molecules (*e.g.* glycogen or starch) that can be metabolized for energy over long periods of time[1] or as structural components for plant physiology, such as cellulose.[2] Carbohydrates can be utilized in a vast array of physiological applications and intracellular functions owing to the diversity of their constituents, which play an enormous role in carbohydrate recognition events within the mammalian body.[3] With recent developments in biotechnology, the role

RSC Polymer Chemistry Series No. 15
Glycopolymer Code: Synthesis of Glycopolymers and Their Applications
Edited by C. Remzi Becer and Laura Hartmann

Published by the Royal Society of Chemistry, www.rsc.org

of carbohydrate systems at the molecular level are better understood, but the exploitation of carbohydrate-based therapeutics has still not been realized in biomedicine. This could be attributable to some pharmacokinetic drawbacks of carbohydrates, such as low plasma half-life[4,5] and low-affinity binding of carbohydrate-specific interactions.[6] Nonetheless, the fabrication and engineering of highly innovative glyconanoparticles with unique physicochemical properties provide an opportunity to understand better the role of specific carbohydrates in their interactions with specific carbohydrate receptors.

At the molecular level, carbohydrate–protein interactions are governed by hydrogen bonding, van der Waals interactions and hydrophobic stacking.[7–9] Carbohydrate–protein interactions are known to be weak ($k_d \approx 10^{-6}$ M)[10,11] compared with other well-known biosystems, such as the streptavidin–biotin system, which have an extremely high binding affinity ($k_d \approx 10^{-15}$ M).[12] Recently, research has been directed towards multivalency, molecular orientation and multi-functional conjugates to increase specificity and enhance the efficiency of molecular interactions.[13,14] Synthetic and natural polymeric scaffolds present an interesting example for exploiting these unique properties of carbohydrate–protein interactions. Nanoparticles functionalized with glycopolymers are designed to enhance their interactions with specific cell surface receptors, transportation across the biological barrier and maintain high stability in biological media.[13,15,16]

This chapter focuses on recent developments in glyconanotechnology and the pivotal participation of carbohydrates residues in the overall properties of such systems is discussed.

7.2 Synthesis of Glyconanoparticles

7.2.1 Metallic Nanoparticles

Glycopolymer-functionalized nanoparticles such as gold and silver nanoparticles (AuNPs and AgNPs) have been the most often studied nanomaterials in the past decade owing to their stability, ease of preparation and unique optoelectronic properties. The first example of carbohydrate-functionalized AuNPs was reported by Penadés and co-workers in 2001, where the thiol-terminated disaccharide lactose (Galβ(1→4)Glcβ1) or the trisaccharide Lewis X (Galβ(1→4)[Fucα(1→3)]GlcNAcβ1, Le^X) were immobilized on gold nanoparticle surfaces by covalent linkages.[17] In general, the synthesis of glyco-AuNPs follows two major approaches. One is direct synthesis, in which a gold/silver salt ($AuCl_4^-$/Ag^+) is reduced in the presence of a reducing agent ($NaBH_4$) and the desired thiol-functionalized glycopolymers are introduced to the gold/silver surface by covalent interaction.[17–21] A disadvantage of this approach is that the size of the AuNPs can be difficult to control.[22] Scaiano and co-workers reported a photochemical method that permits the fast preparation of unprotected water-soluble AuNPs/AgNPs with controlled particle size.[23,24] Using this strategy, Narain and co-workers[19,25] subsequently developed glyco-AuNPs by the *in situ* photochemical reduction of $HAuCl_4$ and glycopolymers in the presence

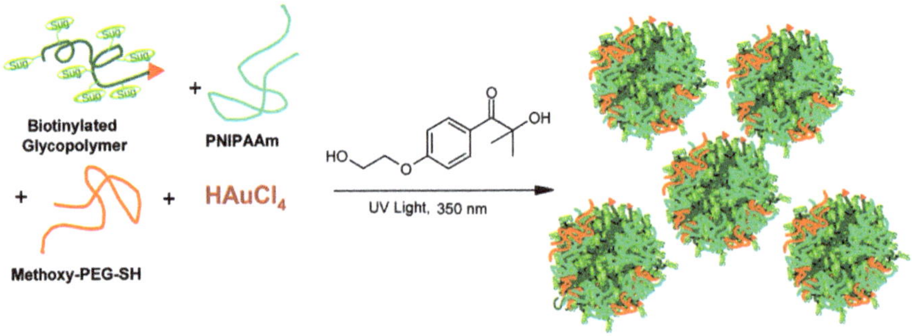

Figure 7.1 Preparation of glycopolymer-functionalized AuNPs by a photochemical method. Reprinted with permission from ref. 19. Copyright 2007 American Chemical Society.

of Irgagure-2959, a water-soluble benzoin. The size of the glyco-AuNPs was controlled under 100 nm with no aggregation noted after several weeks, indicating a very high colloidal stability of the nanoparticles (Figure 7.1).

The other strategy is a multi-step approach in which carbohydrates are conjugated to a previously prepared AuNP. The multi-step approach can be further sub-grouped into 'grafting to' or 'grafting from' methods based on how glycopolymers are immobilized on the AuNP surface. In the 'grafting to' method, glycopolymers and AuNPs are usually prepared separately and the polymers are conjugated to the gold surface by layer-by-layer (LBL) self-assembly,[26] or covalent linkage.[27–29] For example, Davis and co-workers obtained glyco-AuNPs by alternately depositing cationic polyethylenimine (PEI) and anionic poly(*tert*-butyl acrylate-*co*-hydroxyethyl acrylate/galactopyranose) on citrate-stabilized AuNPs (20 nm) (Figure 7.2(a)).[26] Gibson and co-workers presented a strategy for reducing glycopolymers synthesized by the RAFT (reversible addition–fragmentation chain-transfer) process to thiol-terminal products and then grafting them on to the citrate-stabilized AuNP surface (Figure 7.2(b)).[28] Compared with the 'grafting to' method, the polymer-functionalized NPs prepared by the 'grafting from' method showed certain advantages in biological recognition due to their higher polymer density on the material surfaces.[30–33] Yoon *et al.* reported the preparation of glyco-AuNPs by immobilizing a disulfide initiator on the AuNP surface, then growing 2-lactobionamidoethyl methacrylate (LAMA) chains by surface-initiated atom-transfer radical polymerization (SI-ATRP) (Figure 7.2(c)).[32]

7.2.2 Quantum Dots and Magnetic Nanoparticles

7.2.2.1 Quantum Dots

Quantum dots (QDs) are fluorescent semiconductor nanocrystals (CdSe, CdTe, InP, *etc.*) that have attracted enormous attention in recent years owing to their unique size-dependent optical properties.[13,34,35] However, the initial

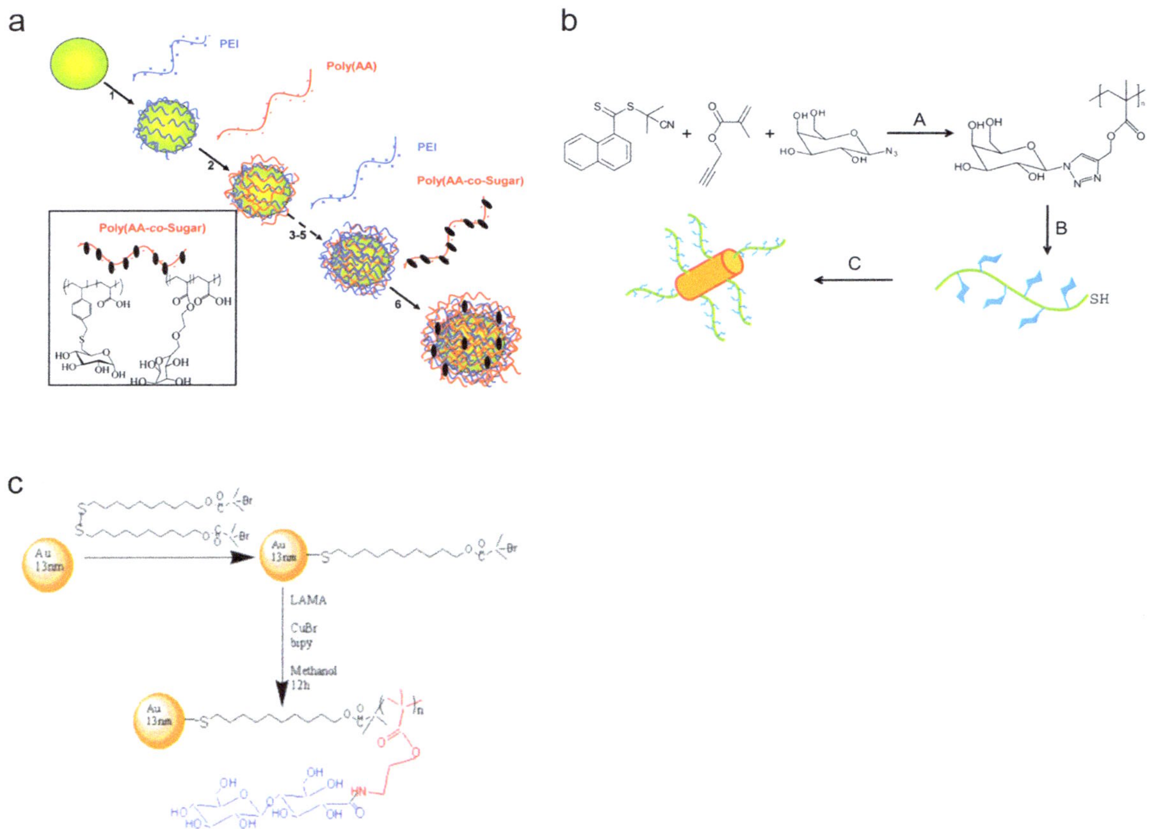

Figure 7.2 Synthesis of glycopolymer-functionalized AuNPs by multi-step approaches: (a) and (b) 'grafting to' method; (c) 'grafting from' method. (a) Reprinted with permission from ref. 26. Copyright 2010 American Chemical Society. (b) Reproduced from ref. 28 with permission from The Royal Society of Chemistry. (c) Reprinted with permission from ref. 32. Copyright 2008 Wiley.

hydrophobic QDs were found to be unsuitable for biological applications owing to their poor solubility in biological systems and high cytotoxicity.[36] Therefore, the functionalization of QDs with proteins,[37] peptides,[38] DNA[39,40] and other small ligands[41] has been exploited to improve the materials' water solubility and biocompatibility while maintaining their photoluminescence and single excitation narrow emission at the same time. Among the various approaches to the functionalization of QDs, more and more studies have been focusing on decorating the QD surface with carbohydrate residues for enhanced stability or for carbohydrate–protein or carbohydrate–carbohydrate recognition processes.[42] However, the individual carbohydrate–carbohydrate interactions are generally weak and multivalent forms of carbohydrate ligands have been used to enhance the interaction through the so-called 'glycoside cluster effect.[14]

Narain and co-workers reported biotinylated glycopolymer-functionalized QDs by coupling a RAFT-synthesized biotinyl-glycopolymer to QD-COOH in the presence of 1-ethyl-3-[3-dimethylaminopropyl]carbodiimide hydrochloride (EDC). Owing to their easy aqueous dispersion and potential low cytoxicity, the glycofunctionalized QDs may act as suitable fluorescent probes in biological systems (Figure 7.3(a)).[43] Tabrizian and co-workers managed to exchange the ligand [trioctylphosphine oxide (TOPO)] on InGaP@ZnS QDs with mercaptohexadecanoic acid (MHA) and then activated with EDC to covalently coupled deacetylated chitosan.[44] Chen and co-workers demonstrated a strategy that directly functionalized ZnS and ZnS:Mn^{2+} QDs with chitosan (CS–ZnS and CS-ZnS:Mn^{2+} QDs). Chitosan-functionalized QDs with a mean size of 4.5 nm were obtained when a mixture of chitosan, $Zn(OAc)_2$, $Mn(OAc)_2$ and $Na_2S_2O_3$ in aqueous solution was irradiated with a 1.1×10^{15} Bq ^{60}Co γ-ray source at room temperature and atmospheric pressure (Figure 7.3(b)).[45] Recently, Jiang *et al.* reported the use of chitosan-QD (CS-QD) hybrid nanospheres (~150 nm) for bioimaging and biolabeling.[46] The hybrid nanospheres were synthesized by a non-solvent-aided counterion complexation method, in which the as-prepared QDs were added to the CS aqueous solution and crosslinked by the addition of glutaraldehyde aqueous solution (25%).

7.2.2.2 *Magnetic Nanoparticles*

Magnetic nanoparticles (MNPs) are nanomaterials that can be manipulated by an external magnetic field. These MNPs commonly consist of magnetic elements such as iron, nickel and cobalt and can potentially be used in clinical diagnosis and therapies owing to their unique physical properties. Taking advantage of the protein–carbohydrate interaction, several glycopolymer-functionalized MNPs have been prepared over recent decades and used as tools for cellular therapy, drug delivery, magnetic resonance imaging (MRI), hyperthermia, magnetic separation and biosensors.[47–51]

Unlike metallic NPs and QDs, most MNPs are synthesized using a multistep approach. However, according to the method of immobilization of glycopolymers on MNPs, the synthesis of glycofunctionalized NMPs can be

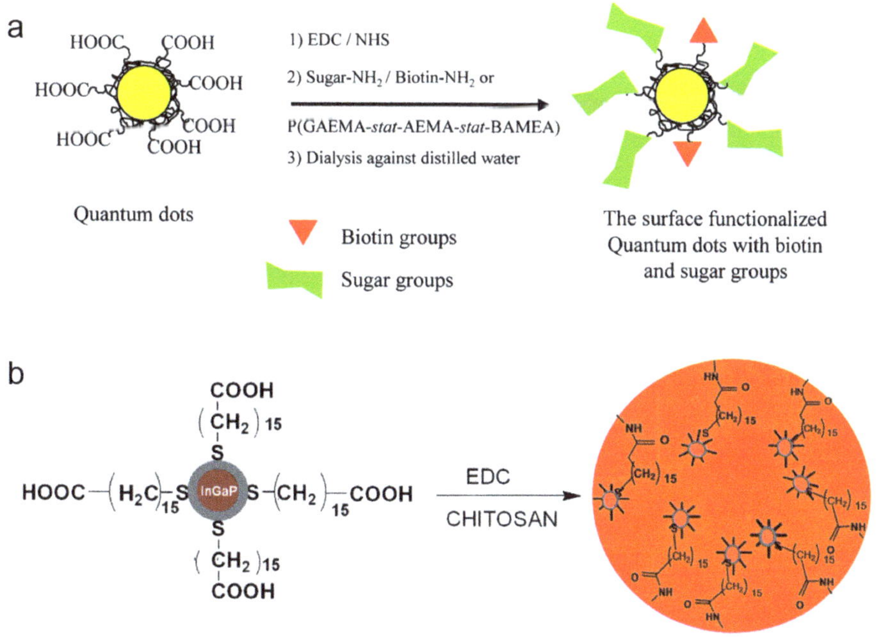

Figure 7.3 (a) Glycopolymer- and (b) chitosan-functionalized QDs. (a) Reprinted with permission from ref. 43. Copyright 2009 American Chemical Society. (b) Reprinted with permission from ref. 45. Copyright 2007 Wiley.

divided into 'grafting to' and 'grafting from' methods. In the 'grafting to' method, glycopolymers and MNPs were usually synthesized separately and the polymers were then immobilized on the MNPs in a covalent or non-covalent fashion. For example, Davis and co-workers functionalized iron oxide nanoparticles (IONPs) with three different glycopolymers (α-D-mannose-, α-D-glucose- and β-D-glucose-bearing glycopolymers) by the 'grafting to' method.[48] To do this, poly[oligo(ethylene glycol) acrylate-*block*-glycidyl acrylate] [P(OEGA)-*b*-GA] diblock copolymers were first synthesized by ATRP and then the sugars were attached to polymer chains by click chemistry. The glycocopolymers with a phosphonic acid terminus were then used to graft the polymer onto IONPs.

In the 'grafting from' method, the synthesized MNPs were usually functionalized with vinyl or bromine groups and glycomonomers were grown from the NP surfaces by either conventional or living free radical polymerization. For example, Muñoz-Bonilla *et al.* recently synthesized glyco-Fe_3O_4 MNPs with an average size of 17 nm by exchanging the oleic acid-functionalized Fe_3O_4 MNPs with either dopamine methacrylamide (DMA) or glycomonomer {2-[(D-glucosamin-2-*N*-yl)carbonyl]oxyethyl methacrylate (HEMAGl)}, then generated the glycopolymer [P(HEMAGl)] coating by conventional radical polymerization (Figure 7.4).[47] Similarly, Muller and co-workers. provided a strategy to produce glycopolymer-functionalized γ-Fe_2O_3 MNPs with fluorescent properties by

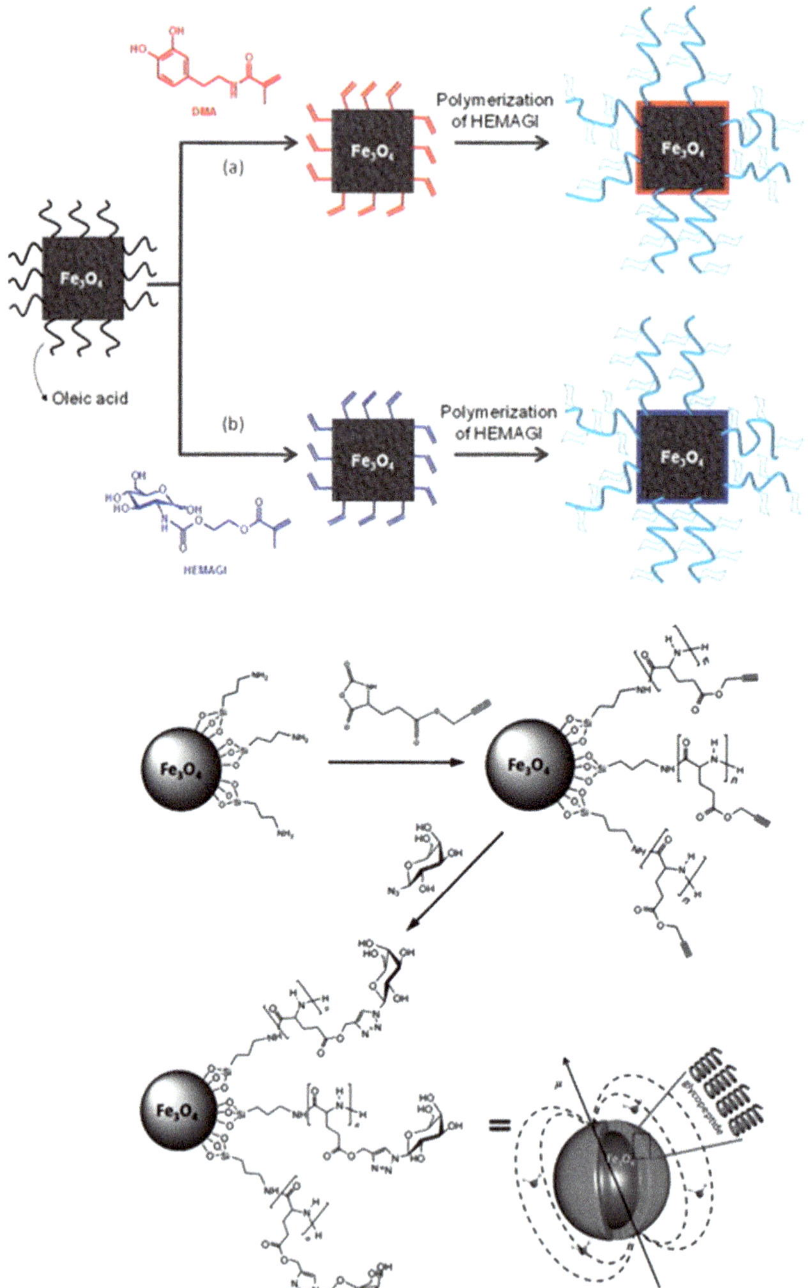

Figure 7.4 Preparation of glycopolymer-functionalized MNPs by the 'grafting from' method. Reprinted with permission from ref. 47 and 50. Copyright 2013 Wiley.

growing polymer from MNPs surface present with double bonds.[49] Recently, the groups of Brougham[50] and Harms[51] described a novel strategy for producing glycopolymer-modified MNPs by combination of a polymer 'grafting from' approach with glycosylation by click chemistry. For example, Brougham and co-workers polymerized *N*-carboxyanhydrides (NCAs) from 3-aminopropyltriethoxysilane (APTS)-functionalized MNPs by ring-opening polymerization and to give clickable alkyne groups. The azide-functionalized galactose was then attached to MNPs and glycol-MNPs were formed by the Huisgen click reaction. Such particles exhibited excellent water dispersion properties, optimal T_1-weighting and selective binding to lectins (Figure 7.4).[50]

7.2.3 Polymeric Nanoparticles

Recently, polymeric glyconanoparticles have attracted major attention for the encapsulation and delivery of drugs and biomacromolecules. Glycopolymers can form nanoparticles that can be covalently immobilized on some polymer-based nanoparticle surfaces,[30,31] chemically or by physically crosslinking[52–54] and self-assembly.[55,56] For example, Müller and co-workers studied the functionalization of polymeric nanoparticles [polydivinylbenzene (PDVB)[30] and polystyrene (PS)[31]] by both 'grafting from' and 'grafting to' methods. The surface modification *via* 'grafting from' was conducted by RAFT or ATRP polymerization from the nanoparticle surfaces, whereas thiol–ene chemistry was used to graft glycopolymer chains onto the particle surface. Their results indicated that the 'grafting from' methods are more advantageous for multivalent biorecognition processes than the 'grafting to' methods owing to the higher grafting density on the nanoparticle surfaces.

Cationic glycopolymers can also self-assemble with nucleic acids to become transformed into nanocomplexes and are used in gene delivery applications. For example, Narain and co-workers synthesized cationic,[57] anionic[58] and zwitterionic[59,60] galactose-based glycopolymers by RAFT polymerization and formed nanoplexes with β-galactosidase plasmid through electrostatic interaction for targeting gene delivery (Figure 7.5). Similarly, Reineke and co-workers reported the synthesis of cationic polymethacrylamidotrehalose (polytrehalose) by RAFT polymerization and subsequent formation of nanocomplexes with siRNA for targeting delivery of nucleic acids to glioblastoma cells.[61] A different strategy for synthesizing glycopolymers through condensation polymerization of esterified carbohydrates and diamines was also reported by the same group. These glycopolymers formed nanocomplexes with nucleic acids and were used for gene delivery purposes.[62–65]

Controlled crosslinking of the glycopolymers can also transform them into glycopolymer-based nanogels or nanoparticles for use in drug delivery. For example, Smeets and co-workers explored a combination of catalytic chain-transfer polymerization (CCTP), thiol–Michael addition chemistry and reductive amination to synthesize block copolymers. A hydrophobic hyperbranched core was formed through CCTP of methyl methacrylate (MMA) and ethylene glycol dimethacrylate (EGDMA). The unsaturated core in the

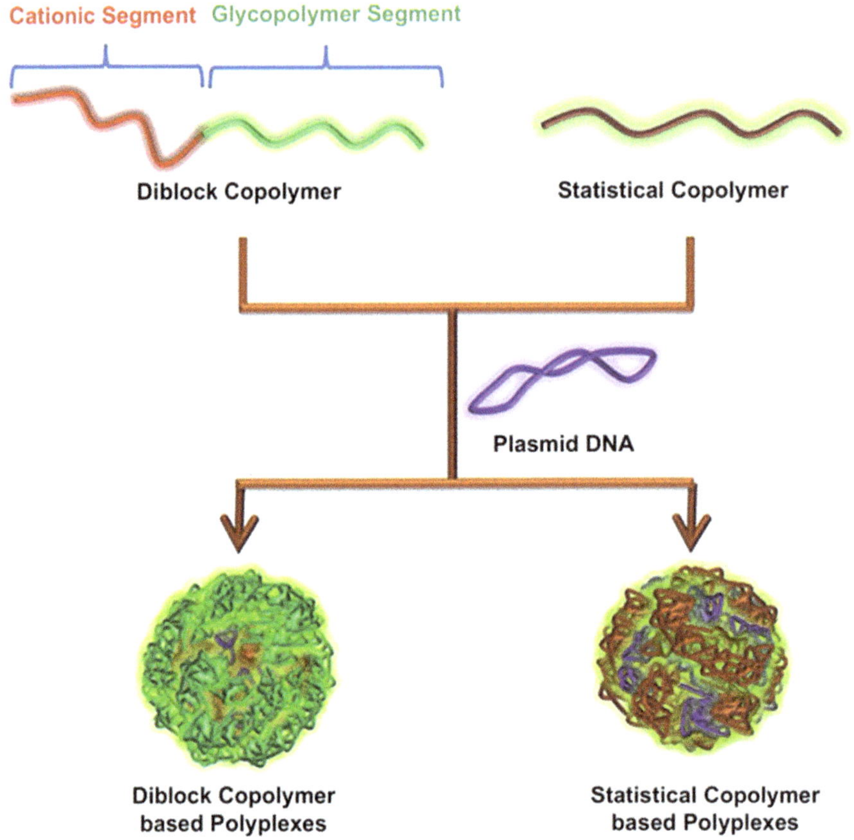

Figure 7.5 Schematic illustration of the glycopolymer-based polyplexes formed by the complexing β-galactosidase plasmid with statistical and diblock cationic glycopolymers. Reprinted with permission from ref. 57. Copyright 2011 Elsevier.

acrylic component was then converted into nucleophilic pendant amines by thiol–Michael addition using cysteamine hydrochloride, so that the polysaccharide shell could be grafted onto the hyperbranched core *via* reductive amination.[66] Narain and co-workers pursued RAFT polymerization in the presence of an acid-degradable crosslinker to synthesize glycopolymer-based nanogels[52,53,67] that had a galactose- or glucose-based glycopolymer shell and a crosslinked thermally sensitive polymer [poly(methoxydiethylene glycol methacrylate)] [poly(MeODEGM)] core (Figure 7.6). At temperatures over the lower critical solubility temperature (LCST) of poly(MeODEGM), the core becomes hydrophobic and collapses so that drugs can be encapsulated by the nanogels *via* hydrophobic interactions.

Boronic acid and its derivatives are known to be capable of reversible interactions with diols, sugars and glycoproteins.[68] Therefore, glyco-NPs can simply be formed by self-assembly if boronic acid-based polymers

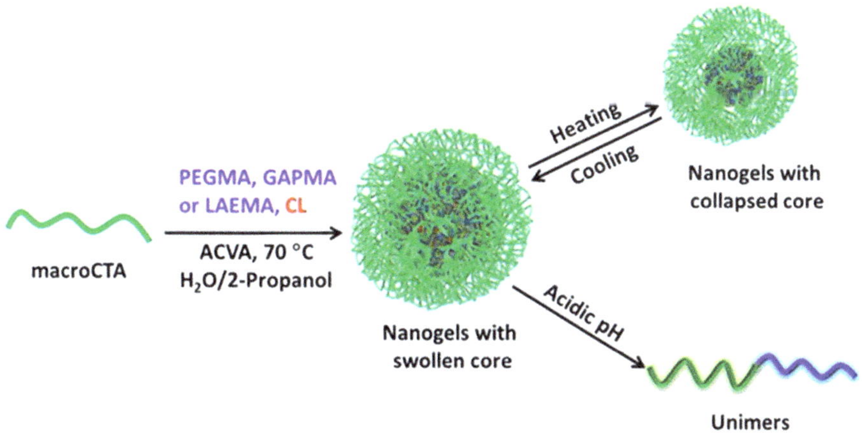

Figure 7.6 Synthesis of glycopolymer-based nanogels and their degradation in an acidic environment. Reprinted with permission from ref. 67. Copyright 2012 American Chemical Society.

[*e.g.* phenylboronic acid (PBA)-containing polymers] are present in the glycopolymer solution.[69,70] Since PBA can only form a stable complex with diols in its tetrahedral anionic state,[71] the interactions between sugar and PBA have been performed under alkaline conditions (usually close to or higher than the pK_a of PBA) in most studies, which is unfavorable for biomedical applications. By introducing secondary or tertiary amines, tetrahedral boronic acids are created at neutral pH, with amines placed proximal to boron centers to facilitate diol complexation.[72] Based on this chemistry, Li and co-workers introduced an aminoglucose-based glycopolymer, poly(D-gluconamidoethyl methacrylate) [poly(GAMA)] to 3-acrylamidophenylboronic acid (AAPBA) by RAFT polymerization to lower the pK_a of AAPBA (Figure 7.7).[54] The NP structure was dissociated at higher glucose concentration, demonstrating its excellent potential in insulin delivery applications. Similarly, Zhang and co-workers reported new insulin delivery vehicles using poly(2-lactobionamidoethyl methacrylate-*r*-3-acrylamidophenylboronic acid) [p(LAMA-*r*-AAPBA)]. The polymer was synthesized by conventional free radical polymerization and the glycopolymer was found to increase the transport of insulin through the respiratory epithelia. After administration of nanoparticle-loaded insulin, the glucose levels were significantly reduced in diabetic rats.[73]

Glyco-NPs can also be prepared by self-assembly of amphiphilic diblock glycocopolymers. For example, Pasparakis and Alexander prepared temperature-sensitive glucose-decorated vesicles from a double-hydrophilic block copolymer obtained by ATRP and RAFT polymerization and found that such vesicles undergo strong interactions with glucose-binding proteins on the bacterial surface.[74] Hedrick and co-workers reported the synthesis of biodegradable polycarbonates displaying either glucose or galactose surface moieties by controlled ring-opening polymerization and formation of micelles *via* self-assembly in aqueous media.[75] Heise and co-workers synthesized

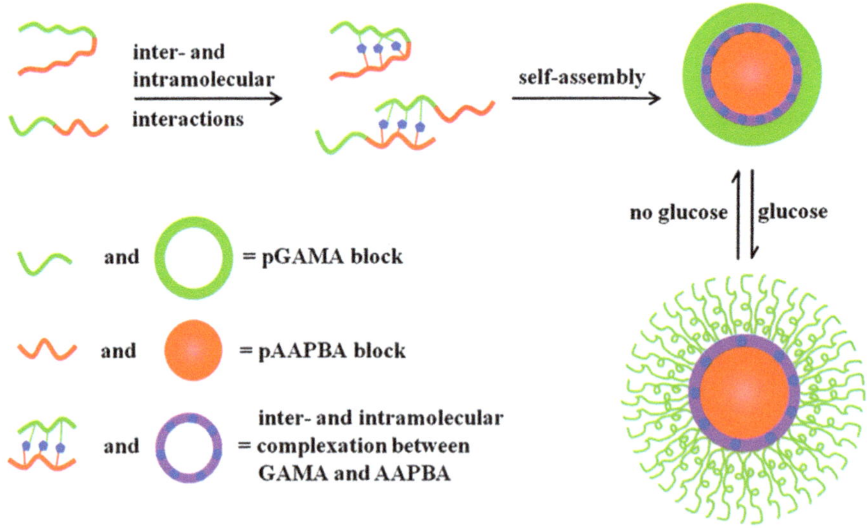

Figure 7.7 Formation of glyco-phenylboronic acid diblock copolymer-based NPs. Reproduced from ref. 54 with permission from The Royal Society of Chemistry.

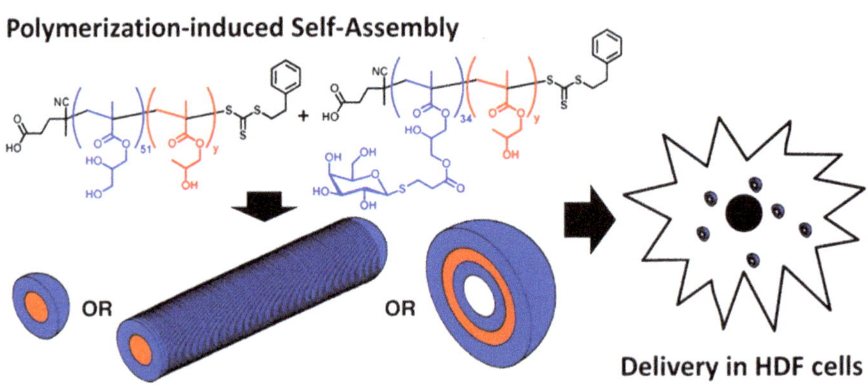

Figure 7.8 Galactose-based glycopolymer self-assembly into NPs with different morphologies. Reprinted with permission from ref. 55. Copyright 2014 American Chemical Society.

glycosylated peptide-based block copolymers by the ring-opening polymerization of *N*-carboxyanhydrides and formed well-defined spherical, worm-like micelles or vesicles in aqueous solution.[76] Armes and co-workers recently developed a method of preparing a range of galactose-functionalized diblock copolymer NPs (spheres, worm-like micelles or vesicles) by a polymerization-induced self-assembly (PISA) approach in concentrated aqueous solution and exploited the relationship between the interaction of galactose-specific lectins with different morphologies of the NPs (Figure 7.8).[55]

7.3 Application of Glyconanoparticles in the Biomedical Fields

7.3.1 Biosensing and Bioimaging

Biotechnological advance over the past decade have focused on improving the efficiency and selectivity of modern biosensors used in the medical community and novel conjugates and nanomaterial scaffolds have created tremendous interest among researchers in this field. The simple and facile detection of infectious pathogens, debilitating bacterial strains and carcinogenic cells is of significant diagnostic importance, as it can provide early prognosis and detection of potential health concerns that can affect humans, animals and the environment.[77] Enhanced research activities in this field have provided the opportunity to engineer incredibly versatile biosensors by conjugation and detection of oligosaccharides,[11] nucleic acids,[78] proteins,[79] pathogens[80] and many more *via* physiochemical methods. The symbiosis of uniquely designed sensors with modern biotechnology has paved a new avenue for dealing with previously undetectable ailments and providing a new scaffold for improving the effectiveness of biomedical theranostics.

One of the most intensively studied applications of glyconanoparticles of the past decade is the use of gold nanoparticle (AuNP) bioconjugates for imaging and targeting purposes. The use of glyco-AuNPs as biosensors and in colorimetric assays has emerged as a prominent detection method owing to their characteristic optical properties, known as surface plasmon band resonance (SPBR), and multifaceted functionalization capabilities.[81] According to the Mie theory, SPBR can be induced by an electromagnetic frequency and results in resonant coherent oscillations at the surface of AuNPs detectable in the visible light spectrum.[82] As a result of enhanced electromagnetic fields at the surface of the nanoparticles, surface-enhanced optical properties can be manipulated using spectroscopic techniques to determine successful binding of biorecognizable molecules to receptors or by colloidal aggregation. These optical properties are strongly correlated with their physical size, shape and aggregation capability, which makes it possible to engineer different geometric shapes for various applications.[83] Originally, the detection of bacterial DNA using AuNPs was pioneered by Mirkin *et al.*, who described a facile method for detecting oligonucleotides at levels as low as 10 fmol by exploiting the aggregation phenomenon of AuNPs.[84] However, recent advances in glyco-AuNPs exploited as biosensors have led to a revolutionary phenomenon for the specific targeting and capture of nanoscale materials *via* carbohydrate–protein interactions.[18,85] Lectin recognition and binding have been an intensively studied field in the past century, since lectin expression can be observed in virtually all cellular architectures of plants and animals, and many studies have been conducted on different types of lectins such as collectins, selectins, lymptocyte lectins and proteoglycans.[3,9,86] Gibson *et al.* recently developed a system that is capable of discriminating between different strains of *Escherichia coli* by multivalent thioglucose-conjugated AuNPs based on their variable expression and binding to lectin receptor FimH on the cell surface.[87] Additional studies by Lim *et al.* led to the development of a colorimetric bioassay that determines the glucose concentration in human blood serum, especially useful for patients with diabetes, without the use of degrading enzymes.[88] Furthermore, similar studies with biosensing glyconanoparticles with SPBR functionality have been carried out with other elemental metallic nanoparticles, such as gadolinium, platinum, aluminum and iron oxides.[89–92]

On the other hand, Zhou *et al.* described the novel synthesis of TEMPO-oxidized chitin nanocrystals labeled with fluorescent imidazoisoquinolinone dye and simultaneously functionalized them with mannose carbohydrate ligands (Figure 7.9).[93] These studies showed that the carbohydrate-functionalized nanocrystals selectively bind to lectins and bacteria *via* carbohydrate–protein interactions and could be visualized by optical, fluorescence, scanning transmission electron microscopy (STEM) and TEM imaging techniques. Similar explorations in this direction by Ma *et al.* have also demonstrated the synthesis of glucose-responsive self-assembly micelles of phenylboronic acid (PBA)-containing block copolymer PEG-*b*-P(AA-*co*-AAPBA) and glycopolymer P(AA-*co*-AGA) that formed by complexation of PBA with glycosyl groups.[94] Recently, Veerapandian *et al.* investigated a new class of functionalized graphene oxide (FGO) nanosheet-based amperometric glucose biosensor.[95] The FGO nanosheet was composed of metalloid polymer hybrid (MPH) nanoparticles and a gold-printed circuit board electrode on the surface of a graphene oxide nanosheet. Astonishingly, the FGO nanosheet has a characteristic response time of 7 s, a correlation efficiency of 0.9981 and a wide linear range of binding capacity up to 55.5 mM, and also showed long-term stability and close similarity to current commercial glucose sensors.

Quantum dots (QDs) also offer a model system that contains various features that are beneficial for applications in biosensors, such as their relatively small size (less than 100 nm) and versatile surface chemistry functionalization, which make them useful for designing and engineering nanocarrier platforms.[35,96] However, most QDs consist of highly toxic elements

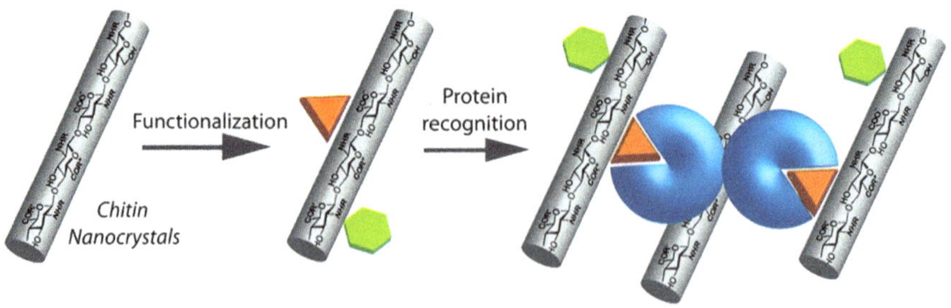

Figure 7.9 Schematic illustration of colorimetric mannose–lectin binding detection with TEMPO-oxidized chitin nanocrystals functionalized with mannose. Reprinted with permission from ref. 93. Copyright 2014 American Chemical Society.

such as cadmium, tellurium, zinc and selenium, which cause long-term toxicity problems when being used in biomedical applications.[97] Owing to the nature of the QDs, research on conjugating ligands that can increase solubility in aqueous solvents and provide biocompatibility and functionality has become a topic of interest. Along with these properties, glyco-QDs also provide the unique ability to monitor drug pharmacokinetics in real time and by fluorescence resonance energy transfer.[98] Recently, Shinchi *et al.* described the facile development of cadmium-free sugar-chain-immobilized fluorescent nanoparticles containing low-toxicity $ZnS–AgInS_2$ cores for probing lectins and cells.[99] Interestingly, their study provided a comprehensive study of the versatility of this low-toxicity $ZnS–AgInS_2$ core by conjugating a variety of different sugar chains such as mannose, galactose, glucose, sialic acids and *N*-acetylglucosamine by simple ligand exchange for specific lectin-binding applications. In comparison with other studies, the $ZnS–AgInS_2$ cores were significantly less toxic compared with cadmium-based QDs and the versatility of carbohydrate conjugation can be utilized in various biosensing applications. Other studies on QDs conducted by Savla *et al.* demonstrated the efficacy of an acid-degradable linker of doxorubicin (DOX) conjugated to a glyco-QD targeted towards Muc1 overexpressing ovarian cancer (Figure 7.10).[100] The glyco-QD–MUC1–DOX complex was stable in neutral and slight basic pH environments and rapidly hydrolyzed inside the tumor endosome. These recent innovations in hybrid carbohydrate scaffolds demonstrate the benefits of biosensing versatility for a multitude of applications in the biomedical industry.

7.3.2 Pathogen Inhibitions

In order to study the modes of pathological recognition, many researchers have elucidated the effects of surface carbohydrate interactions between extracellular binding receptors with their respective ligands.[101] Ligand polyvalency has been an intensively studied field, as most bacterial and viral infections are caused by glycan ligand–protein interactions, and blocking these

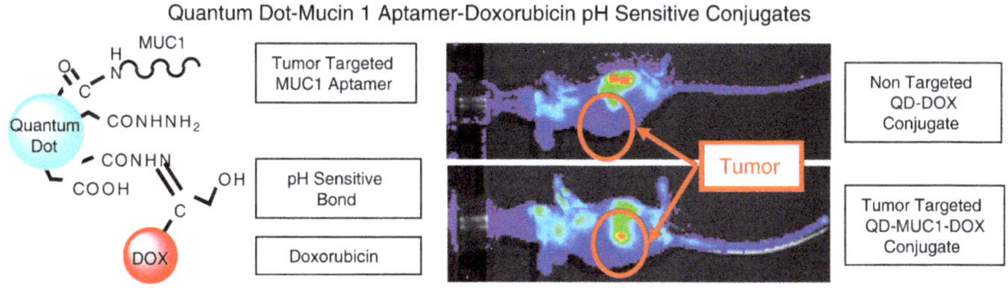

Figure 7.10 Applications of quantum dot-MUC1 aptamer doxorubicin (QD-MUC1-DOX) conjugate and organ tumor content of unmodified QDs and tumor-targeted by MUC1 aptamer QD (QD-MUC1) conjugate. Reprinted with permission from ref. 100. Copyright 2011 Elsevier.

interactions in a facile and efficient manner has been a topic of interest.[102–104] Recently, Davis and co-workers established a ground-breaking method for designing hypervalent glycodendrimers with highly branched glycodendrimer constructs up to 1620 glycans.[102] Their constructs were synthesized up to 32 nm, designed to mimic pathogens both in size and their highly glycosylated cell surfaces, and showed highly efficient models for inhibiting cellular entry of Ebola virus in T-lymphocytes and human dendritic cells. Similar studies by Luczkowiak *et al.* indicated that glycofullerenes decorated with up to 36 mannose or galactose moieties were able to bind and inhibit DC-SIGN-dependent cell entry of Ebola virus competitively (Figure 7.11).[105] These studies demonstrated the 'proof of concept' of glyco-multivalency in blocking viral infectivity, which could potentially be used as a tool for developing more efficient models for inhibiting bacterial and viral infection. Further studies by Davis's group led to the synthesis of a multivalent glycopolymer-stabilized AuNP as a potential synthetic cancer vaccine.[106] It was shown that the development of an AuNP decorated with tumor-associated antigen glycans can generate a significant immune response upon binding to mucin-1 glycoproteins expressed on breast cancer cell lines. Most importantly, these results showed promising indications that the generated antibody response from B-cells is capable to recognizing mammalian mucin

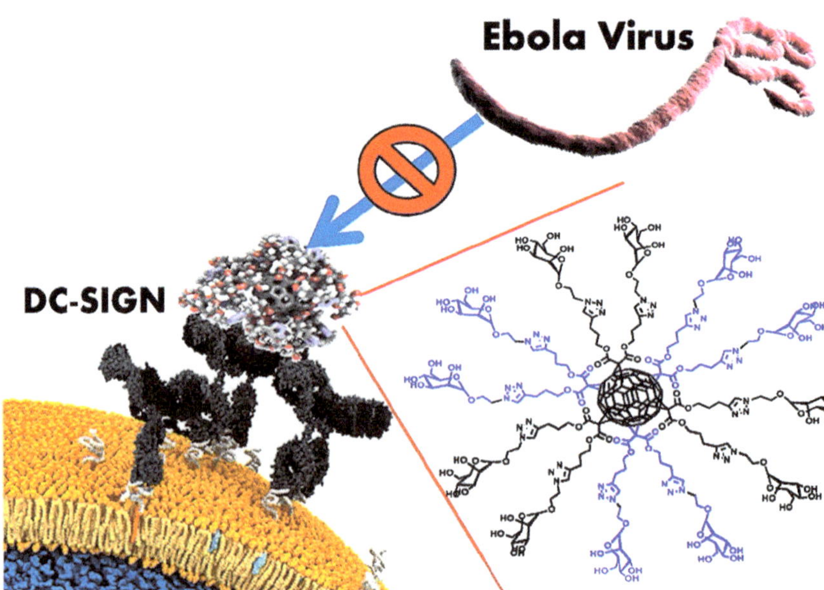

Figure 7.11 Overview of approach to inhibit Ebola virus entry to cells *via* blocking DC-SIGN-mediated entry with mannose-decorated glycofullerene. Reprinted with permission from ref. 105. Copyright 2013 American Chemical Society.

glycoproteins and demonstrated a 'proof of concept' of using peptide- and protein-free synthetic glycoconjugates as efficient vaccines.

Although many multivalent inhibitors have been developed for bacterial and viral infections, glyco-NPs have emerged at the forefront of producing the most efficient and facile methods for developing novel therapeutics.[107] An interesting study by Ogata *et al.* reported the design of artificial sulfated sialoglycopolypeptides carrying Neu5Acα2,3Galβ1,4(6-sulfo)GlcNAc and Neu5Acα2,6Galβ1,4(6-sulfo)GlcNAc residues in the side chain that were designed as hemagglutinin inhibitors against influenza viruses.[104] They used a hemagglutination assay that showed that the binding affinity to influenza virus was approximately 1.2×10^3-fold higher than a fetuin control. Other studies by Garcia-Vallejo *et al.* have also shown the inhibitory effects of PAMAM dendrimer competitive binding with dendritic cell-specific intercellular adhesion molecule-3-grabbing non-integrin (DC-SIGN) with HIV gp120 envelope protein binding with high avidity.[108] They observed a more successful binding of HIV gp120 to DC-SIGN receptor in larger molecular diameter glycodendrimers, and the geometry was pertinent in determining functionality when competing for CD4 T-cell binding. Kouki *et al.* also showed that (Gal1–4Gal)-containing oligosaccharide dendrimers were able to inhibit cell wall anchoring of *Streptococcus suis* at nanomolar concentrations.[109]

Gillies and co-workers described the synthesis of poly(ethylene oxide)–polycaprolactone polymersome designed to interact and inhibit influenza virus at two different stages of infectivity (Figure 7.12).[110] At the primary level, the polymersome is conjugated with sialic acid *N*-acetylneuraminic acid (Neu5Ac) multivalent dendrons, which is designed to bind to viral hemagglutinin gp120 protein to prevent entry into host cells. At the secondary level, a neuraminidase inhibitor, zanamivir, is encapsulated within the polymersome core, which is designed to be released upon aggregation and prevent progeny virion release that further decreases host cell interaction and replication potential. The results showed that the sialodendron had a 17-fold (per sialoside) enhancement, whereas the dendritic sialopolymersome exhibited a 2000-fold increase in binding affinity. Further studies showed that the encapsulated zanamivir sustained controlled release over 4 days and was able to be encapsulated at ~35% w/w. These multifunctional studies of dendritic scaffolds and polymersomes as potential nanocarriers showed the effectiveness of glycopolymeric engineering and its applications in the biomedical field.

7.3.3 Drug and Gene Delivery

The enhanced permeability and retention (EPR) effect is a unique property in which small-sized nanoparticles can accumulate in higher densities in tumors than normal tissue. This is generally due to larger pore sizes in neovascular tissues and poor lymphatic clearance of nanoparticles from the tumor.[111] However, on the other hand, owing to their small size and surface

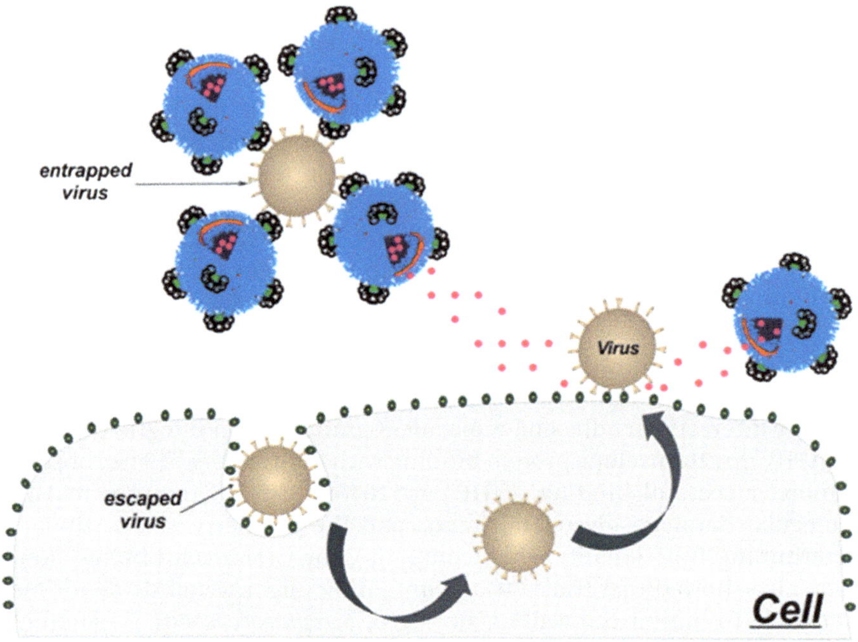

Figure 7.12 Aggregation of poly(ethylene oxide)–polycaprolactone polymersome with influenza virus and release of neuraminidase inhibitor zanamivir to prevent progeny release. Reprinted with permission from ref. 110. Copyright 2014 American Chemical Society.

charges, nanoparticles loaded with therapeutics might also be taken up by normal cells or healthy tissues and therefore result in side effects.[112,113] A wide range of polymeric nanocarriers have therefore been investigated for their targeted delivery efficacies as non-viral vectors for gene and drug delivery applications.[57,114–116] Currently, gene therapy with small targeting molecule and triggered drug release holds great promise for the treatment of many debilitating diseases and malignant tumors, such as certain metastatic cancers, neurofibromatosis, cystic fibrosis, dysplasia and other conditions.[117–120] Furthermore, common chemotherapeutic reagents used to treat cancer, such as cisplatin, doxorubicin, methotrexate and cytarabine, are generally associated with non-localized systemic toxicity.[121] Therefore, the use of more biocompatible carriers with high loading capacity, low overall toxicity and receptor-mediated endocytosis specificity is required.

While many cancer types are known to have heterogeneous subpopulations and show variations in phenotypic expression, one of the most studied and best documented cancers is hepatocellular carcinoma. The asialoglycoprotein receptor (ASGPR) is a lectin expressed on the cell surface which binds galactose analogs and other asialoglycoprotein derivatives to undergo endocytosis, and has been found to be overexpressed in numerous liver carcinomas.[122,123] Drug delivery vesicles designed for targeting the ASGPR has been a topic of long-term investigation by many research groups.[57,115,124] Wang *et al.* carried out a series of novel studies on the interactions of glycopolymers with reducible ionic cores for hepatocyte-targeted delivery for DOX.[115] They described a strategy for encapsulation of DOX into galactose-decorated reduction-sensitive cystamine crosslinked micelles that permits the rapid release of the drug in a reductive cancer environment, as evidenced by the stronger fluorescence intensity from DOX compared with micelles lacking galactose molecules. Furthermore, their studies showed that toxicity had higher selectivity for HepG2 cells than NIH 3T3 cells owing to the overexpression of ASGPR receptor on the cell surface of HepG2 cells, which was in accordance with other published findings.[125,126] Studies by Yu *et al.* showed similar results with galactosylated cholesterol modified-glycol chitosan micelles in the delivery of DOX in HepG2 cells,[127] and similar work by Detampel *et al.* indicated the efficacy of PEGylated asialofetuin-conjugated liposomes in targeted ASGPR binding.[128] With the advances in nanotechnology and external stimuli-responsive polymers, many new and improved polymeric complexes have demonstrated much higher efficacy compared with current methods of drug delivery. Li *et al.* reported the synthesis of pH-sensitive pullulan–DOX conjugated nanoparticles that have a release profile of more than 75% of DOX within 2 h at pH 5 and with less than 15% release after 12 h at pH 7.4.[129] Additional studies by Lai *et al.* demonstrated the efficiency of targeted drug delivery of paclitaxel–galactose functionalized fluorescent SiO_2 nanoparticle in HepG2 cells.[130] The presence of a fluorescent probe conjugated to the nanoparticles allowed real-time cell imaging, which also indicated that the cell-killing effect was much improved compared with free paclitaxel at the same concentration.

Gene delivery has introduced a variety of molecular techniques for manipulating viral and non-viral vectors to transform mRNA and DNA into host cells.[131] Narain and co-workers demonstrated enhanced DNA transfection with 40 nm gold nanoparticles conjugated with cationic galactose-based glycopolymers compared with commercially available transfection reagents.[120] Other studies by the same group achieved the synthesis of thermoresponsive degradable cationic glyco-nanogels for delivery of siRNA in knockdown expression of EGFR in HeLa cells by Janus green cell viability assay.[52] Furthermore, by mixing siRNA with cationic polytrehalose and forming stable nanocomplexes in high salt and serum protein conditions, Reineke and co-workers successfully delivered siRNA into glioblastoma cells (Figure 7.13).[61] Using confocal microscopy, the trehalose–polycomplex was found to undergo cytoplasmic trafficking and siRNA-induced downregulation of the target gene was achieved at an IC_{50} of 19 nM. Ragelle *et al.* further demonstrated the biocompatibility, stability and uncompromised functionality of PEGylated chitosan-conjugated nanoparticles of siRNA for gene silencing in B16 melanoma cells.[132] These studies showed the biocompatibility of carbohydrate-coated nanoplexes, their ability to retain function even under harsh conditions and their multiple applications in biomedical therapeutics.

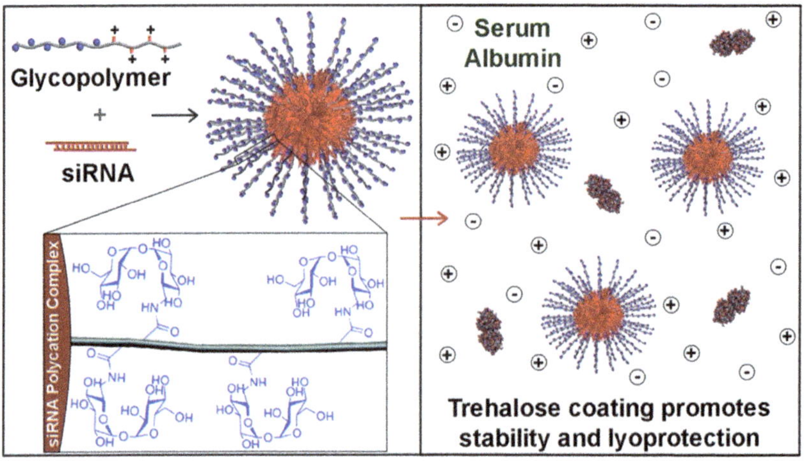

Figure 7.13 Schematic illustration of polyplex formation with siRNA and diblock cationic glycopolymers, cellular internalization and gene knockdown in glioblastoma cells. Reprinted with permission from ref. 61. Copyright 2013 American Chemical Society.

7.4 Conclusion and Future Perspective

In this chapter, we have summarized recent studies on the synthesis of glyconanoparticles and their applications in nanomedicine. Although the use of glyconanoparticles has shown great promise in drug and gene drug delivery, as inhibitors of pathogen and toxin adhesion on host cells and as probes in fluorescence and MRI, considerable work remains to be carried out towards the design and synthesis of more complex and biologically relevant carbohydrate mimics in order to gain a better understanding of carbohydrate–carbohydrate and carbohydrate–protein interactions. The full potential of these nanoparticle systems can be achieved once the vital participation of the carbohydrates in biological systems is fully understood.

References

1. J. F. Robyt, Polysaccharides: energy storage, in: *eLS*, Wiley, New York, 2001.
2. S. C. Mueller and R. M. Brown, *J. Cell Biol.*, 1980, **84**, 315–326.
3. S. K. Podder, A. Surolia and B. K. Bachhawat, *Eur. J. Biochem.*, 1974, **44**, 151–160.
4. H. E. Savaki, L. Davidsen, C. Smith and L. Sokoloff, *J. Neurochem.*, 1980, **35**, 495–502.
5. J. I. Weitz, *Circulation*, 2004, **110**, I-19–I-26.
6. L. Cipolla, A. C. Araújo, D. Bini, L. Gabrielli, L. Russo and N. Shaikh, *Expert Opin. Drug Discovery*, 2010, **5**, 721–737.
7. D. R. Bundle and N. M. Young, *Curr. Opin. Struct. Biol.*, 1992, **2**, 666–673.

8. J. Holgersson, A. Gustafsson and M. E. Breimer, *Immunol. Cell Biol.*, 2005, **83**, 694–708.
9. W. I. Weis and K. Drickamer, *Annu. Rev. Biochem.*, 1996, **65**, 441–473.
10. M. J. Linman, J. D. Taylor, H. Yu, X. Chen and Q. Cheng, *Anal. Chem.*, 2008, **80**, 4007–4013.
11. E. A. Smith, W. D. Thomas, L. L. Kiessling and R. M. Corn, *J. Am. Chem. Soc.*, 2003, **125**, 6140–6148.
12. O. Livnah, E. A. Bayer, M. Wilchek and J. L. Sussman, *Proc. Natl. Acad. Sci. U. S. A.*, 1993, **90**, 5076–5080.
13. M. Marradi, F. Chiodo, I. Garcia and S. Penades, *Chem. Soc. Rev.*, 2013, **42**, 4728–4745.
14. Y. C. Lee and R. T. Lee, *Acc. Chem. Res.*, 1995, **28**, 321–327.
15. N. Kottari, Y. Chabre, R. Sharma, R. Roy, in: *Multifaceted Development and Application of Biopolymers for Biology, Biomedicine and Nanotechnology*, ed. P. K. Dutta, J. Dutta, Springer, Berlin, 2013, vol. 254, pp. 297–341.
16. C. Fasting, C. A. Schalley, M. Weber, O. Seitz, S. Hecht, B. Koksch, J. Dernedde, C. Graf, E.-W. Knapp and R. Haag, *Angew. Chem. Int. Ed.*, 2012, **51**, 10472–10498.
17. J. M. de la Fuente, A. G. Barrientos, T. C. Rojas, J. Rojo, J. Cañada, A. Fernández and S. Penadés, *Angew. Chem. Int. Ed.*, 2001, **40**, 2257–2261.
18. A. L. Parry, N. A. Clemson, J. Ellis, S. S. Bernhard, B. G. Davis and N. R. Cameron, *J. Am. Chem. Soc.*, 2013, **135**, 9362–9365.
19. R. Narain, A. Housni, G. Gody, P. Boullanger, M.-T. Charreyre and T. Delair, *Langmuir*, 2007, **23**, 12835–12841.
20. W. Lu, W. Ma, J. Lu, X. Li, Y. Zhao, G. Chen, *Macromol. Rapid Commun.*, 2014, **35**, 827–833.
21. N. K. Vishwakarma, A. K. Mishra, A. Mishra, T. Paira, V. K. Patel, C. S. Biswas, T. K. Mandal, P. Maiti and B. Ray, *J. Appl. Polym. Sci.*, 2013, **128**, 1369–1380.
22. A. Sundgren and J. J. Barchi Jr., *Carbohydr. Res.*, 2008, **343**, 1594–1604.
23. K. L. McGilvray, M. R. Decan, D. Wang and J. C. Scaiano, *J. Am. Chem. Soc.*, 2006, **128**, 15980–15981.
24. J. C. Scaiano, C. Aliaga, S. Maguire and D. Wang, *J. Phys. Chem. B*, 2006, **110**, 12856–12859.
25. F. B. Lollmahomed and R. Narain, *Langmuir*, 2011, **27**, 12642–12649.
26. C. Boyer, A. Bousquet, J. Rondolo, M. R. Whittaker, M. H. Stenzel and T. P. Davis, *Macromolecules*, 2010, **43**, 3775–3784.
27. Z. Deng, S. Li, X. Jiang and R. Narain, *Macromolecules*, 2009, **42**, 6393–6405.
28. J. Lu, W. Zhang, S.-J. Richards, M. I. Gibson and G. Chen, *Polym. Chem.*, 2014, **5**, 2326–2332.
29. M. Takara, M. Toyoshima, H. Seto, Y. Hoshino and Y. Miura, *Polym. Chem.*, 2014, **5**, 931–939.
30. A. Pfaff, L. Barner, A. H. E. Müller and A. M. Granville, *Eur. Polym. J.*, 2011, **47**, 805–815.

31. H. Arslan, A. Pfaff, Y. Lu, P. Stepanek and A. H. E. Müller, *Macromol. Biosci.*, 2014, **14**, 81–91.
32. K. R. Yoon, S. M. Lee, B. Ramaraj and D.-P. Kim, *Surf. Interface Anal.*, 2008, **40**, 1139–1143.
33. M. K. Corbierre, N. S. Cameron, M. Sutton, K. Laaziri and R. B. Lennox, *Langmuir*, 2005, **21**, 6063–6072.
34. R. Kikkeri, B. Lepenies, A. Adibekian, P. Laurino and P. H. Seeberger, *J. Am. Chem. Soc.*, 2009, **131**, 2110–2112.
35. A. P. Alivisatos, *Science*, 1996, **271**, 933–937.
36. K. Susumu, H. T. Uyeda, I. L. Medintz, T. Pons, J. B. Delehanty and H. Mattoussi, *J. Am. Chem. Soc.*, 2007, **129**, 13987–13996.
37. M. A. Ali, S. Srivastava, M. K. Pandey, V. V. Agrawal, R. John and B. D. Malhotra, *Anal. Chem.*, 2014, **86**, 1710–1718.
38. F. Pinaud, D. King, H.-P. Moore and S. Weiss, *J. Am. Chem. Soc.*, 2004, **126**, 6115–6123.
39. H. Peng, L. Zhang, T. H. M. Kjällman and C. Soeller, *J. Am. Chem. Soc.*, 2007, **129**, 3048–3049.
40. A. Samanta, Z. Deng, Y. Liu, *Nanoscale*, 2014, **6**, 4486–4490.
41. P. Babu, S. Sinha and A. Surolia, *Bioconjugate Chem.*, 2006, **18**, 146–151.
42. M. Ahmed, R. Narain, in: *Engineered Carbohydrate-Based Materials for Biomedical Applications*, ed. R. Narain, Wiley, Hoboken, NJ, 2011, pp. 167–188.
43. X. Jiang, M. Ahmed, Z. Deng and R. Narain, *Bioconjugate Chem.*, 2009, **20**, 994–1001.
44. M. G. Sandros, M. Behrendt, D. Maysinger and M. Tabrizian, *Adv. Funct. Mater.*, 2007, **17**, 3724–3730.
45. S.-Q. Chang, B. Kang, Y.-D. Dai, H.-X. Zhang and D. Chen, *Nanoscale Res. Lett.*, 2011, **6**, 591.
46. Y. Lin, L. Zhang, W. Yao, H. Qian, D. Ding, W. Wu and X. Jiang, *ACS Appl. Mater. Interfaces*, 2011, **3**, 995–1002.
47. A. Muñoz-Bonilla, G. Marcelo, C. Casado, F. J. Teran and M. Fernández-García, *J. Polym. Sci., Part A: Polym. Chem.*, 2012, **50**, 5087–5096.
48. J. S. Basuki, L. Esser, H. T. T. Duong, Q. Zhang, P. Wilson, M. R. Whittaker, D. M. Haddleton, C. Boyer and T. P. Davis, *Chem. Sci.*, 2014, **5**, 715–726.
49. A. Pfaff, A. Schallon, T. M. Ruhland, A. P. Majewski, H. Schmalz, R. Freitag and A. H. E. Müller, *Biomacromolecules*, 2011, **12**, 3805–3811.
50. T. Borase, T. Ninjbadgar, A. Kapetanakis, S. Roche, R. O'Connor, C. Kerskens, A. Heise and D. F. Brougham, *Angew. Chem. Int. Ed.*, 2013, **52**, 3164–3167.
51. T. D. Farr, C.-H. Lai, D. Grünstein, G. Orts-Gil, C.-C. Wang, P. Boehm-Sturm, P. H. Seeberger, C. Harms, *Nano Lett.*, 2014, **14**, 2130–2134.
52. M. Ahmed, P. Wattanaarsakit and R. Narain, *Polym. Chem.*, 2013, **4**, 3829–3836.
53. R. Sunasee, P. Wattanaarsakit, M. Ahmed, F. B. Lollmahomed and R. Narain, *Bioconjugate Chem.*, 2012, **23**, 1925–1933.

54. Q. Guo, Z. Wu, X. Zhang, L. Sun and C. Li, *Soft Matter*, 2014, **10**, 911–920.
55. V. Ladmiral, M. Semsarilar, I. Canton and S. P. Armes, *J. Am. Chem. Soc.*, 2013, **135**, 13574–13581.
56. V. Percec, P. Leowanawat, H.-J. Sun, O. Kulikov, C. D. Nusbaum, T. M. Tran, A. Bertin, D. A. Wilson, M. Peterca, S. Zhang, N. P. Kamat, K. Vargo, D. Moock, E. D. Johnston, D. A. Hammer, D. J. Pochan, Y. Chen, Y. M. Chabre, T. C. Shiao, M. Bergeron-Brlek, S. André, R. Roy, H.-J. Gabius and P. A. Heiney, *J. Am. Chem. Soc.*, 2013, **135**, 9055–9077.
57. M. Ahmed and R. Narain, *Biomaterials*, 2011, **32**, 5279–5290.
58. M. Ahmed and R. Narain, *Biomaterials*, 2013, **34**, 4368–4376.
59. M. Ahmed, M. Jawanda, K. Ishihara and R. Narain, *Biomaterials*, 2012, **33**, 7858–7870.
60. M. Ahmed, K. Ishihara and R. Narain, *Chem. Commun.*, 2014, **50**, 2943–2946.
61. A. Sizovs, L. Xue, Z. P. Tolstyka, N. P. Ingle, Y. Wu, M. Cortez and T. M. Reineke, *J. Am. Chem. Soc.*, 2013, **135**, 15417–15424.
62. Y. Liu and T. M. Reineke, *J. Am. Chem. Soc.*, 2005, **127**, 3004–3015.
63. V. P. Taori, H. Lu and T. M. Reineke, *Biomacromolecules*, 2011, **12**, 2055–2063.
64. K. M. Fichter, N. P. Ingle, P. M. McLendon and T. M. Reineke, *ACS Nano*, 2012, **7**, 347–364.
65. M. Tranter, Y. Liu, S. He, J. Gulick, X. Ren, J. Robbins, W. K. Jones and T. M. Reineke, *Mol. Ther.*, 2012, **20**, 601–608.
66. D. J. Krasznai, T. F. L. McKenna, M. F. Cunningham, P. Champagne and N. M. B. Smeets, *Polym. Chem.*, 2012, **3**, 992–1001.
67. M. Ahmed and R. Narain, *Biomaterials*, 2012, **33**, 3990–4001.
68. S. D. Bull, M. G. Davidson, J. M. H. van den Elsen, J. S. Fossey, A. T. A. Jenkins, Y.-B. Jiang, Y. Kubo, F. Marken, K. Sakurai, J. Zhao and T. D. James, *Acc. Chem. Res.*, 2012, **46**, 312–326.
69. Y. Wang, X. Zhang, Y. Han, C. Cheng and C. Li, *Carbohydr. Polym.*, 2012, **89**, 124–131.
70. Y. Kotsuchibashi, R. V. C. Agustin, J.-Y. Lu, D. G. Hall and R. Narain, *ACS Macro Lett.*, 2013, **2**, 260–264.
71. Q. Wu, L. Wang, H. Yu, J. Wang and Z. Chen, *Chem. Rev.*, 2011, **111**, 7855–7875.
72. S. L. Wiskur, J. J. Lavigne, H. Ait-Haddou, V. Lynch, Y. H. Chiu, J. W. Canary and E. V. Anslyn, *Org. Lett.*, 2001, **3**, 1311–1314.
73. C. Zheng, Q. Guo, Z. Wu, L. Sun, Z. Zhang, C. Li and X. Zhang, *Eur. J. Pharm. Sci.*, 2013, **49**, 474–482.
74. G. Pasparakis and C. Alexander, *Angew. Chem. Int. Ed.*, 2008, **47**, 4847–4850.
75. F. Suriano, R. Pratt, J. P. K. Tan, N. Wiradharma, A. Nelson, Y.-Y. Yang, P. Dubois and J. L. Hedrick, *Biomaterials*, 2010, **31**, 2637–2645.
76. J. Huang, C. Bonduelle, J. Thévenot, S. Lecommandoux and A. Heise, *J. Am. Chem. Soc.*, 2011, **134**, 119–122.

77. J. Wang, G. Chen, H. Jiang, Z. Li and X. Wang, *Analyst*, 2013, **138**, 4427–4435.
78. J. Liu, Z. Cao and Y. Lu, *Chem. Rev.*, 2009, **109**, 1948–1998.
79. K. J. Cash, F. Ricci and K. W. Plaxco, *J. Am. Chem. Soc.*, 2009, **131**, 6955–6957.
80. M. D. Disney and P. H. Seeberger, *Chem. Biol.*, 2004, **11**, 1701–1707.
81. L. Dykman and N. Khlebtsov, *Chem. Soc. Rev.*, 2012, **41**, 2256–2282.
82. Y. Du, L. Shi, M. Hong, H. Li, D. Li and M. Liu, *Opt. Commun.*, 2013, **298**, 232–236.
83. S. Balamurugan, K. M. Mayer, S. Lee, S. A. Soper, J. H. Hafner and D. A. Spivak, *J. Mol. Recognit.*, 2013, **26**, 402–407.
84. R. Elghanian, J. Storhoff, R. Mucic, R. Letsinger and C. Mirkin, *Science*, 1997, **277**, 1078–1081.
85. L. Otten, S.-J. Richards, E. Fullam, G. S. Besra and M. I. Gibson, *J. Mater. Chem. B*, 2013, **1**, 2665–2672.
86. A. G. Barrientos, J. M. Fuente, M. Jiménez, D. Solís, F. J. Cañada, M. Martín-Lomas and S. Penadés, *Carbohydr. Res.*, 2009, **344**, 1474–1478.
87. M. I. Gibson, E. Fullam, G. S. Besra, S.-J. Richards, *J. Mater. Chem. B*, 2014, **2**, 1490–1498.
88. K. R. Lim, J.-M. Park, H. N. Choi and W.-Y. Lee, *Microchem. J.*, 2013, **106**, 154–159.
89. D. Granot and E. M. Shapiro, *Magn. Reson. Med.*, 2011, **65**, 1253–1259.
90. H. Wu, J. Wang, X. Kang, C. Wang, D. Wang, J. Liu, I. A. Aksay and Y. Lin, *Talanta*, 2009, **80**, 403–406.
91. P. Takmakov, I. Vlassiouk and S. Smirnov, *Anal. Bioanal.Chem.*, 2006, **385**, 954–958.
92. A. Kaushik, R. Khan, P. R. Solanki, P. Pandey, J. Alam, S. Ahmad and B. Malhotra, *Biosens. Bioelectron.*, 2008, **24**, 676–683.
93. J. Zhou, N. Butchosa, H. S. N. Jayawardena, Q. Zhou, M. Yan, O. Ramström, *Bioconjugate Chem.*, 2014, **25**, 640–643.
94. R. Ma, H. Yang, Z. Li, G. Liu, X. Sun, X. Liu, Y. An and L. Shi, *Biomacromolecules*, 2012, **13**, 3409–3417.
95. M. Veerapandian, H. Y. Kim, Y.-T. Seo, K.-N. Lee, K. Yun and M.-H. Lee, *Mater. Res. Bull.*, 2014, **49**, 593–600.
96. C.-Y. Zhang, H.-C. Yeh, M. T. Kuroki and T.-H. Wang, *Nat. Mater.*, 2005, **4**, 826–831.
97. R. Hardman, *Environ. Health Perspect.*, 2006, **114**, 165–172.
98. R. Wilson, D. G. Spiller, A. Beckett, I. A. Prior and V. Sée, *Chem. Mater.*, 2010, **22**, 6361–6369.
99. H. Shinchi, M. Wakao, N. Nagata, M. Sakamoto, E. Mochizuki, T. Uematsu, S. Kuwabata, Y. Suda, *Bioconjugate Chem.*, 2014, **25**, 286–295.
100. R. Savla, O. Taratula, O. Garbuzenko and T. Minko, *J. Controlled Release*, 2011, **153**, 16–22.
101. J. Arnaud, A. Audfray and A. Imberty, *Chem. Soc. Rev.*, 2013, **42**, 4798–4813.

102. R. Ribeiro-Viana, M. Sánchez-Navarro, J. Luczkowiak, J. R. Koeppe, R. Delgado, J. Rojo and B. G. Davis, *Nat. Commun.*, 2012, **3**, 1303.
103. S.-J. Richards, M. W. Jones, M. Hunaban, D. M. Haddleton and M. I. Gibson, *Angew. Chem. Int. Ed.*, 2012, **51**, 7812–7816.
104. M. Ogata, H. Uzawa, K. I. P. J. Hidari, T. Suzuki, E. Y. Park and T. Usui, *J. Appl. Glycosci.*, 2014, **61**, 1–7.
105. J. Luczkowiak, A. Muñoz, M. Sánchez-Navarro, R. Ribeiro-Viana, A. Ginieis, B. M. Illescas, N. Martín, R. Delgado and J. Rojo, *Biomacromolecules*, 2013, **14**, 431–437.
106. A. L. Parry, N. A. Clemson, J. Ellis, S. S. Bernhard, B. G. Davis and N. R. Cameron, *J. Am. Chem. Soc.*, 2013, **135**, 9362–9365.
107. M. Khanal, T. Vausselin, A. Barras, O. Bande, K. Turcheniuk, M. Benazza, V. Zaitsev, C. M. Teodorescu, R. Boukherroub and A. Siriwardena, *ACS Appl. Mater. Interfaces*, 2013, **5**, 12488–12498.
108. J. J. Garcia-Vallejo, N. Koning, M. Ambrosini, H. Kalay, I. Vuist, R. Sarrami-Forooshani, T. B. Geijtenbeek and Y. van Kooyk, *Int. Immunol.*, 2013, **25**, 221–233.
109. A. Kouki, R. J. Pieters, U. J. Nilsson, V. Loimaranta, J. Finne and S. Haataja, *Biology*, 2013, **2**, 918–935.
110. A. Nazemi, S. M. Haeryfar and E. R. Gillies, *Langmuir*, 2013, **29**, 6420–6428.
111. K. Cho, X. Wang, S. Nie, Z. Chen and D. M. Shin, *Clin. Cancer Res.*, 2008, **14**, 1310–1316.
112. C. Huang, Y. Zhang, H. Yuan, H. Gao and S. Zhang, *Nano Lett.*, 2013, **13**, 4546–4550.
113. T.-G. Iversen, T. Skotland and K. Sandvig, *Nano Today*, 2011, **6**, 176–185.
114. O. Onaca, R. Enea, D. W. Hughes and W. Meier, *Macromol. Biosci.*, 2009, **9**, 129–139.
115. Y. Wang, X. Zhang, P. Yu and C. Li, *Int. J. Pharm.*, 2013, **441**, 170–180.
116. R. Mo, T. Jiang, R. DiSanto, W. Tai, Z. Gu, *Nat. Commun.*, 2014, **5**, 3364.
117. R. Sinha, G. J. Kim, S. Nie and D. M. Shin, *Mol. Cancer Ther.*, 2006, **5**, 1909–1917.
118. T. Laitala-Leinonen, *Recent Pat. DNA Gene Sequences*, 2010, **4**, 113–121.
119. N. Vij, T. Min, R. Marasigan, C. N. Belcher, S. Mazur, H. Ding, K.-T. Yong and I. Roy, *J. Nanobiotechnol.*, 2010, **8**, 1–18.
120. M. Ahmed, Z. Deng, S. Liu, R. Lafrenie, A. Kumar and R. Narain, *Bioconjugate Chem.*, 2009, **20**, 2169–2176.
121. V. T. Huynh, S. Binauld, P. L. de Souza and M. H. Stenzel, *Chem. Mater.*, 2012, **24**, 3197–3211.
122. U. Treichel, z. B. K. Meyer, R. J. Stockert, T. Poralla and G. Gerken, *J. Gen. Virol.*, 1994, **75**, 3021–3029.
123. E. I. Rigopoulou, D. Roggenbuck, D. S. Smyk, C. Liaskos, M. G. Mytilinaiou, E. Feist, K. Conrad and D. P. Bogdanos, *Autoimmun. Rev.*, 2012, **12**, 260–269.
124. D. W. Lim, Y. I. Yeom and T. G. Park, *Bioconjugate Chem.*, 2000, **11**, 688–695.

125. L. Seymour, K. Ulbrich, S. Wedge, I. Hume, J. Strohalm and R. Duncan, *Br. J. Cancer*, 1991, **63**, 859–866.
126. Y. Cao, Y. Gu, H. Ma, J. Bai, L. Liu, P. Zhao and H. He, *Int. J. Biol. Macromol.*, 2010, **46**, 245–249.
127. J.-M. Yu, W.-D. Li, L. Lu, X.-Y. Zhou, D.-Y. Wang, H.-M. Li, X.-Y. Xu and J. Chen, *J. Mater. Sci. Mater. Med.*, 2013, **25**, 691–701.
128. P. Detampel, D. Witzigmann, S. Krähenbühl and J. Huwyler, *J. Drug Target.*, 2013, 1–10.
129. H. Li, S. Bian, Y. Huang, J. Liang, Y. Fan and X. Zhang, *J. Biomed. Mater. Res., Part A*, 2014, **102**, 150–159.
130. C.-H. Lai, T.-C. Chang, Y.-J. Chuang, D.-L. Tzou and C.-C. Lin, *Bioconjugate Chem.*, 2013, **24**, 1698–1709.
131. A. J. Mellott, M. L. Forrest and M. S. Detamore, *Ann. Biomed. Eng.*, 2013, **41**, 446–468.
132. H. Ragelle, R. Riva, G. Vandermeulen, B. Naeye, V. Pourcelle, C. Le Duff, C. D'Haese, B. Nysten, K. Braeckmans and S. De Smedt, *J. Controlled Release*, 2014, **176**, 54–63.

CHAPTER 8

Hybrid Glycopolymer Materials and Their Expansive Applications

MARTA FERNÁNDEZ-GARCÍA*[a] AND ALEXANDRA MUÑOZ-BONILLA*[a]

[a]Instituto de Ciencia y Tecnología de Polímeros (ICTP-CSIC), C/Juan de la Cierva 3, 28006 Madrid, Spain
*E-mail: martafg@ictp.csic.es, sbonilla@ictp.csic.es

8.1 Introduction

The fabrication of hybrid glycopolymer materials, particularly nanomaterials, has considerable importance for biomedical uses such as sensors, drug delivery systems, imaging agents and diagnostic platforms. The incorporation of glycopolymers improves the properties of nanomaterials and expands their utility, reducing the risk of cytotoxicity, conferring solubility/dispersability and leading to biorecognition.

Hybrid glycopolymer materials are normally fabricated by direct modification of the surface of certain substrates with glycomonomers or glycopolymers. Glycopolymer-functionalized substrates can be easily prepared, for instance by self-assembly processes, Langmuir–Blodgett techniques and spin coating, directed by relatively weak interactions. However, as the glycopolymers are physically absorbed, the resulting coating is relatively fragile

RSC Polymer Chemistry Series No. 15
Glycopolymer Code: Synthesis of Glycopolymers and Their Applications
Edited by C. Remzi Becer and Laura Hartmann

Published by the Royal Society of Chemistry, www.rsc.org

and also the control of the density is poor. Alternatively, layer-by-layer (LbL) and grafting approaches allow more robust attachment, which is required for certain purposes. Therefore, all these strategies yield very different coatings, *i.e.* polymer brushes, monolayer or multilayers of polymer chains, *etc.*, where the surface structure and bond strength vary considerably and accordingly so do the properties of the final material. In order to attach glycopolymers to the surface of a material, covalent bond formation by a grafting process is the most widely used approach to construct hybrid glycomaterials. Typically in the 'grafting to' approach, preformed end-functionalized or side-functionalized polymers directly anchor to a suitable surface, whereas the 'grafting from' approach involves surface-initiated polymerization. 'Grafting through' comprises copolymerization of polymerizable groups attached to the surface. In all these cases, a polymer brush layer of glycopolymer is created at the surface of the substrate. Although the 'grafting to' technique is apparently simple and its efficiency has been significantly enhanced with the development of 'click' chemistry, high grafting densities are normally achieved with the 'grafting from' approach. Moreover, both techniques exert major control in comparison with the 'grafting through' process.

This chapter deals with the preparation of hybrid glycopolymer materials by using different approaches and also discusses their potential applications in the biomedical field. Of the variety of materials/nanomaterials that have been functionalized with glycopolymers, including quantum dots,[1] silk[2] and polymer membranes,[3] this chapter discusses in greater detail the most promising and studied materials for advanced applications, in particular gold surfaces and nanoparticles (NPs), magnetic NPs, silica and carbon-based NPs.

8.2 Gold Nanoparticles and Surfaces

Gold nanoparticles (AuNPs) have attracted much attention in recent decades as innovative nanomaterials with unique properties in the fields of electronics, catalysis, biomedicine and so on,[4,5] because of their quantum-size effects, that is, the resulting physical properties are neither those of bulk metal nor those of molecular compounds, but they depend strongly on the particle size, interparticle distance, nature of the protecting organic shell and shape of the nanoparticles.[5] Surface modification of AuNPs is essential for enhancing their functionality and versatility, hence great efforts have been dedicated to studies towards the synthesis of AuNPs and modification of their surfaces with a variety of functional molecules.

Carbohydrates have become a key aspect because of their specific molecular characteristics and actions in living systems for surface modification of AuNPs. Many types of carbohydrate-related molecules have been used for conjugation with gold stabilized with thiolates in two- and three-dimensional (2D and 3D) arrays (see Figure 8.1).[6–10] These systems are built using Brust's method, which consists in growing the metallic clusters with the simultaneous attachment of self-assembled thiol monolayers on the growing nuclei.[5,11] The advantage of using this strategy on a 3D scale is that highly

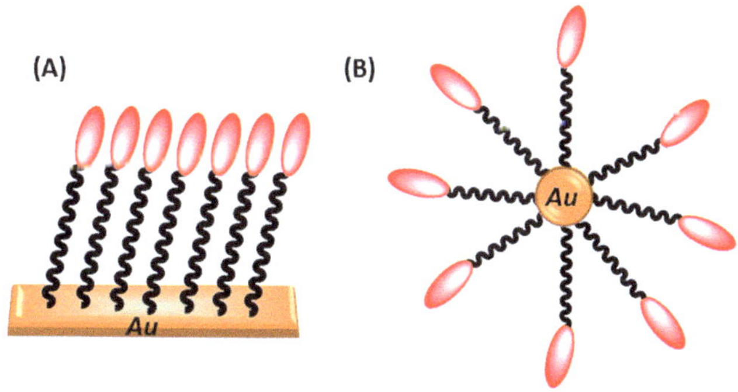

Figure 8.1 (A) Two- and (B) three-dimensional gold model substrates conjugated with carbohydrates.

polyvalent water-soluble carbohydrate surfaces with globular shapes can be tailored, which allows the analysis of the influence of carbohydrate clustering and orientation effects on their interactions with specific receptors.

The first attempt to functionalize gold glyconanoparticles *in situ* was reported by de la Fuente and co-workers[7,12] using thiol-derivatized neoglycoconjugates of two biologically significant oligosaccharides, the disaccharide lactose [Galβ(1→4)Glcβ1] and the Lewis trisaccharide Le^X {Galβ(1→4)[Fucα(1→3)]GlcNAcβ1}. In that work, the enzymatic hydrolysis of their corresponding nanoclusters with β-galactosidase of *Escherichia coli* bacteria was shown. They demonstrated the selective ability of the Le^X epitope for self-recognition in aqueous solution, supporting the proposal that specific interactions between carbohydrates are possible and may represent an additional mechanism for cell adhesion and recognition. This group also synthesized different glyconanoparticles functionalized with the tetrasaccharide Le^Y {[Fucα(1→2)]Galβ(1→4)[Fucα(1→2)]GlcNAcβ1}, the disaccharides lactose [Galβ(1→4)Glcβ1] and maltose [Glcα(1→4)Glcβ1] and the monosaccharide glucose, among others[10,13–15] (see Figure 8.2). Svarovsky *et al.*[16] also synthesized glyconanoparticles bearing Thomsen–Friedenreich tumor-associated carbohydrate antigen disaccharide [Galβ(1→3)GalNAc-α-*O*-Ser/Thr] as a model to develop therapeutic agents that inhibit protein–carbohydrate interactions.

Around the same time, Kataoka and co-workers,[17] seeking a convenient method to construct various colloidal sensor systems currently applied in bioassays and biorecognition, synthesized heterobifunctional poly(ethylene glycol) (PEG) containing acetal and thiol terminal groups, acetal-PEG-SH, and the acetal group was immediately converted into a reactive aldehyde by simple treatment with dilute acid. Subsequently, lactose was successfully introduced at the distal end of the PEG chain to induce lectin interaction with *Ricinus communis* agglutinin (RCA_{120}) in a quantitative and reversible manner, producing a simultaneous color change in the visible spectra (pinkish red → purple → pinkish-red) (see Figure 8.3).

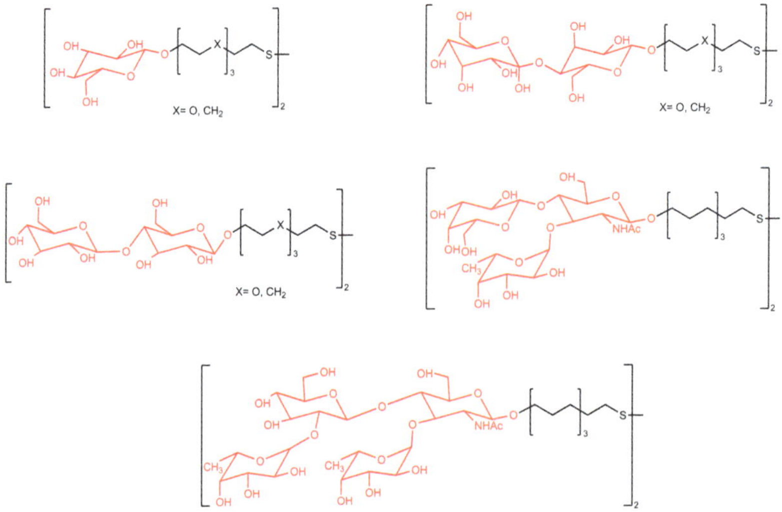

Figure 8.2 Structures of saccharides with thiol end-group.

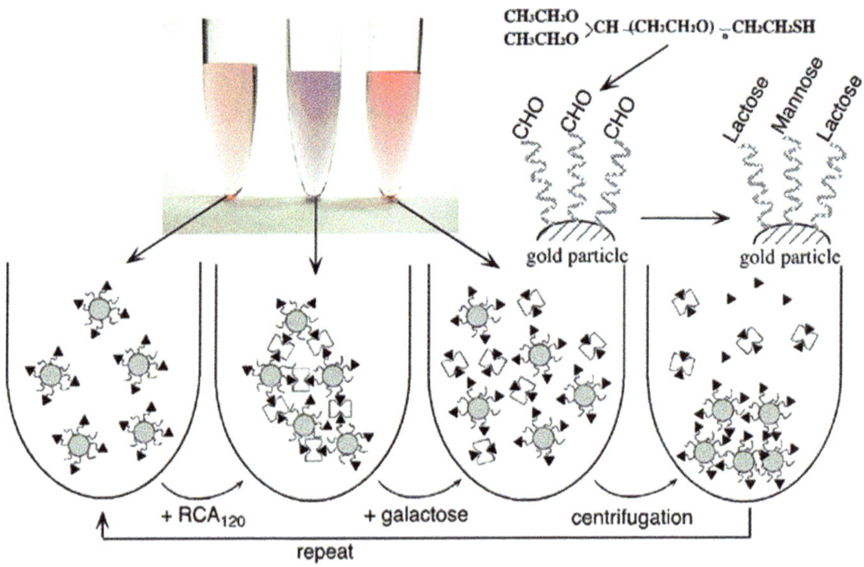

Figure 8.3 Schematic representation of the reversible aggregation–dispersion behavior of Lac-PEGylated gold nanoparticles by sequential addition of RCA_{120} lectin and galactose. Reprinted with permission from ref. 17. Copyright 2001 American Chemical Society.

They also analyzed the effects of ligand density on the aggregation behavior of ligand-bearing PEGylated gold nanoparticles through the evaluation of the interactions between RCA_{120} lectin and lactosyl-PEGylated gold nanoparticles with regulated density of functional groups. The control of lactose density on the PEGylated gold nanoparticles was performed by immobilization of mixtures of oxidized dimers of acetal-PEG-SH and lactosyl-PEG-SH in different ratios. Interestingly, they found a critical lactose density (>20%) that induces detectable particle aggregation, indicating that the interaction between the particles is triggered by the multimolecular bridging *via* lectin molecules.[18]

Kitaoka's prolific group further prepared cellulose nanolayers with parallel chain alignment by chemoselective modification of the reducing ends of cellulose with thiosemicarbazide (TSC) and subsequent self-assembly chemisorption on a gold surface.[19] This approach, named vectorial chain immobilization, utilizes *S*-derivatization of one terminus (the reducing end) of poly/oligosaccharides and spontaneous thiol–gold interaction to anchor the carbohydrates unidirectionally on the gold surface. The method was applied to a variety of carbohydrates, *e.g.* chitohexaose, cellohexaose, galactosyllactose, cellobiose and hyaluronan,[19–26] focusing on the development of practical cell culture scaffolds. These interface systems are able to adhere preferentially rat liver cells (IAR-20), human hepatocellular carcinoma cells (HepG2) and mouse fibroblasts (NIH-3T3). There are various reports based on the fact that gold or silver substrates can be easily covered with polysaccharides, improving surface-enhanced Raman scattering (SERS). The SERS effect is several orders of magnitude greater than that shown by the same substance in the pure state or in solution. This is the case with glass slides dipped rapidly in a mixed solution of methylcellulose, a semi-synthetic carbohydrate-based polymer and $HAuCl_4$. After photoirradiation to crosslink the hybrid material, the resulting gold films were able to detect different SERS-active analytes, *i.e.* 1-aminopyrene, nicotinic acid, nicotinamide, nicotine, Cresyl Fast Violet (CFV), Brilliant Cresyl Blue and a DNA derivatized with CFV.[27]

A hyperbranched polyethylenimine core with a glycidol, gluconolactone or lactonionic acid shell (Figure 8.4) has been also used as support material for metal nanoparticles (*i.e.* Cu, Ag, Au and Pt) in water.[28] In this study, various parameters such as pH, concentration, [metal ions]-to-[polymer] ratio and the nature of the metal precursors ($HAuCl_4$, $AgNO_3$, $CuSO_4$ and H_2PtCl_6) were optimized to obtain stable nanoparticle systems in water. In the case of stabilized Pt nanoparticles, the authors investigated the catalytic activity in the hydrogenation of isophorone by monitoring the disappearance of isophorone by gas chromatography. The recycling of the catalyst (by ultracentrifugation and isolation as a powder) did not lead to a significant decrease in activity after its redispersion in water.

Uzawa *et al.*[29] prepared synthetic glycopolymers carrying the key carbohydrate epitopes of α-D-galactobioside (Gb2), β-lactoside and α-D-mannopyranoside. The Gb2 epitope prepared from penta-*O*-acetyl-D-galactopyranose, which serves as a ligand of bacterial toxins and adhesion proteins, was coupled with poly(ethylene-*alt*-maleic anhydride) to afford a Gb2-embedded glycopolymer having also carboxylate (COOK) polyanions on the side chain. The polyanionic glycopolymer was then applied to the preparation of sugar-coated gold electrodes, which involves alternating layer-by-layer adsorption based on electrostatic interactions (Figure 8.5). These glycosyl arrays

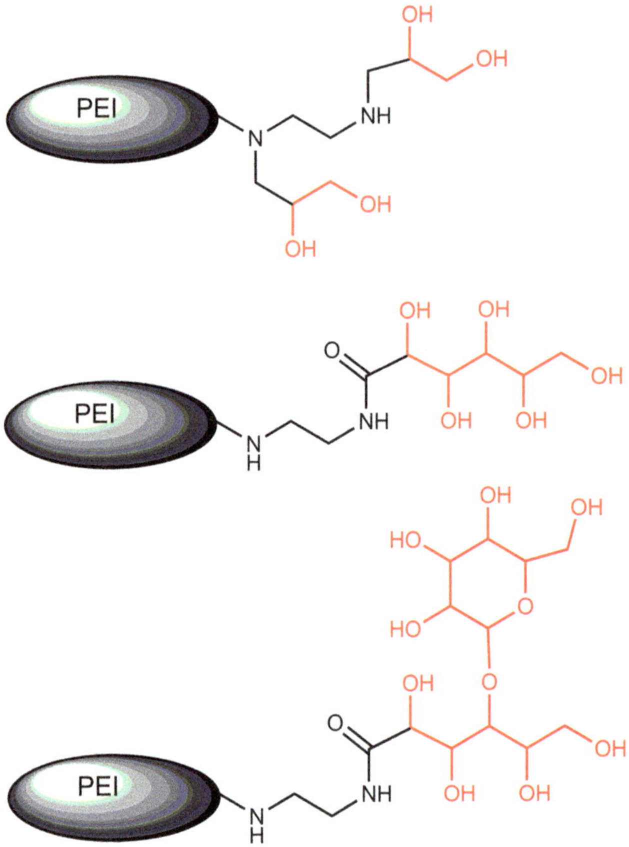

Figure 8.4 Structure of hyperbranched glycopolymers.

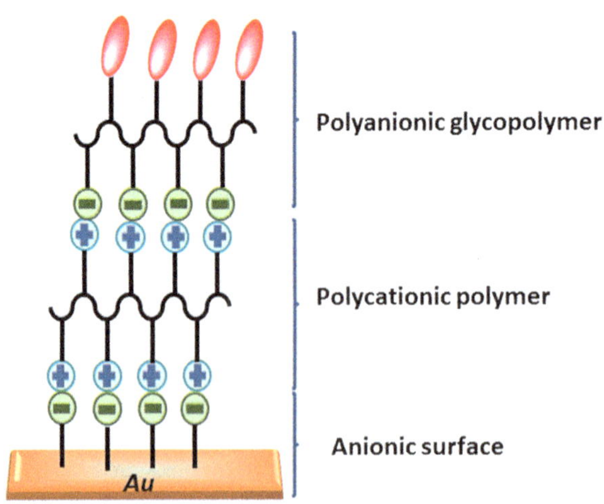

Figure 8.5 Schematic diagram of alternating layer-by-layer membranes on the gold surface.

or glycosyl chips for the surface plasmon resonance (SPR) or quartz crystal microbalance (QCM) technique are functional for detecting Shiga toxins, other pathogenic toxins and viruses.[30,31]

In a different approach, instead of thiol- or disulfide-derivatized carbohydrate structures, Pei *et al.*[32] used aryl azide click chemistry to obtain glycopolymer-coated gold surfaces on QCM crystals for carbohydrate–lectin interaction (Figure 8.6). Gold surfaces, functionalized with azide groups, are covalently coated with thin layers of PEG, to which photoprobe-derivatized carbohydrate structures (α-D-mannopyranoside, β-D-galactopyranoside and α-L-fucopyranoside) are covalently attached by a nitrene photoinsertion reaction. The resulting surfaces applied to a flow-through QCM instrument are able to analyze carbohydrate–protein interactions [*i.e.* concanavalin A (ConA), *Ulex europaeus* agglutinin-I (UEA), *Pisum sativum* agglutinin (PSA) and *Viscum album* agglutinin (VAA)] in real time. These platforms allow the study of the kinetics and thermodynamics of the interactions between carbohydrates and lectins. The results show high specificity, where the surface with α-D-mannopyranoside residues is selective for ConA and PSA, β-D-galactopyranoside is only selective for VAA and α-L-fucopyranoside only for VAA (see Figure 8.7).

The same group used the same approach to synthesize and functionalize AuNPs with perfluorophenyl azide by a ligand-exchange reaction. Subsequently, monosaccharides (D-mannose, D-glucose, D-galactose) and disaccharides (maltose, sucrose, α-1,3-galactobiose) were subsequently immobilized by a fast light activation.[33] The resulting particles demonstrated strong interactions with their corresponding carbohydrate-binding proteins, *i.e.* ConA, PNA, *Griffonia simplicifolia* (GS II) or soy bean agglutinin (SBA).

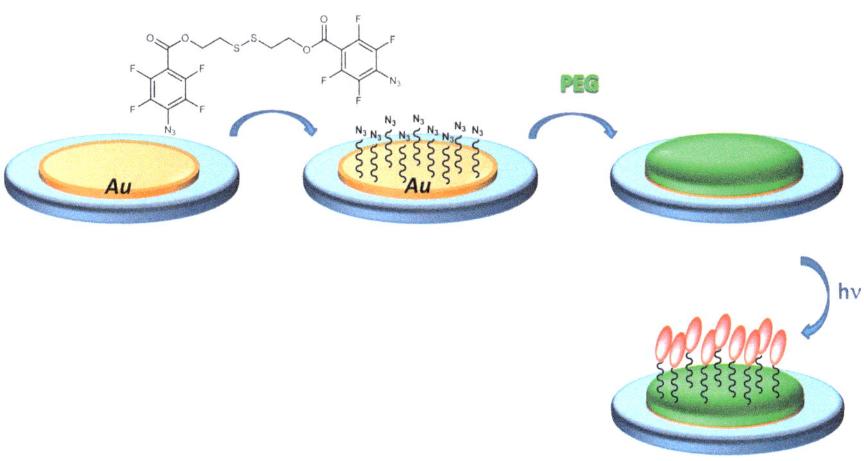

Figure 8.6 Double surface ligation of carbohydrate structures to gold-plated QCM crystals. The gold electrode is initially derivatized with 2,2′-dithioethyl bis(4-azido-2,3,5,6-tetrafluorobenzoate), yielding a perfluorophenyl azide-presenting surface. A PEG monolayer is subsequently covalently attached to the surface by azide thermo- or photolysis. Finally, carbohydrate is coupled to the PEG layer by the nitrene photoinsertion reaction.

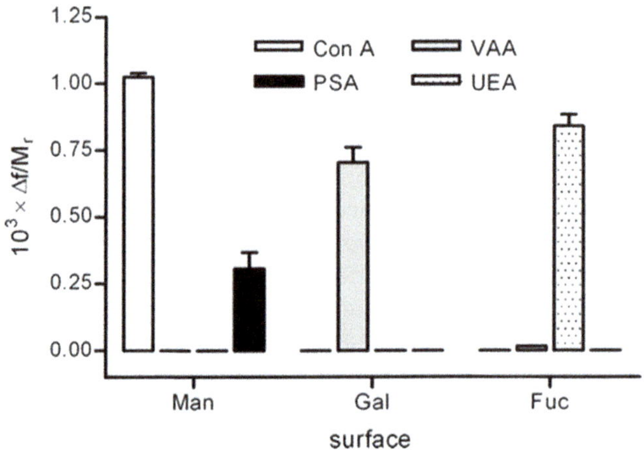

Figure 8.7 Determination of binding selectivity. Residual binding (columns) for all four lectins (ConA, PSA, VAA, UEA) recorded for all three carbohydrate-derivatized surfaces; where Δf is the frequency difference at a certain concentration and M_r the molecular weight of the respective lectin. Reprinted with permission from ref. 32. Copyright 2007 American Chemical Society.

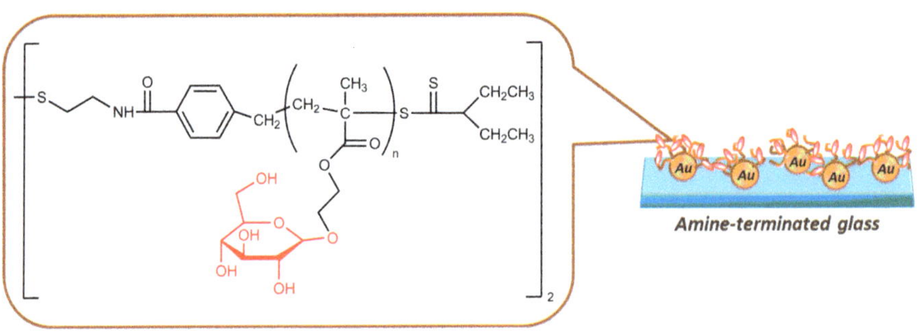

Figure 8.8 Preparation of polymer with many pendant glucose residues using the disulfide-carrying iniferter (Cys-BDC): probable structure of the prepared biofunctionalized sensor chip.

On the other hand, a glycopolymer based on 2-methacryloyloxyethyl-D-glucopyranoside was synthesized by the iniferter (*ini*tiation, chain trans*fer* and *ter*mination) technique, and was easily accumulated on colloidal Au attached to a glass substrate (Figure 8.8). The glycopolymer-coated sensor chip showed very high stability and detectability against ConA lectin, [ConA] = 1.9 nM.[34]

A glycopolymer brush on cap-shaped gold particles deposited on a vacuum-evaporated gold film has been used in another procedure for sensing glycoproteins.[35] A monolayer of polystyrene (PS) latex particles was deposited on a solid substrate covered with a thin gold film. After vacuum evaporation of gold on the latex particles, the poly(2-methacryloyloxyethyl-D-mannopyranoside), which had been prepared with an ATRP initiator, was deposited as a

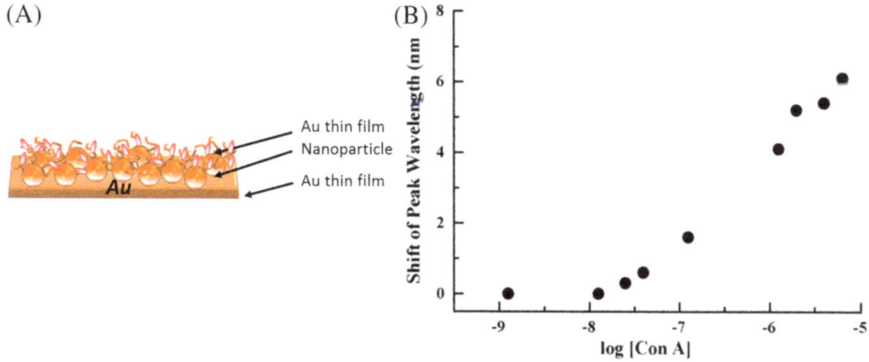

Figure 8.9 (A) Probable structure of the biofunctionalized sensor cells. (B) Shift of peak wavelength, 30 min after the immersion, as a function of the logarithm of the ConA concentration. [ConA] = 1.2×10^{-9}, 1.2×10^{-8}, 2.3×10^{-8}, 3.9×10^{-8}, 1.2×10^{-7}, 1.2×10^{-6}, 1.9×10^{-6}, 3.9×10^{-6}, 5.7×10^{-6} and 1.2×10^{-5} M. Reprinted with permission from ref. 35. Copyright 2007 Elsevier.

polymer brush on the cap-shaped gold film. The sensor coated with the glycopolymer was tested in interactions with ConA using a reflection-type visible spectrophotometer. This polymer-coated sensor cell showed very high stability and detectability ([ConA] = 12 nM; see Figure 8.9(B)).

Moreover, a thin layer of crosslinked *N*-acryloylaminophenyl-α-D-mannopyranosyl glycopolymer showed high binding sensitivity and selectivity to ConA at a concentration as low as 5×10^{-10} M.[36] The response was proportional to the ConA concentration up to 10^{-7} M in phosphate-buffered saline. The use of crosslinked polymer decreased the flexibility of the polymer backbone between the carbohydrate binding sites. Therefore, the cost of conformational entropy for multivalent binding was minimized. The binding constants of the so-prepared crosslinked polymer with ConA were found to be significantly larger than those obtained in a carbohydrate self-assembled monolayer and linear glycopolymers.[37,38] In addition to the carbohydrate–lectin recognition, additional selectivity can be achieved by controlling the degree of crosslinking. The same group recently described functionalized mannosylated polyaniline (Manno-PANI) as an electrochemical platform to study carbohydrate–ConA lectin interactions by exploiting the conductivity change of a glycopolymer (see Figure 8.10).[39] The detection limit reached in this case was also 1.2 nM with a linear range of 3–15 nM. In addition, it is very selective to ConA and non-selective *versus* peanut agglutinin (PNA), elderberry lectin (SNA), UEA, wheat germ agglutinin (WGA) and PSA.

A sugar-containing polymer was grown on a gold surface by surface-initiated ATRP (SI-ATRP) of a methacrylate derivative, 3-*O*-methacryloyl-1,2:5,6-di-*O*-isopropylidene-α-D-glucofuranoside.[40] In this case, there was no mention of the ability for biomolecular recognition.

A glycosylated surface on a poly[acrylonitrile-*co*-(hydroxyethyl methacrylate)] (PANCHEMA) film has been built by chemical modification.[41] The

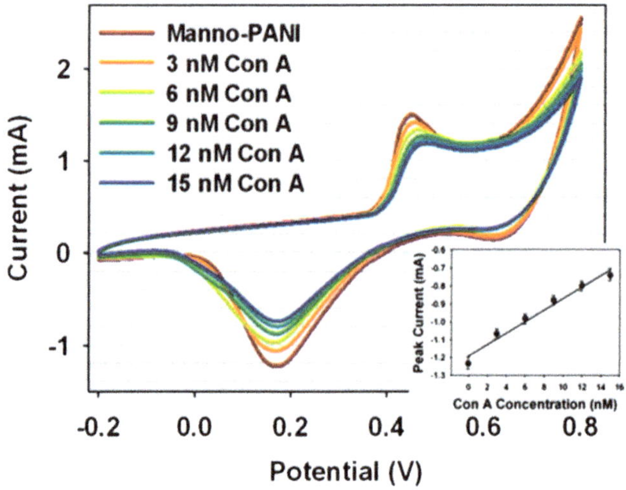

Figure 8.10 Cyclic voltammograms of Manno-PANI film in the absence and presence of ConA at different concentrations. Reprinted with permission from ref. 39. Copyright 2013 Elsevier.

binding affinity indicates the presence of strong and selective multivalent interactions between ConA and the glucose residues, but was resistant to the non-specific adsorption of BSA. The association saturation constant was determined as $2.86 \times 10^6\ M^{-1}$ by QCM and $4.09 \times 10^6\ M^{-1}$ by SPR.

Spain *et al.*[42] employed glycopolymers synthesized by reversible addition–fragmentation chain-transfer (RAFT) polymerization of 2-(β-D-galactosyloxy) ethyl methacrylate glycomonomer for immobilization on AuNP surfaces (Figure 8.11). The RAFT methodology utilizes thiocarbonylthio compounds, such as dithioesters, as mediating agents and these dithioester moieties at the chain ends of the polymers can easily be converted to thiols, allowing the polymers to be anchored to gold surfaces. The specificity of the glyconanoparticle with lectin was proved by the agglomeration of peanut agglutinin (PNA)-coated agarose beads with the particles. In Figure 8.11 it is easily observed that the PNA beads are well dispersed without the gold glycoparticles, but after their addition the agglomeration is evident and it is almost complete after 30 min. The incorporation of galactoside induces rapid redispersion of the beads. In addition, a glucose derivative was utilized under the same conditions and no agglomeration occurred, confirming that binding between the nanoparticles and lectin is carbohydrate specific. In a recent study, Cameron and co-workers[43] also synthesized well-defined glycopolymers by RAFT polymerization and conjugated them to gold nanoparticles. In this case, they used over-expressed mucin-related carbohydrates, providing a disease-specific target for cancer immunotherapy (Figure 8.12). The immunological studies indicated that these nanomaterials generated strong and long-lasting production of antibodies that are selective to the tumor-associated (Tn) antigen glycan and cross-reactive towards mucin proteins displaying Tn.

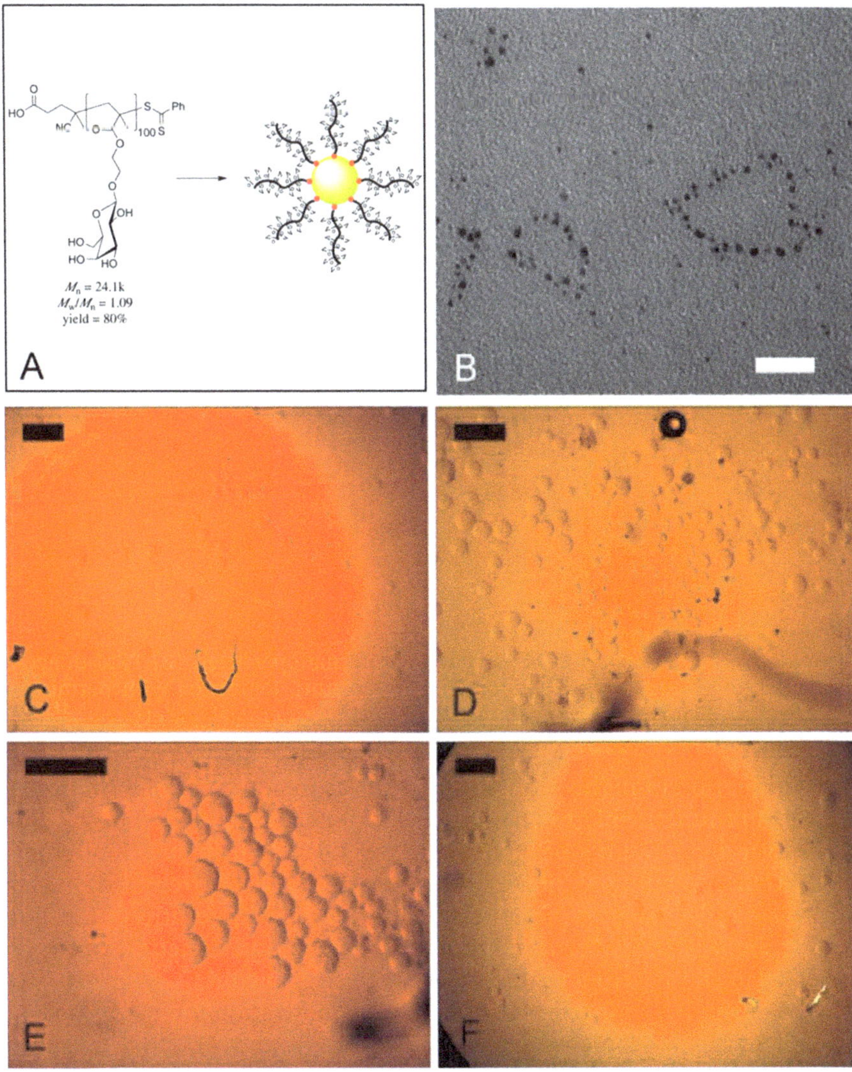

Figure 8.11 (A) Schematic diagram of gold glyconanoparticle. (B) TEM image (scale bar represents 20 nm) of gold glyconanoparticles. (C)–(F) Optical micrographs of PNA-coated agarose beads: (C) before, (D) 5 min after and (E) 30 min after addition of gold glyconanoparticles; (F) after addition of methyl β-D-galactopyranoside to suspension shown in (E). Scale bars in (C)–(F) represent 500 mm. Reprinted with permission from ref. 42. Copyright 2006 The Royal Society of Chemistry.

Narain's group synthesized well-defined glycopolymers based on D-gluconamidoethyl methacrylate (GAMA) and 2-lactobionamidoethyl methacrylate (LAMA) by RAFT polymerization. The glycopolymers were also used in the synthesis of stabilized glyconanoparticles in aqueous solution. The facile *in situ* reduction of the RAFT-prepared glycopolymers and $HAuCl_4$ resulted

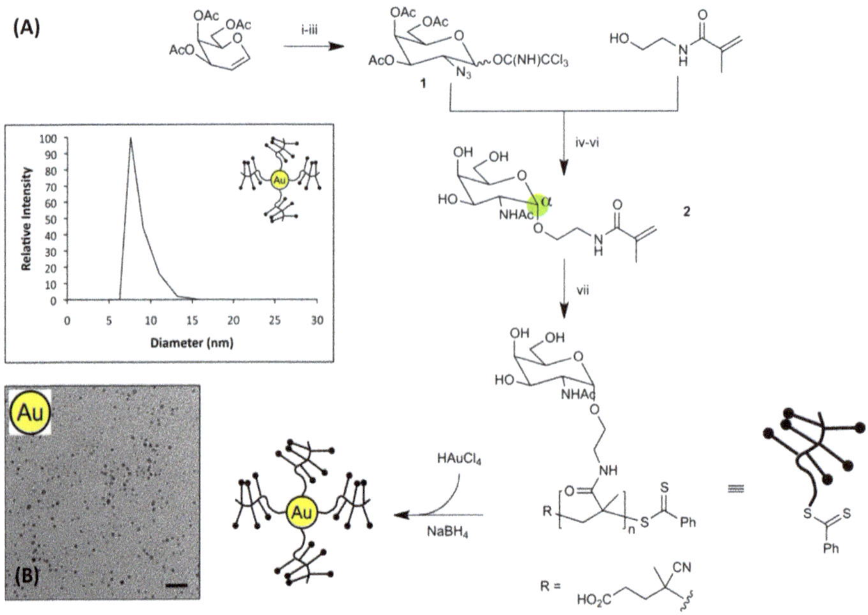

Figure 8.12 (A) Preparation of Tn antigen gold nanoparticles. (B) TEM image of glyconanoparticles (scale bar = 20 nm). Reprinted with permission from ref. 43. Copyright 2013 American Chemical Society.

in the formation of highly stable gold particles in aqueous solution (Figure 8.13).[44] Multifunctional glyconanoparticles were also generated in the presence of varying amounts of biotinylated PEG having terminal thiol groups (bio-PEG-SH) as a model bioactive domain. Addition of streptavidin to the biotinylated glyconanoparticles resulted in particle aggregation, thus confirming the accessibility and recognition of the biotin moieties, which are subjected to steric factors towards binding to streptavidin.

Glycopolymers synthesized by RAFT polymerization of mannose and *N*-acetylglucosamine methacrylamide derivatives and PEG having terminal thiol groups (PEG-SH) were mixed in an aqueous solution containing $HAuCl_4$ and the photoinitiator Irgacure-2959.[45] Further, the biotinylated glyconanoparticles were immobilized on an avidin-coated diffraction optics technology (DOT) sensor chip because of the well-known strong affinity between biotin and avidin (Figure 8.14). The specific recognitions for different carbohydrates and lectins, *i.e.* α-D-mannoside with ConA and *N*-acetyl-β-D-glucosaminoside with WGA, were confirmed using a dotLab biosensing device, which shows a response when specific lectin and almost no response when non-specific lectin is injected in the DOT equipment, *i.e.* α-D-mannoside with WGA and *N*-acetyl-β-D-glucosaminoside with ConA (Figure 8.14(b) plots).

Narain's group also used cationic glyconanoparticles as gene delivery systems.[46–48] First, they synthesized by the RAFT process diblock copolymers of

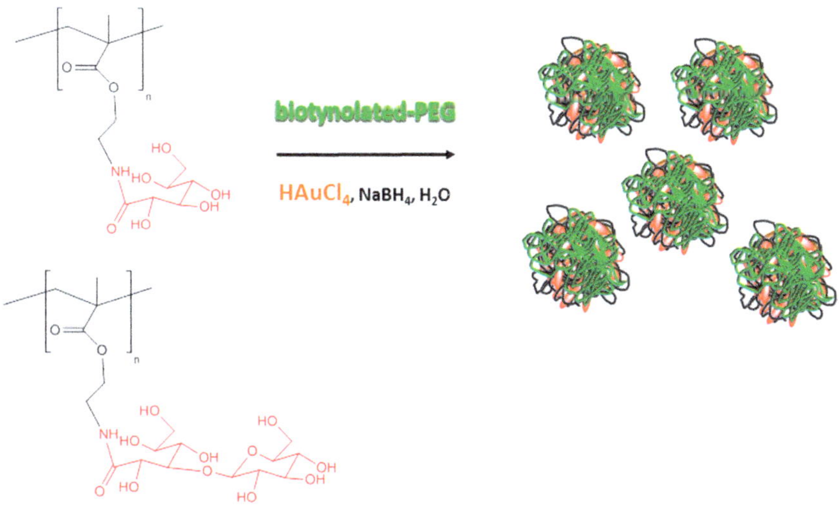

Figure 8.13 Preparation of biotinylated gold glyconanoparticles.

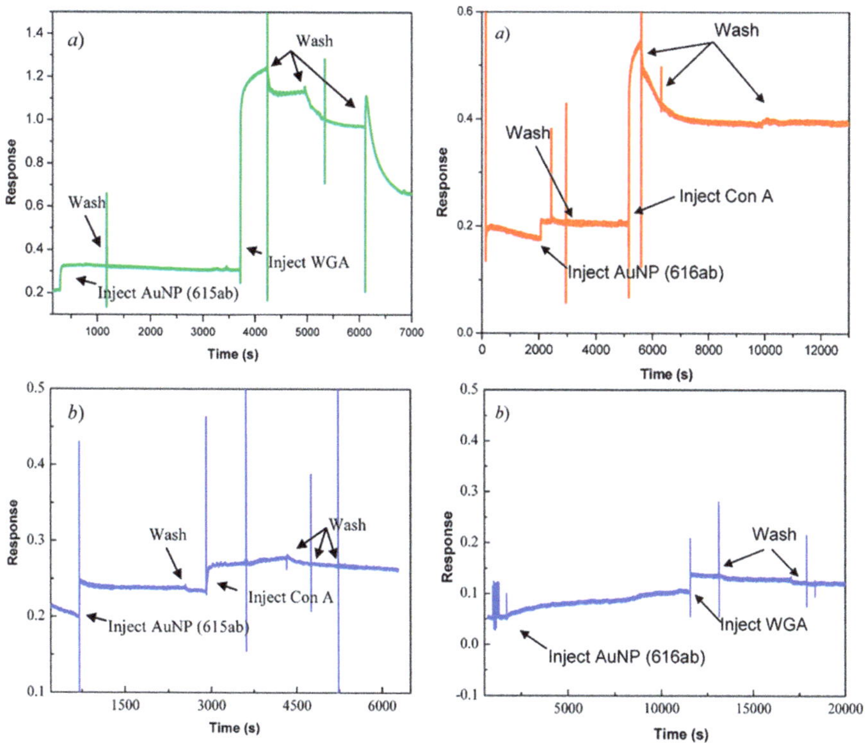

Figure 8.14 Diffraction optics responses of biotinylated gold glyconanoparticles. Left plots: (615ab) prepared from the biotin-P(NAM225-*co*-GlcNAcAm51) and right plots: (616ab) prepared from the biotin-P(NAM40-*co*-ManAm7) glycopolymer with Con A (a) and WGA (b) lectins. Reprinted with permission from ref. 45. Copyright 2013 American Chemical Society.

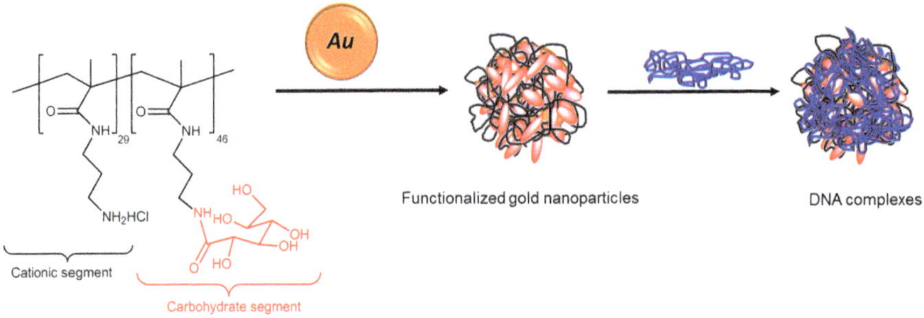

Figure 8.15 Preparation of cationic gold glyconanoparticles and their complexation with DNA.

3-aminopropylmethacrylamide hydrochloride (APMA) with 3-gluconamidopropylmethacrylamide (GAPMA). In a subsequent step, $HAuCl_4$ was added to an aqueous solution of the block copolymers and then Irgacure-2959. UV photoirradiation of the reaction mixture at different times was carried out to obtain various sized glycopolymer-coated AuNPs. These glycoparticles do not aggregate in physiological conditions and undergo receptor-mediated endocytosis following vesicular leakage, owing to their cationic character. The transfection efficacy of these systems in complexation with DNA plasmid was demonstrated by agarose gel electrophoresis experiments (Figure 8.15).[46] These processes are described in more detail in Chapter 7.

Further, they also synthesized block copolymers of 2-lactobionamidoethylmethacrylamide (LAEMA) with APMA, which were further modified with biotinyl-*N*-hydroxysuccinimide ester for anchoring to gold nanoparticles. These glyconanoparticles were able to interact specifically with RCA_{120} lectin and avidin.[49]

In a different approach and to analyze the antitumor activity of organo-gold compounds, glycopolymer-based dithiocarbamate (DTC) conjugates were obtained by RAFT, and subsequently modified with gold(I) phosphine (Figure 8.16). In this study, both APMA and GAPMA monomers were copolymerized and the resulting copolymers were modified by a Makovnikov addition reaction to yield DTC-derived polymeric chains.[50] The ratio of DTC/DTC-Au to glycopolymer was demonstrated to be critical in order to maintain the anticancer activity.

Toyoshima and Miura[51] also performed RAFT polymerization of *p*-(*N*-acrylamido)phenyl α-mannoside and *p*-(*N*-acrylamido)phenyl *N*-acetyl-β-glucosamine with acrylamide to obtain thiol-terminated glycopolymers with a range of molecular weights and saccharide densities for modification of AuNPs (Figure 8.17). The hybrid glycoparticles, glycopolymer@AuNPs, showed strong, specific molecular recognition capabilities with ConA and WGA lectins and *E. coli* bacteria. They also synthesized glycopolymers with α-Gal, which displayed, in addition to RCA_{120} lectin aggregation, a strong interaction with Shiga toxins according to SPR

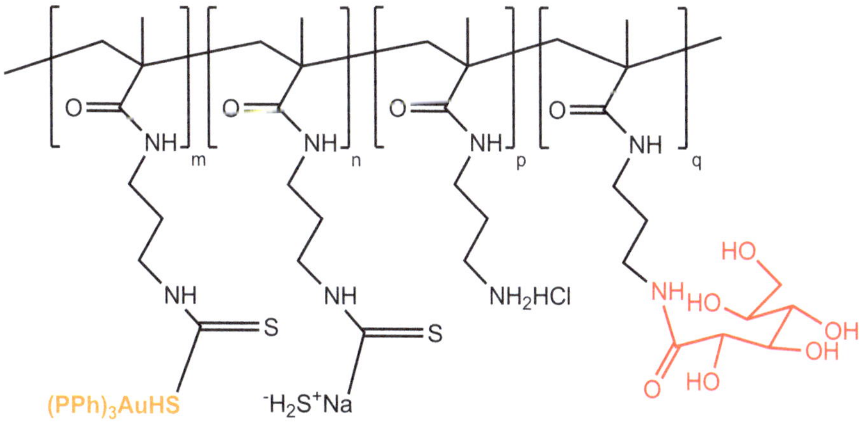

Figure 8.16 Synthesis of glycopolymer–dithiocarbamate gold conjugates.

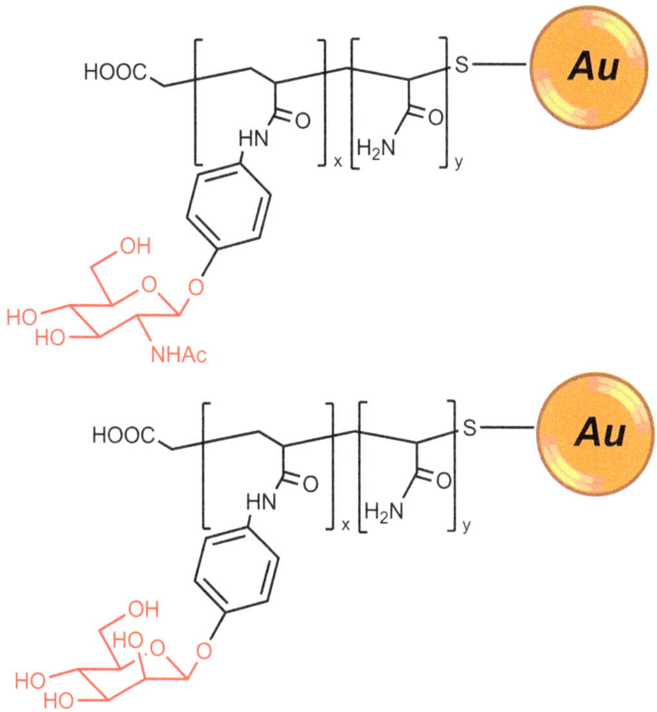

Figure 8.17 Glycopolymer@Au nanoparticles.

measurements, suggesting a possible application in biosensing.[52,53] In the case of mannose derivatives, they were also used in lateral flow,[54] electrochemical[55] and immunochromatographic[56] assays of saccharide–protein interactions for the detection of ConA. These procedures are explained in more detail in Chapter 9.

Miura and co-workers also prepared glycodendrimers by a combination of click chemistry and the convergent method.[57] These glycodendrimers were grafted to an alkyne-modified gold surface by Huisgen 1,3-dipolar cycloaddition (Figure 8.18). The molecular recognition abilities of the saccharide-immobilized substrates were evaluated by SPR with ConA, WGA and

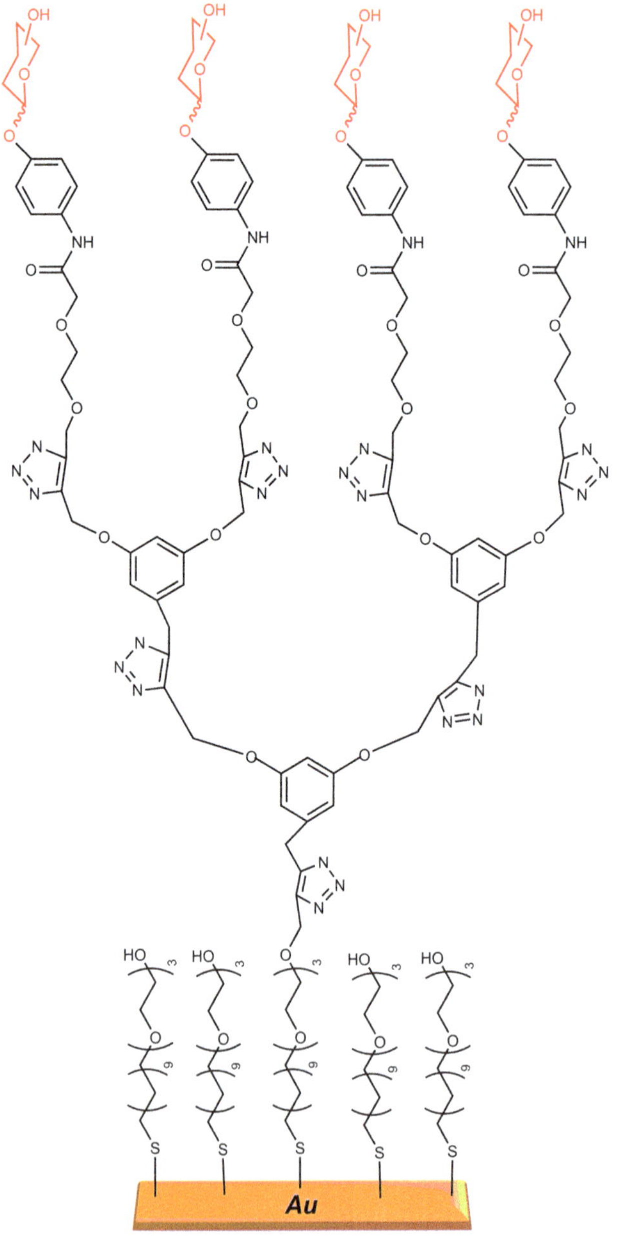

Figure 8.18 Representation of immobilized glycodendrimer in gold surfaces.

RCA_{120} lectins, calculating the association constants (K_A) and kinetic rate constants (k_a and k_d). The K_A values between carbohydrates and specific lectins increased with increasing density of the saccharide, demonstrating the multivalent effect of saccharide–protein interactions.

Recently, Borsali's group used the ability of block glycopolymers to self-assemble into micellar structures encapsulating gold nanoparticles (Figure 8.19).[58] In particular, they used maltoheptaose-*b*-polystyrene diblock glycopolymer to form micelles by nanoprecipitation of the copolymer solution in a mixture of tetrahydrofuran (THF) and water, where the copolymer molecules exist as well-swollen single chains, into a large volume of water in which the carbohydrate segment is soluble. It is important to mention that gold nanoparticles are well dispersed in THF, since thiol-terminated PS is used to anchor the metallic particles, conferring stability.

Instead of encapsulation, Boyer *et al.*[59] used copolymers that assembled onto positively charged gold nanoparticle surfaces using a layer-by-layer

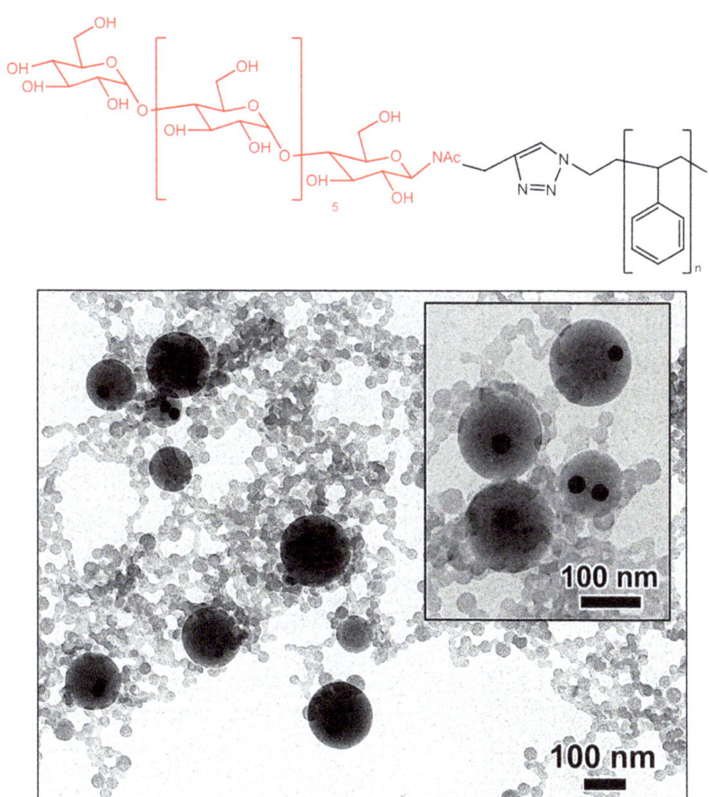

Figure 8.19 Block glycopolymer structure and TEM images of the self-assembled block glycopolymer nanoparticles in the presence of 20 nm PS–AuNPs (revealed as dark dots). Inset: magnified view of the particles. Reprinted with permission from ref. 58. Copyright 2013 American Chemical Society.

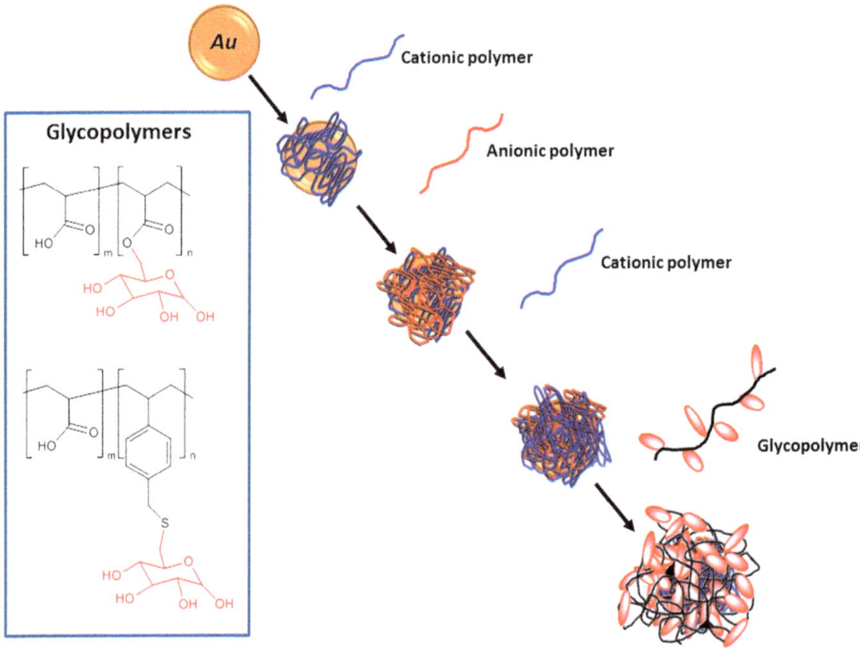

Figure 8.20 Schematic illustration of the functionalization of gold nanoparticles using layer-by-layer methodology.

methodology (Figure 8.20). Specifically, they synthesized two copolymers, poly(*tert*-butyl acrylate-*co*-chloromethylstyrene) [P(*t*BA-*co*-ClMeS)] and poly(*tert*-butyl acrylate-*co*-hydroxyethyl acrylate) [P(*t*BA-*co*-HEA)], by RAFT polymerization. P(*t*BA-*co*-ClMeS) copolymer was subsequently modified with thioglucose using a thio–halogen click nucleophilic substitution reaction and P(*t*BA-*co*-HEA) copolymer was subsequently modified with *p*-toluenesulfonyl chloride, followed by sugar functionalization (galactose) *via* a nucleophilic substitution reaction. Then, deprotection of the *tert*-butyl groups was carried out to form carboxylic acid functionality. The glucose-coated nanoparticles bind ConA specifically whereas galactose-coated nanoparticles do not interact.

Stenzel and co-workers[60] prepared polymeric types of deacetylated auranofin, a gold complex with a sugar ligand, by post-modifying RAFT block glycopolymers. The micellization of a block copolymer containing pendant Au(I) units produced nanoparticles with an increased antiproliferative effect against OVCAR-3 human ovarian carcinoma cells compared with its small molecules analog, overcoming the problems of stability and toxicity of gold nanoparticles (see Figure 8.21).

Recently, gold nanoparticles containing multiple long, flexible linkers decorated with both lactose and β-cyclodextrin (β-CD) have been prepared (Figure 8.22).[61] It was demonstrated that they can function as targeted drug

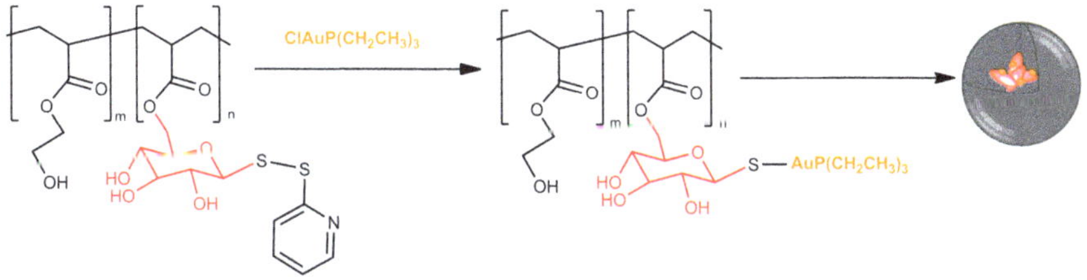

Figure 8.21 Schematic representation of glycopolymer–gold complex assembly.

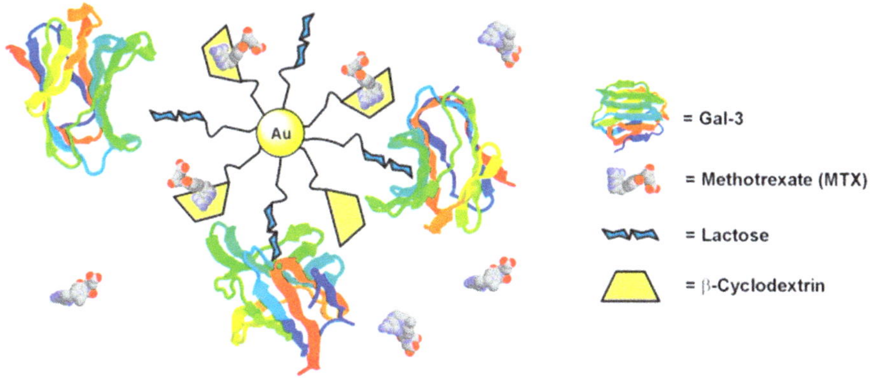

Figure 8.22 Representation of β-cyclodextrin-bearing gold glyconanoparticles. Reprinted with permission from ref. 61. Copyright 2013 American Chemical Society.

delivery systems for anticancer drugs since these particles are able to recognize PNA and human galectin-3 (Gal-3) lectins. This procedure allowed the estimation of their loading capability towards the anticancer drug methotrexate. The results showed that the presence of CD moieties on the nanoparticles clearly enhanced their ability to load this anticancer drug; therefore, they can be used as site-specific delivery systems.

An interesting procedure to form hybrid glycosystems was reported by Pfaff *et al.*,[62] in which spherical sugar-containing polymer brushes consisting of PS cores onto which chains of glycopolymers have been grafted were used as carriers for catalytically active gold nanoparticles (Figure 8.23). The crosslinking reaction of styrene and the methacrylic inimer 2-(2-bromoisobutyryloxy)ethyl methacrylate in aqueous dispersion led to the formation of spherical PS–inimer cores. Then, ATRP was used to graft poly(*N*-acetylglucosamine) arms, which, after deprotection, led to water-soluble particles. The preparation of Au nanoparticles in an aqueous solution was carried out by the chemical reduction of the metal salt–brush mixture with sodium borohydride and the resulting hybrid particles were able to catalyze the reduction of nitrophenol to aminophenol. In addition, these glycopolymer chains showed a high affinity to adsorb WGA whereas

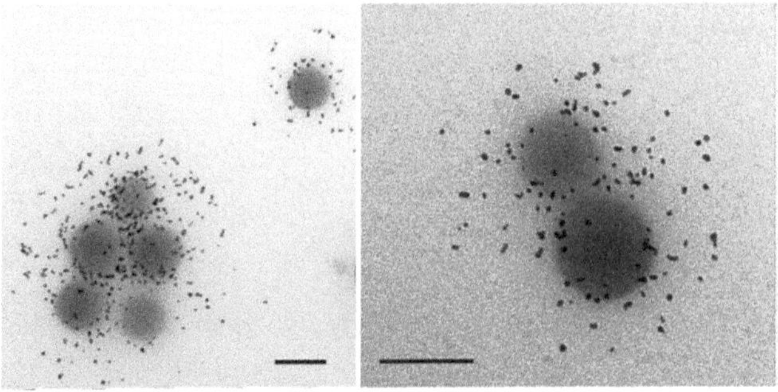

Figure 8.23 TEM images of gold nanoparticles immobilized on glucosamine brushes. The scale bars represent 100 nm. Reprinted with permission from ref. 62. Copyright 2011 Wiley-VCH.

no binding to BSA or PNA was detected, that is, they showed strong and selective binding.

8.3 Magnetic Nanoparticles

Magnetic nanoparticles based on iron oxide have generated great interest in recent years in many fields as storage devices, for catalysis and in biomedical applications.[63] They are especially attractive for the latest applications including protein and cell separation, biosensors, stem cell tracking, hyperthermia treatment, magnetic resonance imaging (MRI)[64] and drug delivery as they are approved for clinical use. This is due to their low toxicity and biocompatibility as they are easily degraded and metabolized *in vivo*. Most biomedical applications additionally require a controlled size, narrow size distribution and uniform morphology as they exhibit unique nanoscale size-dependent physicochemical properties. The major advantages of using magnetic nanoparticles of small size is the high surface area, long sedimentation times in the absence of an applied magnetic field and enhanced tissue diffusion. Moreover, below a critical particle size (generally <15 nm), each particle becomes a single magnetic domain showing superparamagnetic behavior,[65,66] so when the applied magnetic field is removed they do not exhibit coercivity and remanence. This is important in order to avoid aggregation during clinical applications. Superparamagnetic iron oxide NPs (SPIONs) have been extensively investigated in MRI, producing an enhanced proton relaxation in comparison with paramagnetic particles and consequently smaller amounts of particles are needed.[67] Magnetic hyperthermia is a promising therapy for tumor treatment based on the capacity of SPIONs to heat local tumor regions through Néel relaxation loss of the magnetic NPs under an external alternating magnetic field.[68] For these uses and in general for *in vivo* biomedical applications, the SPIONs should also have a high magnetization to ensure their movement in the blood to the targeted tissue controlled with clinically acceptable magnetic field gradients.

However, a weak magnetic response to applied magnetic fields with reduction in sizes has also been demonstrated.[69] Therefore, the size of the magnetic nanoparticles is critical for practical uses and the materials have to be carefully designed. Many efforts have also been made to design SPIONs with improved properties in order to increase the circulation time, enhance vectorization and decrease concentration. This requires better control of the size, shape and chemical composition of the magnetic NPs and also of the surface functionalization of the NPs. In the last respect, numerous studies have been performed to stabilize the iron oxide NPs with different biocompatible ligands including polymers,[70] peptides[71] and nucleic acid molecules.[72] The incorporation of carbohydrate moieties on the surface of the magnetic nanoparticles is particularly interesting as it not only imparts stability in aqueous media, preventing agglomeration, but additionally provides selective recognition and the ability to target specific cells. Huang and co-workers[73] developed magnetic glyconanoparticles bearing carbohydrates for MRI able to detect and differentiate cancer cells based on their membrane receptors. Further, carbohydrate-functionalized iron oxide nanoparticles have been utilized in MRI studies, where the carbohydrate molecules reduced non-specific adsorption and suppressed immunogenicity, while providing specific targeting.[74,75] In addition to their use as contrast agents for MRI, sugar-coated iron oxide NPs can be designed as magnetic fluid hyperthermia heat mediators to achieve targeted treatment of tumor cells.[76] Moreover, it is reported that carbohydrate binding specificity imparted to magnetic nanoparticles also allows the detection and magnetic separation of proteins[73] and pathogen bacteria such as *E. coli*.[77] Focusing on glycopolymers, their immobilization on the surface of magnetic iron oxide nanoparticles has been extensively explored since these carbohydrate–lectin interactions responsible for the target properties can be enhanced by exploiting the multivalent effect of densely packed saccharides along a polymer backbone. The design of the glycopolymer in terms of the architecture, molecular weight, type of sugar, density and distribution of the pendant sugar, is important as it strongly influences the binding effect. In spite of their promising significance, there are not many examples of synthetic glycopolymers being applied as coatings of magnetic nanoparticles. For instance, SPIONs have been functionalized with poly(vinylbenzyl-*O*-β-D-galactopyranosyl-D-gluconamide) that can be recognized by asialoglycoprotein receptors on hepatocytes.[78] MRI experiments revealed that the coated magnetic nanoparticles are specifically accumulated in rat liver compared with controls, thus having potential as a contrast liver-targeting MRI contrast agent. In this study, the glycopolymer chains were incorporated on previously synthesized magnetic nanoparticles due to their amphiphilic structures, thus acting as an emulsifier. However, grafting methods rather than physically adsorbed methods are generally used to attach the glycopolymers on the magnetic nanoparticle surface. Biotin end-terminated glycopolymer (Figure 8.24) was grafted onto the surface of streptavidin-derivatized magnetic particles.[79] The glycosylated magnetic particles obtained were evaluated for

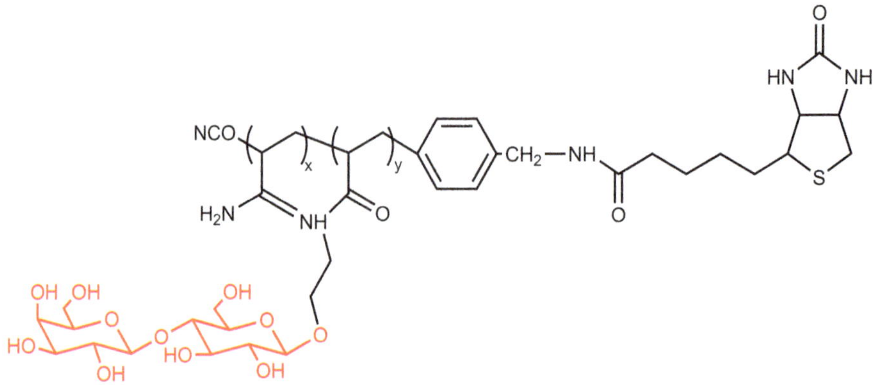

Figure 8.24 Structure of biotin end-terminated glycopolymer.

the specific and rapid isolation of lectins that can be further released and identified with an excess of free biotin.

Control of the glycopolymer structure is crucial for optimizing the binding ability and is also required for reproducibility of the system. Davis and co-workers[80] synthesized well-defined block glycopolymers using a combination of controlled/living radical polymerization and click chemistry to attach α-D-mannose, α-D-glucose and β-D-glucose to iron oxide nanoparticles. Initially the designed block copolymers consisted of an oligo(ethylene glycol) acrylate (OEGA) segment and glycidyl acrylate (GA) block for subsequent 'click' modifications. The block copolymer was synthesized from a phosphonic ester-bearing initiator in order to cleave the copolymer to the SPIONs in the subsequent step. Two different approaches were followed to incorporate sugars into the copolymers, thiol–epoxide addition and azide–alkyne cycloaddition (Figure 8.25). Then, the MRI responses of the resulting magnetic nanoparticles coated with the block glycopolymers were evaluated and also the effect of lectin concentration on the magnetic properties. It was found that the incorporation of sugar moieties significantly increases the relaxivity as a result of the enhanced stabilization of the SPIONs in water, reducing the aggregation. More important, the r_2/r_1 relaxivity is also higher, thus showing great potential as a negative contrast agent in clinical T_2-weighted MRI. On the basis of the specific lectin binding capability of the functionalized SPIONs, an aggregation process occurring with increasing concentration of lectin, and consequently a decrease in their transverse relaxation time, were expected. Effectively, the reported results demonstrated that specific binding events can be quantitatively detected using MRI measurement, which is important in targeting and in diagnostic applications.

Babiuch *et al.*[81] also functionalized iron oxide nanoparticles with well-defined glycopolymers, in this case prepared *via* nitroxide-mediated radical polymerization of glycomonomer bearing glucose (see Figure 8.26). The

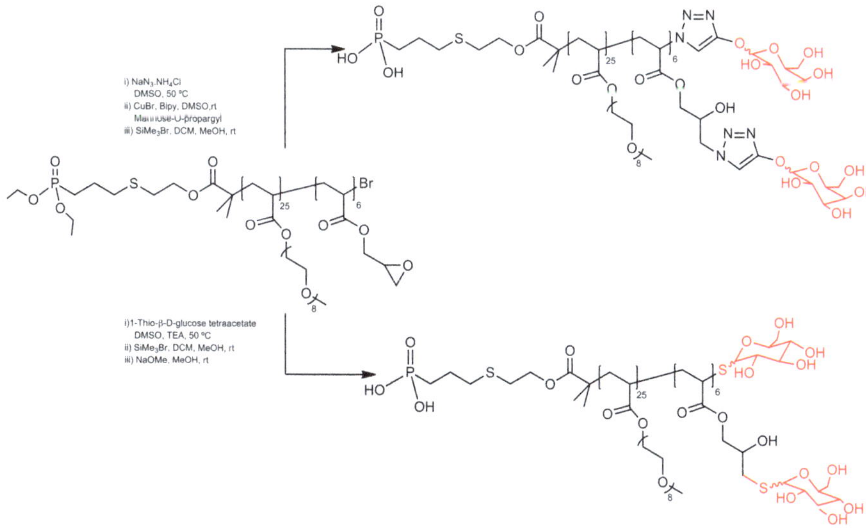

Figure 8.25 Schematic representation of the synthesis of glycopolymers bearing mannose or glucose by click reactions from POEGA-*b*-PGA block copolymers containing a phosphonic ester initiator to cleave the copolymer on the SPIONs in the subsequent step.

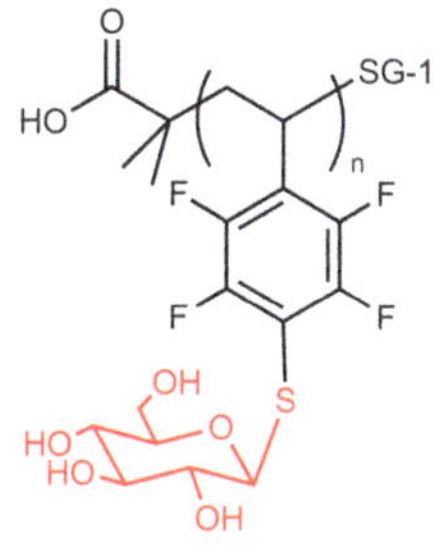

Figure 8.26 Structure of a glycopolymer bearing glucose prepared *via* nitroxide-mediated radical polymerization.

glycomonomer was synthesized by a click reaction, coupling acetylated β-D-thioglucopyranose *via* a nucleophilic substitution of *para*-fluorine of pentafluorostyrene and subsequent deacetylation. The corresponding glycopolymers were then prepared from a β-phosphonylated alkoxyamine initiator that was subsequently used to attach the glycopolymer to the magnetic iron oxide nanoparticles. It was also demonstrated that the nanoparticles obtained show no cytotoxicity towards 3T3 mouse fibroblasts, which is a prerequisite for their application as the surrounding and sane cells must to be preserved.

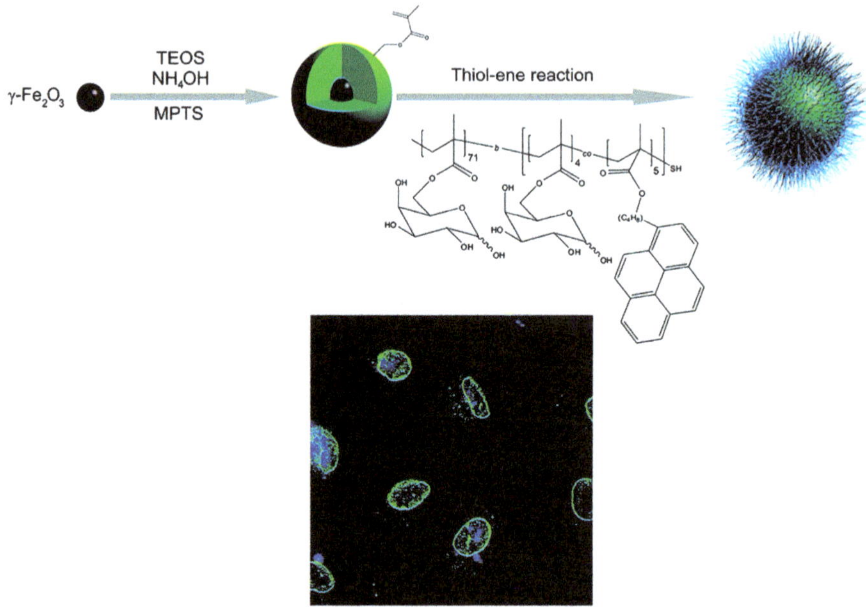

Figure 8.27 Scheme of the synthesis of fluorescent, magnetic, glycopolymer-grafted nanoparticles. Reprinted with permission from ref. 83. Copyright 2011 American Chemical Society.

This thiol–*para*-fluorine click reaction was also employed to attach carbohydrates covalently on magnetite–polymer composite materials.[82] Particles of divinylbenzene-*co*-pentafluorostyrene were prepared by precipitation polymerization in the presence of magnetite nanoparticles functionalized with vinyl groups. Such magnetic particles containing a large number of accessible active *para*-fluorine groups can be simply modified, introducing glucose moieties. Biorecognition experiments corroborated the ability of the resulting glycoparticles to interact specifically with ConA with the additional advantage of being easily separated from the medium due to their response to an external magnetic field. Another strategy that has been reported for attaching glycopolymers to magnetic nanoparticles consisted in the previous encapsulation of the magnetic particles with silicon oxide.[83] With this approach, further functionalization of the silica surface is possible *via* 3-(trimethoxysilyl)propyl methacrylate, providing the particles with vinyl groups available for the incorporation of the glycopolymer chains by thiol–ene reactions (Figure 8.27). In particular, the glycopolymer grafted onto the nanoparticles consisted of 6-*O*-methacryloylgalactopyranose and a fluorescent 4-(pyrenyl)butyl methacrylate [PMAGal-*b*-(PMAGal-*co*-PPyMA)]. Magnetization measurements of the particles obtained indicated that the encapsulation of the iron oxide nanoparticles and the grafting of the glycopolymers had almost no effect on the superparamagnetic properties, thus having potential application in MRI and in hyperthermia. Additionally, the

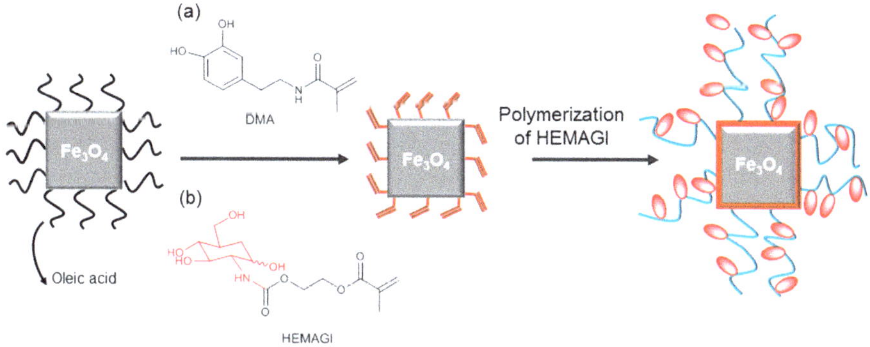

Figure 8.28 Schematic representation of the vinyl modification of Fe_3O_4 NPs with a) DMA or b) HEMAGl and posterior 'graft through' polymerization of HEMAGl.

incorporation of fluorescent pyrene groups allows the detection of the particles within mammalian cells, showing preferential nuclear localization probably due to the galactose moieties on the nanoparticles surface.

As described above, most of the approaches used to incorporate glycopolymers on magnetic nanoparticles are based on 'grafting onto' methods. A different strategy was employed by Muñoz-Bonilla *et al.*,[84] consisting in the polymerization of 2-{[(D-glucosamin-2*N*-yl)carbonyl]oxy}ethyl methacrylate glycomonomer (HEMAGl) at the interface of vinyl-functionalized SPIONs. First, oleic acid-functionalized magnetite nanoparticles were surface modified to introduce active vinyl groups *via* ligand exchange with either dopamine methacrylamide (DMA), a monomer with properties inspired by the adhesive capacity of mussels, or unprotected HEMAGl glycomonomer. Then, the modified NPs were used as seeds of the radical polymerization of the glycomonomer bearing glucose moieties to grow glycopolymer chains from the magnetite surface, therefore providing a targeting function to be recognized for specific proteins (Figure 8.28). The biomolecular recognition ability against FITC-conjugated ConA indicated that the lectin–NP interactions increase with increasing glycopolymer percentage. Regarding their magnetic behavior, the nanoparticles maintained their superparamagnetic properties after the incorporation of the glycopolymer chains. In addition, the synthesized glycopolymer-coated magnetite nanoparticles were evaluated for hyperthermia treatment. The hybrid glycoparticles exhibit a suitable heat dissipation power when exposed to alternating magnetic field conditions, showing a specific absorption rate (SAR) variation between 10 and 50 W g^{-1} when the frequency varies from 40 to 200 kHz at a constant field amplitude of 25 mT.

8.4 Silica Particles

Silica-based hybrid materials possess very attractive properties, such as rigidity, stability and lack of toxicity, and are very attractive for a variety of applications, including purification processes, chemical synthesis, drug delivery and many others.[85] The versatile chemistry of silica-based materials in sol–gel processes facilitates the preparation of hybrid materials. In addition, silica surfaces can be facilely post-modified for introducing multiple functionalities *via* silanization, which also allows the development of hybrid materials. The immobilization of sugar molecules in the preparation of silica-based hybrid glycomaterials has been carried out using different synthetic strategies, including click chemistry reactions,[86] and has been mainly proposed for applications such as protein detection,[87] protein purification,[88] boron removal[89] and even photodynamic therapy.[90] However, up to now, only a few examples have been reported concerning the functionalization of silica materials with glycopolymers. Most of these strategies require a previous silanization step for introducing initiator molecules or active groups at the silica surface to graft the glycopolymer chains further. One approach consists in the introduction of double bonds on the surface of silica particles by reaction of, for instance, methacryloxypropyltrimethoxysilane (MPTMS) with the hydroxyl groups of silica particles. Guo *et al.*[91] reported the free-radical graft polymerization of 2-*O*-methacryloyloxyethoxyl(2,3,4,6-tetra-*O*-acetyl-β-D-galactopyranosyl)-(1–4)-2,3,6-tri-*O*-acetyl-β-D-glucopyranoside (MAEL) onto silica gel particles previously functionalized with vinyl groups (see Figure 8.29). The potential of the resulting glycopolymer-grafted silica particles as chromatographic supports for the isolation of biomolecules that are difficult to separate, such as flavonoid compounds of quercetin and isorhamnetin, was also evaluated. The particular affinity is based, in principle, on the high hydrophilicity and strong hydrogen bonding ability of the glycopolymer chains. Indeed, HPLC measurements showed that modified silica particles are able to separate flavonoid compounds effectively compared with a pristine silica gel column. Moreover, well-defined glycopolymers of MAEL can be grafted onto silica particles following a different strategy.[92] First,

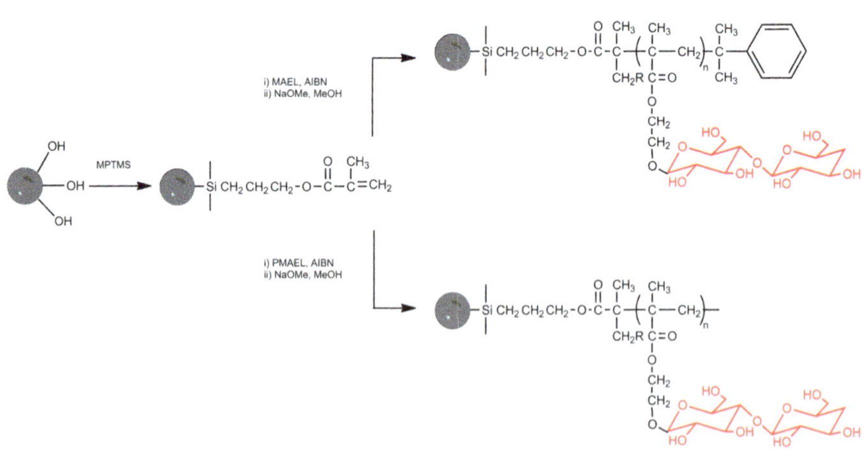

Figure 8.29 Schematic illustration of the preparation of glycopolymer-coated silica particles *via* either a 'grafting from' or a 'grafting to' approach.

the glycopolymer is synthesized by RAFT polymerization and then grafted onto the MPTMS-modified silica gel through its dithioester residues using azobisisobutyronitrile.

Following the 'grafting onto' approaches, *O*-cyanate chain-end functionalized glycopolymers permit attachment to a commercially available amine-functionalized silica gel *via* isourea bond formation.[93] An *O*-cyanate chain-end functionalized glycopolymer containing acrylaminoethyl lactoside and acrylamide was synthesized *via* cyanoxyl free-radical polymerization varying the sugar content. Then, the glycopolymer chains can be easily immobilized onto the 3-aminopropyl-functionalized silica gel at room temperature. Again, the most promising application of these glycopolymer-grafted silica particles is their use as a chromatographic support, in this case to isolate lectins. The study was carried out against β-galactose-specific and non-specific lectins, namely *Arachis hypogea* and ConA, respectively. The results demonstrated the specific capture of *A. hypogea*, which can also be released after elution with free lactose (Figure 8.30).

In addition to spherical silica particles, glycopolymer chains have been anchored to flat surfaces. Most of the studies reported used SI-ATRP to graft glycopolymers from silicon substrates. In one example, the initiator 2-(4-chlorosulfonylphenyl)ethyltrimethoxysilane is immobilized by the Langmuir–Blodgett technique. Subsequently, the glycomonomer, *i.e.* 3-*O*-methacryloyl-1,2 : 5,6-di-*O*-isopropylidene-D-glucofuranose, is polymerized from the surfaces *via* ATPR in the presence of a sacrificial free initiator.[94] The thickness of the resulting graft layer was analyzed by ellipsometry and atomic force microscopy, showing that the thickness increases with reaction time and with molecular weight, confirming the control of the polymerization. Further, unprotected glycomonomers containing different carbohydrate residues, 2′-acrylamidoethyl-α-D-mannopyranoside, 2′-acrylamidoethyl-β-D-glucopyranoside and 2′-acrylamidoethyl-β-D-galactopyranoside, were synthesized and directly grafted onto silicon wafers by SI-ATRP through either ester- or amido-functionalized initiators, 11-(2-bromo-2-methylpropionyloxy)-undecyltrichlorosilane or 11-(2′-chloropropionamido)undecyltrichlorosilane,

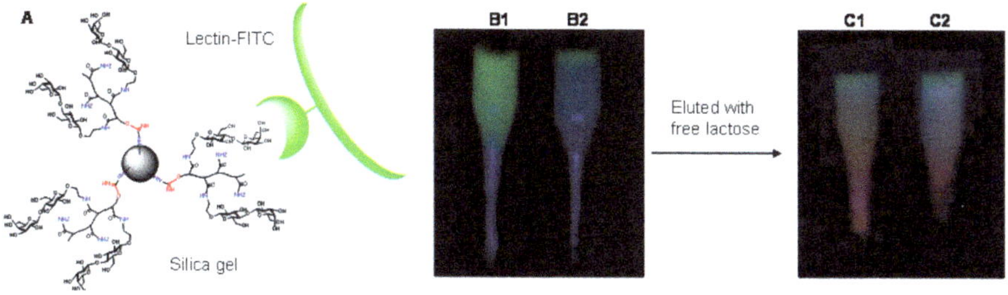

Figure 8.30 Representation of glycopolymer immobilization on silica gel particles and its lectin capture (A). Lectin capture on the glycopolymer–silica gel column (B1) and raw silica gel (B2), after elution with free lactose (C1 and C2). Reprinted with permission from ref. 93. Copyright 2011 The Royal Society of Chemistry.

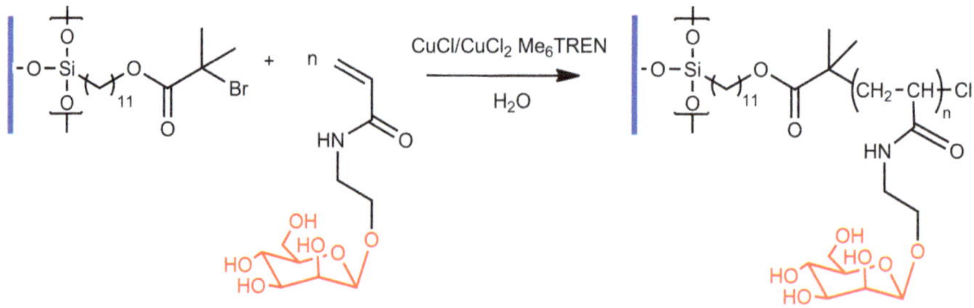

Figure 8.31 Illustration of immobilization of a glycopolymer chain on a silicon wafer by SI-ATRP through the ester initiator 11-(2-bromo-2-methyl) propionyloxy)undecyltrichlorosilane.

respectively (see Figure 8.31).[95] This study analyzed in detail the optimal polymerization conditions and also characterized the thickness of the layer formed and the grafting density. The resulting surfaces were highly hydrophilic, showing antifouling properties against proteins such as bovine serum albumin and fibrinogen, and showed specific interactions with lectins.

In another elegant study, a glycopolymer bearing α-D-mannose and also trimethoxysilane groups as matrix-binding sites was synthesized for further surface modification of porous glass membranes in order to fabricate devices for the selective separation of lectins.[96] First, lectin recognition experiments with ConA were carried out by QCM using a deacetylated sensor cell. The resulting modified membranes were able to bind almost double the amount of lectin compared with unmodified cells. Moreover, as the hydrophilicity of the surfaces increases with the glycopolymer functionalization, the non-specific interactions of, for instance, BSA are almost prevented. This biorecognition ability of the membrane obtained was then evaluated by analyzing the protein adsorption during permeation of a protein solution. The results were in agreement with those obtained by QCM, hence these hybrid porous materials show great potential for purification processes and even for the separation of pathogens.

8.5 Carbon-Based Particles

Carbon-based materials, especially carbon nanotubes (CNTs) and graphene, have gained great importance in recent years for a wide variety of applications in nanotechnology, such as supercapacitors, batteries, polymer composites, solar cells and biosensing, owing to their unique mechanical, electrical, thermal and optical properties.[97,98] In spite of their fascinating properties, these nanomaterials show some limitations and probably their main demerit is their high tendency to undergo aggregation. One of the solutions to overcome this drawback might be their hybridization, for instance, introducing functional groups or polymers at the surface of the nanomaterial. The surface modification not only enhances the stability and dispersability in a medium but also extends their utility. In biomedical applications, the use of CNTs and graphene is highly dependent on surface chemistry, that is, water solubility, biocompatibility, biorecognition capability and gene transfer ability. For example, in the development of gene delivery vectors, CNTs are postulated to be a promising candidate to mediate endocytosis and much effort has been focused on the fabrication of cationic polymer-grafted CNTs that show high gene transfer ability while at the same time being non-toxic towards living tissues. In this regard, an interesting approach consists in the incorporation of biocompatible cationic glycopolymers on the surface of CNTs. In particular, two distinct diblock glycopolymers were designed, namely poly(3-aminopropylmethacrylamide)-*b*-poly(3-gluconamidopropylmethacrylamide) (PAPMA-*b*-PGAPMA) and poly(3-aminopropylmethacrylamide)-*b*-poly(2-lactobionamidoethylmethacrylamide) (PAPMA-*b*-LAEMA).[99] The block copolymers were synthesized *via* RAFT polymerization excluding, in this way, the possible toxic effects of residual metals that arise in other polymerization methods such as ATRP. The block glycopolymers obtained can be attached on single-walled CNTs *via* the 'grafting to' approach by coupling the carboxyl groups at the surface of the CNTs and the amine groups of APMA units using 1-ethyl-3-(3-dimethylaminopropyl)carbodiimide (EDC)-*N*-hydroxysuccinimide (NHS).

To study the transfection efficiency, various amounts of glycopolymer-functionalized CNTs were complexed with green fluorescent protein encoded and evaluated in HeLa cells. It was found that the transfection capability increases with the net positive charge of the complex, hence it is dependent on the cationic block. On the other hand, toxicity assays confirmed that the function of the glycopolymer is to provide biocompatibility and it also participates in cell-mediated interactions, decreasing the toxicity as the glycopolymer segment augments. Hence the enormous potential of glycopolymer immobilization in hybrid gene delivery vehicles has been demonstrated.

Other approaches have been employed for the surface functionalization of CNTs with glycopolymers, most of them using SI-ATRP. The general strategy for grafting polymers on the CNTs involves the previous immobilization of an ATRP initiator as illustrated in Figure 8.32. The procedure entails the following steps: (1) reaction of thionyl chloride with carboxylic acid-functionalized CNTs (previously oxidized with HNO_3), (2) introduction of hydroxyl groups by reaction with glycol or amino alcohols, (3) formation of the ATRP initiating sites (CNT-Br) by reacting the hydroxyl groups with initiator precursors such as 2-bromo-2-methylpropionyl bromide, (4) ATRP polymerization of glycomonomer from the CNT-Br and (5) in some cases a deprotection step.

Using this strategy, linear and hyperbranched glycopolymers of poly(3-*O*-methacryloyl-α,β-D-glucopyranose) were effectively grown from the surface of multi-walled CNTs.[100] Further, the percentage of glycopolymer grafted on CNTs can be easily controlled by varying the reaction time, and thus the monomer conversion. As displayed in Figure 8.33, a core–shell structure is formed due to the uniform and high grafting density.

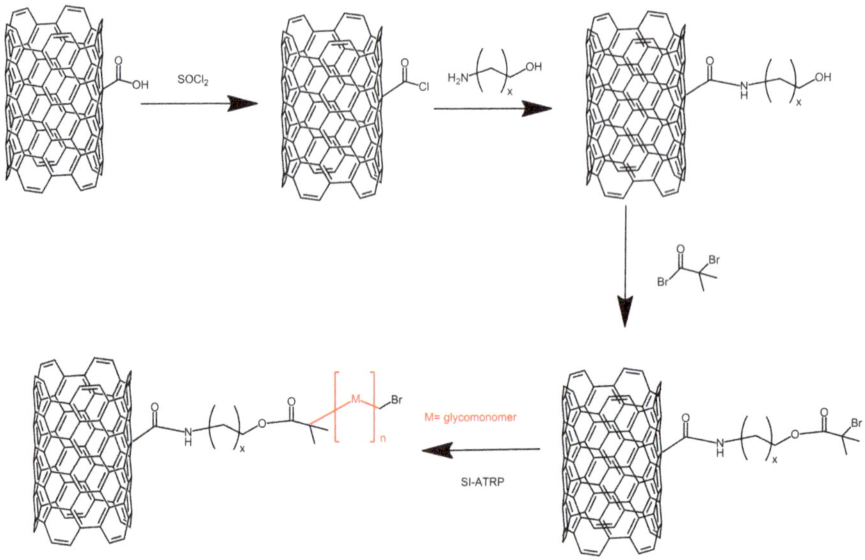

Figure 8.32 Scheme of the preparation of glycopolymer-functionalized CNTs.

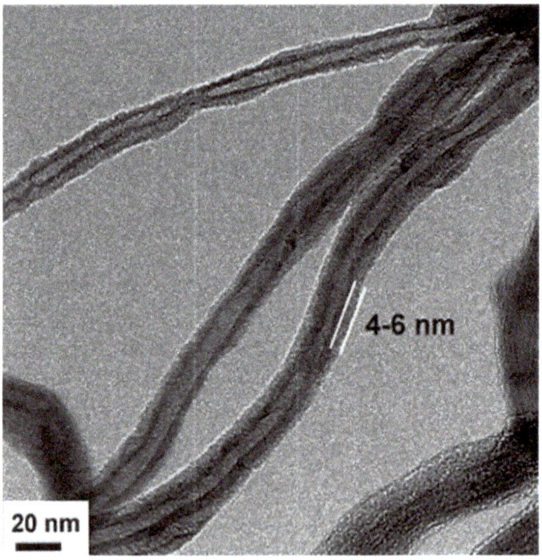

Figure 8.33 TEM image of multi-walled CNTs functionalized with linear glycopolymer. Reprinted with permission from ref. 100. Copyright 2007 American Chemical Society.

Moreover, both multi-walled[101] and single-walled[102] CNTs have been functionalized with glycopolymers by SI-ATRP of LAMA that does not need protection/deprotection steps, an advantage over other glycomonomers. The resulting glyco-functionalized CNTs were dispersible in water and in principle biocompatible.

The 'grafting to' approach has also been explored to functionalize CNTs with glycopolymers by taking advantage of the capacity of aromatic groups to trap radicals. In particular, azo-type radical macroinitiators of poly(2-methacryloyloxyethyl D-glucopyranoside)-(4,4′-azobiscyanovaleric acid) conjugates were simply cleaved to the CNTs in a single step by incubation at 70 °C for 24 h.[103] By varying the concentration, the glycopolymer chains introduced at the surface can be controlled. It was also demonstrated that the specific interaction with lectins increases as the content of glycopolymer increases. This interaction could be further increased, diminishing the hydrophobic surface of the CNTs in contact with the aqueous medium, for instance using amphiphilic macromolecules.

Finally, graphene oxide generated from the oxidation of graphene is emerging as a potential hybrid material for biological uses, mainly as a consequence of its facility to interact with biomolecules through both ionic and aromatic domains. In addition, graphene oxide exhibits a high fluorescence quenching capability induced by its sp^2 aromatic domains *via* fluorescence resonance energy transfer (FRET) or non-radiative dipole–dipole coupling,[104] and consequently can be interestingly utilized for light-up biosensing. Making use of the biorecognition ability of carbohydrates, hybrid probes based on graphene oxide and conjugated oligomers containing mannose moieties have been examined for light-up sensing of lectin and *E. coli*.[105] The authors developed a conjugate oligomer, (4,7-bis(9,9-bis(2-(2-(2-(2,3,4,5,6-pentahydroxyhexanal)-ethoxy)ethyl)fluorenyl)benzothiadiazole, with a high density of α-mannose as a fluorescent macromolecule that interacts strongly with graphene oxide *via* π–π interactions, leading to a high quenching efficiency. Such interactions are inhibited in the presence of ConA or *E. coli* as a result of the specific protein–carbohydrate interactions, hence the quenched fluorescence is recovered. Compared with other bioassays, where the non-specific interactions induce background fluorescence, this method possesses higher sensitivity. This hybrid probe shows a detection limit in the nanomolar range and a high capacity for discrimination between different proteins and even between different bacterial strains.

8.6 Conclusion

In this chapter, we have summarized the most attractive hybrid glycomaterials, including gold surfaces, magnetic NPs and silica- and carbon-based NPs. The key synthetic approaches and their promising applications have been discussed. Currently, various hybrid glycomaterials, especially nanomaterials, have been introduced in many biomedical applications ranging from diagnosis to imaging. The incorporation of glycopolymer chains on the surface of these materials expands and complements their utility, providing selectivity and targeting properties, stability and biocompatibility. In addition, the understanding of carbohydrate-mediated biological events is still an open area of research that needs to be addressed, and surely these hybrid systems will provide interesting platforms to study these biological events.

Acknowledgements

MINECO is gratefully acknowledged for financial support (MAT-2010-17016). A. Muñoz-Bonilla also thanks MINECO for her Juan de la Cierva postdoctoral contract.

References

1. R. Kikkeri, B. Lepenies, A. Adibekian, P. Laurino and P. H. Seeberger, *J. Am. Chem. Soc.*, 2009, **131**, 2110–2112.
2. S. Das, D. Pati, N. Tiwari, A. Nisal and S. Sen Gupta, *Biomacromolecules*, 2012, **13**, 3695–3702.
3. Q. Yang, M.-X. Hu, Z.-W. Dai, J. Tian and Z.-K. Xu, *Langmuir*, 2006, **22**, 9345–9349.
4. M.-C. Daniel and D. Astruc, *Chem. Rev.*, 2003, **104**, 293–346.
5. M. Brust and C. J. Kiely, *Colloids Surf., A*, 2002, **202**, 175–186.
6. M. Mrksich, *Chem. Soc. Rev.*, 2000, **29**, 267–273.
7. J. M. de la Fuente, A. G. Barrientos, T. C. Rojas, J. Rojo, J. Cañada, A. Fernández and S. Penadés, *Angew. Chem.*, 2001, **113**, 2317–2321.
8. X.-L. Meng, Y. Fang, L.-S. Wan, X.-J. Huang and Z.-K. Xu, *Langmuir*, 2012, **28**, 13616–13623.
9. B. T. Houseman and M. Mrksich, *Chem. Biol.*, 2002, **9**, 443–454.
10. Á. G. Barrientos, J. M. de la Fuente, T. C. Rojas, A. Fernández and S. Penadés, *Chem. Eur. J.*, 2003, **9**, 1909–1921.
11. M. Brust, M. Walker, D. Bethell, D. J. Schiffrin and R. Whyman, *J. Chem. Soc., Chem. Commun.*, 1994, 801–802.
12. J. M. de la Fuente, A. G. Barrientos, T. C. Rojas, J. Rojo, J. Cañada, A. Fernández and S. Penadés, *Angew. Chem., Int. Ed.*, 2001, **40**, 2257–2261.
13. J. M. de la Fuente and S. Penadés, *Biochim. Biophys. Acta*, 2006, **1760**, 636–651.
14. J.-L. de Paz, R. Ojeda, Á. G. Barrientos, S. Penadés and M. Martín-Lomas, *Tetrahedron: Asymmetry*, 2005, **16**, 149–158.
15. J. de la Fuente and S. Penadés, *Glycoconjugate J.*, 2004, **21**, 149–163.
16. S. A. Svarovsky, Z. Szekely and J. J. Barchi, *Tetrahedron: Asymmetry*, 2005, **16**, 587–598.
17. H. Otsuka, Y. Akiyama, Y. Nagasaki and K. Kataoka, *J. Am. Chem. Soc.*, 2001, **123**, 8226–8230.
18. S. Takae, Y. Akiyama, H. Otsuka, T. Nakamura, Y. Nagasaki and K. Kataoka, *Biomacromolecules*, 2005, **6**, 818–824.
19. S. Yokota, T. Kitaoka, J. Sugiyama and H. Wariishi, *Adv. Mater.*, 2007, **19**, 3368–3370.
20. S. Yokota, T. Kitaoka, M. Opietnik, T. Rosenau and H. Wariishi, *Angew. Chem., Int. Ed.*, 2008, **47**, 9866–9869.
21. S. Yokota, T. Kitaoka and H. Wariishi, *Carbohydr. Polym.*, 2008, **74**, 666–672.

22. Y. Yoshiike, S. Yokota, N. Tanaka, T. Kitaoka and H. Wariishi, *Carbohydr. Polym.*, 2010, **82**, 21–27.
23. N. Tanaka, Y. Yoshiike, C. Yoshiyama and T. Kitaoka, *Carbohydr. Polym.*, 2010, **82**, 100–105.
24. T. Kitaoka, S. Yokota, M. Opietnik and T. Rosenau, *Mater. Sci. Eng. C*, 2011, **31**, 1221–1229.
25. Y. Yoshiike and T. Kitaoka, *J. Mater. Chem.*, 2011, **21**, 11150–11158.
26. T. Kitaoka, C. Yoshiyama and F. Uemura, *Carbohydr. Polym.*, 2013, **92**, 374–379.
27. A. Pal, D. L. Stokes and T. Vo-Dinh, *Curr. Sci.*, 2004, **87**, 486–491.
28. M. Krämer, N. Pérignon, R. Haag, J.-D. Marty, R. Thomann, N. Lauth-de Viguerie and C. Mingotaud, *Macromolecules*, 2005, **38**, 8308–8315.
29. H. Uzawa, H. Ito, M. Izumi, H. Tokuhisa, K. Taguchi and N. Minoura, *Tetrahedron*, 2005, **61**, 5895–5905.
30. H. Dohi, Y. Nishida, M. Mizuno, M. Shinkai, T. Kobayashi, T. Takeda, H. Uzawa and K. Kobayashi, *Biorg. Med. Chem.*, 1999, **7**, 2053–2062.
31. H. Uzawa, S. Kamiya, N. Minoura, H. Dohi, Y. Nishida, K. Taguchi, S.-i. Yokoyama, H. Mori, T. Shimizu and K. Kobayashi, *Biomacromolecules*, 2002, **3**, 411–414.
32. Y. Pei, H. Yu, Z. Pei, M. Theurer, C. Ammer, S. André, H.-J. Gabius, M. Yan and O. Ramström, *Anal. Chem.*, 2007, **79**, 6897–6902.
33. X. Wang, O. Ramstrom and M. Yan, *J. Mater. Chem.*, 2009, **19**, 8944–8949.
34. S. Morokoshi, K. Ohhori, K. Mizukami and H. Kitano, *Langmuir*, 2004, **20**, 8897–8902.
35. Y. Anraku, Y. Takahashi, H. Kitano and M. Hakari, *Colloids Surf., B*, 2007, **57**, 61–68.
36. L. Yu, M. Huang, P. G. Wang and X. Zeng, *Anal. Chem.*, 2007, **79**, 8979–8986.
37. M. Huang, Z. Shen, Y. Zhang, X. Zeng and P. G. Wang, *Bioorg. Med. Chem. Lett.*, 2007, **17**, 5379–5383.
38. Y. Zhang, S. Luo, Y. Tang, L. Yu, K.-Y. Hou, J.-P. Cheng, X. Zeng and P. G. Wang, *Anal. Chem.*, 2006, **78**, 2001–2008.
39. Z. Wang, C. Sun, G. Vegesna, H. Liu, Y. Liu, J. Li and X. Zeng, *Biosens. Bioelectron.*, 2013, **46**, 183–189.
40. K. R. Yoon, B. Ramaraj, S. Lee, J.-S. Yu and I. S. Choi, *J. Biomed. Mater. Res. A*, 2009, **88**, 735–740.
41. A.-F. Che, X.-J. Huang and Z.-K. Xu, *Macromol. Biosci.*, 2010, **10**, 955–962.
42. S. G. Spain, L. Albertin and N. R. Cameron, *Chem. Commun.*, 2006, 4198–4200.
43. A. L. Parry, N. A. Clemson, J. Ellis, S. S. R. Bernhard, B. G. Davis and N. R. Cameron, *J. Am. Chem. Soc.*, 2013, **135**, 9362–9365.
44. A. Housni, H. Cai, S. Liu, S. H. Pun and R. Narain, *Langmuir*, 2007, **23**, 5056–5061.
45. X. Jiang, A. Housni, G. Gody, P. Boullanger, M.-T. Charreyre, T. Delair and R. Narain, *Bioconjugate Chem.*, 2010, **21**, 521–530.

46. M. Ahmed, Z. Deng, S. Liu, R. Lafrenie, A. Kumar and R. Narain, *Bioconjugate Chem.*, 2009, **20**, 2169–2176.
47. M. Ahmed, Z. Deng and R. Narain, *ACS Appl. Mater. Interfaces*, 2009, **1**, 1980–1987.
48. A. Marya and N. Ravin, *ACS Symp. Ser.*, 2012, **1119**, 81–105.
49. Z. Deng, S. Li, X. Jiang and R. Narain, *Macromolecules*, 2009, **42**, 6393–6405.
50. M. Ahmed, S. Mamba, X.-H. Yang, J. Darkwa, P. Kumar and R. Narain, *Bioconjugate Chem.*, 2013, **24**, 979–986.
51. M. Toyoshima and Y. Miura, *J. Polym. Sci. Part A: Polym. Chem.*, 2009, **47**, 1412–1421.
52. M. Toyoshima, T. Oura, T. Fukuda, E. Matsumoto and Y. Miura, *Polym. J.*, 2009, **42**, 172–178.
53. Y. Miura, *Polym. J.*, 2012, **44**, 679–689.
54. J. Ishii, M. Toyoshima, M. Chikae, Y. Takamura and Y. Miura, *Bull. Chem. Soc. Jpn.*, 2011, **84**, 466–470.
55. J. Ishii, M. Chikae, M. Toyoshima, Y. Ukita, Y. Miura and Y. Takamura, *Electrochem. Commun.*, 2011, **13**, 830–833.
56. M. Takara, M. Toyoshima, H. Seto, Y. Hoshino and Y. Miura, *Polym. Chem.*, 2014, **5**, 931–939.
57. T. Fukuda, S. Onogi and Y. Miura, *Thin Solid Films*, 2009, **518**, 880–888.
58. I. Otsuka, M. Osaka, Y. Sakai, C. Travelet, J.-L. Putaux and R. Borsali, *Langmuir*, 2013, **29**, 15224–15230.
59. C. Boyer, A. Bousquet, J. Rondolo, M. R. Whittaker, M. H. Stenzel and T. P. Davis, *Macromolecules*, 2010, **43**, 3775–3784.
60. S. Pearson, W. Scarano and M. H. Stenzel, *Chem. Commun.*, 2012, **48**, 4695–4697.
61. A. Aykaç, M. C. Martos-Maldonado, J. M. Casas-Solvas, I. Quesada-Soriano, F. García-Maroto, L. García-Fuentes and A. Vargas-Berenguel, *Langmuir*, 2013, **30**, 234–242.
62. A. Pfaff, V. S. Shinde, Y. Lu, A. Wittemann, M. Ballauff and A. H. E. Müller, *Macromol. Biosci.*, 2011, **11**, 199–210.
63. N. A. Frey, S. Peng, K. Cheng and S. Sun, *Chem. Soc. Rev.*, 2009, **38**, 2532–2542.
64. J. Gallo, N. J. Long and E. O. Aboagye, *Chem. Soc. Rev.*, 2013, **42**, 7816–7833.
65. U. Jeong, X. Teng, Y. Wang, H. Yang and Y. Xia, *Adv. Mater.*, 2007, **19**, 33–60.
66. K. M. Krishnan, A. B. Pakhomov, Y. Bao, P. Blomqvist, Y. Chun, M. Gonzales, K. Griffin, X. Ji and B. K. Roberts, *J. Mater. Sci.*, 2006, **41**, 793–815.
67. F. Yu, L. Zhang, Y. Huang, K. Sun, A. E. David and V. C. Yang, *Biomaterials*, 2010, **31**, 5842–5848.
68. L. Zhang, W.-F. Dong and H.-B. Sun, *Nanoscale*, 2013, **5**, 7664–7684.
69. C. T. Yavuz, J. T. Mayo, W. W. Yu, A. Prakash, J. C. Falkner, S. Yean, L. Cong, H. J. Shipley, A. Kan, M. Tomson, D. Natelson and V. L. Colvin, *Science*, 2006, **314**, 964–967.

70. C. C. Berry, S. Wells, S. Charles, G. Aitchison and A. S. G. Curtis, *Biomaterials*, 2004, **25**, 5405–5413.
71. J. Xie, K. Chen, H.-Y. Lee, C. Xu, A. R. Hsu, S. Peng, X. Chen and S. Sun, *J. Am. Chem. Soc.*, 2008, **130**, 7542–7543.
72. M. V. Yigit, D. Mazumdar and Y. Lu, *Bioconjugate Chem.*, 2008, **19**, 412–417.
73. K. El-Boubbou, D. C. Zhu, C. Vasileiou, B. Borhan, D. Prosperi, W. Li and X. Huang, *J. Am. Chem. Soc.*, 2010, **132**, 4490–4499.
74. I. García, J. Gallo, N. Genicio, D. Padro and S. Penadés, *Bioconjugate Chem.*, 2011, **22**, 264–273.
75. J. Gallo, I. García, N. Genicio, D. Padro and S. Penadés, *Biomaterials*, 2011, **32**, 9818–9825.
76. L. Lartigue, C. Innocenti, T. Kalaivani, A. Awwad, M. d. M. Sanchez Duque, Y. Guari, J. Larionova, C. Guérin, J.-L. G. Montero, V. Barragan-Montero, P. Arosio, A. Lascialfari, D. Gatteschi and C. Sangregorio, *J. Am. Chem. Soc.*, 2011, **133**, 10459–10472.
77. K. El-Boubbou, C. Gruden and X. Huang, *J. Am. Chem. Soc.*, 2007, **129**, 13392–13393.
78. M. K. Yoo, I. Y. Kim, E. M. Kim, H. J. Jeong, C. M. Lee, Y. Y. Jeong, T. Akaike and C. S. Cho, *J. Biomed. Biotechnol.*, 2007.
79. X.-L. Sun, W. Cui, C. Haller and E. L. Chaikof, *ChemBioChem*, 2004, **5**, 1593–1596.
80. J. S. Basuki, L. Esser, H. T. T. Duong, Q. Zhang, P. Wilson, M. R. Whittaker, D. M. Haddleton, C. Boyer and T. P. Davis, *Chem. Sci.*, 2014, **5**, 715–726.
81. K. Babiuch, R. Wyrwa, K. Wagner, T. Seemann, S. Hoeppener, C. R. Becer, R. Linke, M. Gottschaldt, J. r. Weisser, M. Schnabelrauch and U. S. Schubert, *Biomacromolecules*, 2011, **12**, 681–691.
82. M. Alvarez-Paino, G. Marcelo, A. Munoz-Bonilla, J. Rodriguez-Hernandez and M. Fernandez-Garcia, *Polym. Chem.*, 2013, **4**, 986–995.
83. A. Pfaff, A. Schallon, T. M. Ruhland, A. P. Majewski, H. Schmalz, R. Freitag and A. H. E. Müller, *Biomacromolecules*, 2011, **12**, 3805–3811.
84. A. Muñoz-Bonilla, G. Marcelo, C. Casado, F. J. Teran and M. Fernández-García, *J. Polym. Sci. Part A: Polym. Chem.*, 2012, **50**, 5087–5096.
85. M. Pagliaro, *Silica-based Materials for Advanced Chemical Applications*, Royal Society of Chemistry, Cambridge, 2009.
86. F. Santoyo-Gonzalez and F. Hernandez-Mateo, *Chem. Soc. Rev.*, 2009, **38**, 3449–3462.
87. K. Hatano, T. Yamazaki, K. Yoshino, N. Ohyama, T. Koyama, K. Matsuoka and D. Terunuma, *Tetrahedron Lett.*, 2008, **49**, 5593–5596.
88. M. Ortega-Muñoz, J. Lopez-Jaramillo, F. Hernandez-Mateo and F. Santoyo-Gonzalez, *Adv. Synth. Catal.*, 2006, **348**, 2410–2420.
89. G. Rodriguez-Lopez, M. Dolores Marcos, R. Martinez-Manez, F. Sancenon, J. Soto, L. A. Villaescusa, D. Beltran and P. Amoros, *Chem. Commun.*, 2004, 2198–2199.
90. M. Gary-Bobo, Y. Mir, C. Rouxel, D. Brevet, I. Basile, M. Maynadier, O. Vaillant, O. Mongin, M. Blanchard-Desce, A. Morère, M. Garcia, J.-O. Durand and L. Raehm, *Angew. Chem., Int. Ed.*, 2011, **50**, 11425–11429.

91. T.-Y. Guo, P. Liu, Y.-Q. Xia and M.-D. Song, *J. Appl. Polym. Sci.*, 2010, **116**, 1611–1616.
92. T.-Y. Guo, P. Liu, J.-W. Zhu, M.-D. Song and B.-H. Zhang, *Biomacromolecules*, 2006, 7, 1196–1202.
93. S. N. Narla and X.-L. Sun, *Org. Biomol. Chem.*, 2011, **9**, 845–850.
94. M. Ejaz, K. Ohno, Y. Tsujii and T. Fukuda, *Macromolecules*, 2000, **33**, 2870–2874.
95. K. Yu and J. N. Kizhakkedathu, *Biomacromolecules*, 2010, **11**, 3073–3085.
96. H. Seto, Y. Ogata, T. Murakami, Y. Hoshino and Y. Miura, *ACS Appl. Mater. Interfaces*, 2011, **4**, 411–417.
97. K. S. Novoselov, V. I. Fal'Ko, L. Colombo, P. R. Gellert, M. G. Schwab and K. Kim, *Nature*, 2012, **490**, 192–200.
98. M. F. L. De Volder, S. H. Tawfick, R. H. Baughman and A. J. Hart, *Science*, 2013, **339**, 535–539.
99. M. Ahmed, X. Jiang, Z. Deng and R. Narain, *Bioconjugate Chem.*, 2009, **20**, 2017–2022.
100. C. Gao, S. Muthukrishnan, W. Li, J. Yuan, Y. Xu and A. H. E. Müller, *Macromolecules*, 2007, **40**, 1803–1815.
101. Y.-W. Lee, S. Kang, K. Yoon, Y. Chi, I. Choi, S.-P. Hong, B.-C. Yu, H.-J. Paik and W. Yun, *Macromol. Res.*, 2005, **13**, 356–361.
102. R. Narain, A. Housni and L. Lane, *J. Polym. Sci. Part A: Polym. Chem.*, 2006, **44**, 6558–6568.
103. H. Kitano, K. Tachimoto and Y. Anraku, *J. Colloid Interface Sci.*, 2007, **306**, 28–33.
104. J. Kim, L. J. Cote, F. Kim and J. Huang, *J. Am. Chem. Soc.*, 2009, **132**, 260–267.
105. L. Wang, K.-Y. Pu, J. Li, X. Qi, H. Li, H. Zhang, C. Fan and B. Liu, *Adv. Mater.*, 2011, **23**, 4386–4391.

CHAPTER 9

Glycopolymers in Biosensing Applications

YOSHIKO MIURA*[a] AND HIROKAZU SETO[a]

[a]Department of Chemical Engineering, Graduate School of Engineering, 744 Motooka, Nishi-ku, Fukuoka 819-0395, Japan
*E-mail: miuray@chem-eng.kyushu-u.ac.jp

9.1 Introduction

9.1.1 Sugar–Protein Interactions

Saccharides on cell surfaces play important roles in living systems.[1] Saccharides have molecular recognition abilities and they mediate biological phenomena such as cell adhesion, cell differentiation and fertilization. They are also related to the infection of various pathogens such as viruses and bacteria. Therefore, the molecular recognition ability of saccharides is relevant in the design of functional molecules and materials. *In vivo*, the development of glycopolymers for medicine is a promising research area, and for *in vitro* applications, the molecular recognition ability of saccharides is utilized to create functional materials and biodevices.[2,3] Although saccharide–protein interactions are generally too weak for biosensing applications, they can be amplified *via* the multivalency of glycopolymers.[4]

In this chapter, biosensing with glycopolymers is described. Examples of saccharide recognition proteins are given in Table 9.1. The molecular recognition of saccharides covers various biological systems and the interactions

RSC Polymer Chemistry Series No. 15
Glycopolymer Code: Synthesis of Glycopolymers and Their Applications
Edited by C. Remzi Becer and Laura Hartmann

Published by the Royal Society of Chemistry, www.rsc.org

Table 9.1 Examples of proteins, cells and pathogens for saccharide recognition.

	Target	Saccharide structure
Lectins	Concanavalin A (ConA)	α-Man/α-Glc
	Wheat germ agglutinin (WGA)	GlcNAc, Neu5Ac
	Ricinus communis agglutinin (RCA_{120})	Gal, Galβ(1→4)GlcNAc
Cells	Hepatocyte	β-Gal/β-GalNAc
	Macrophage	α-Man
Pathogens	Shiga toxin (Stx) type 1 and 2 (from *E. coli* O-157, *etc.*)	Gb3 : Galα(1→4)Galβ(1→4)GlcCer
	Cholera toxin	GM1 : Galβ(1→3)[NeuAcα(2→3)] GalNAcβ(1→4)Galβ(1→4)GlcCer
	E. coli type 1	Manα(1→3)Manα(1→6)Man
	Influenza virus type A for humans	Neu5Acα(2→6)Galβ
	Influenza virus type A for birds	Neu5Acα(2→3)Galβ

of saccharides with pathogens are often stronger than those of the usual saccharide recognition proteins (lectins). In a typical biosensor, antibodies are utilized as the molecular recognition unit. Taking the strong interactions of glycopolymers into account, they can also be considered to be suitable for biosensing.

9.1.2 Saccharide–Protein Interactions in Biosensing

Biosensors are composed of a molecular recognition part and a sensing unit (Figure 9.1). The sensing parts have optical and electrical properties, but saccharides do not have these properties. Therefore, the glycopolymer should be conjugated to other materials to provide the required sensing properties.

Interestingly, the saccharide–protein interactions with specific pathogens and proteins show strong affinities. Generally, the lectin–saccharide interaction is modest for a monomeric saccharide, but some pathogen–saccharide interactions are very strong. For example, the saccharide–protein interactions with AB_5 toxins such as cholera and Shiga toxins (Stxs) are fairly strong ($K_d = 10^{-8}$–10^{-12} M)[5–7] owing to the densely packed sugar recognition units, and become comparable to antibody affinities. Influenza viruses also have densely packed sugar recognition units of hemagglutinin and sialidase, which can exhibit strong interactions.[8] In general, lectins have multivalent sugar recognition units.[9]

In a biosensor, not only the affinity to the target protein but also the molecular recognition specificity is important, since the non-specific adsorption of proteins reduces the sensitivity of biosensors. Non-specific adsorption is induced by hydrophobic, electrostatic and hydrogen-bonding interactions.[10] Notably, hydrophobic interactions contribute strongly to non-specific adsorption. In contrast, saccharides are hydrophilic compounds, which inhibit non-specific adsorption through hydrophobic interactions. In early

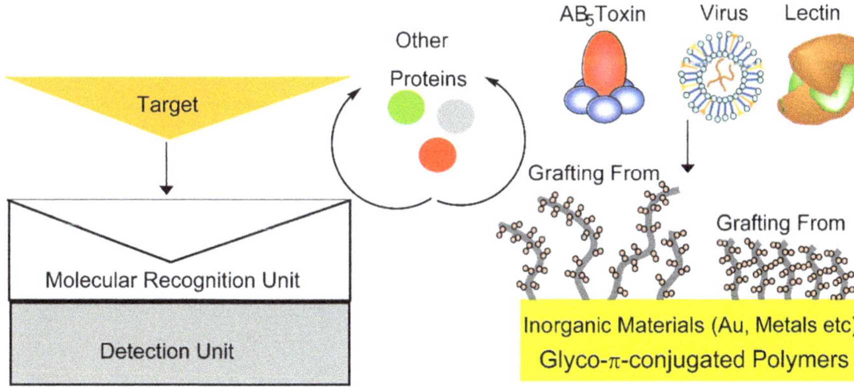

Figure 9.1 Schematic illustration of glycopolymer biosensor. Left, concept of a biosensor; right, concept of a glycopolymer biosensor.

studies of biointerfaces, Whitesides' group reported the inhibition of protein adsorption on saccharide-immobilized interfaces.[11] Biointerfaces with polysaccharides such as dextran are utilized in commercially available biosensors such as the Biacore CM5 chip. Biointerfaces with synthetic glycopolymers also show similar bioinert surface properties to polysaccharide-modified surfaces. We investigated substrates with glycopolymers where the substrates showed interactions with a specific lectin, but inhibited non-specific adsorption, *e.g.* of bovine serum albumin (BSA).[12]

Generally, the non-specific adsorption of proteins is inhibited by the addition of proteins such as BSA and casein.[13,14] However, proteins are larger than 10 nm, so the addition of protein to the interface causes surface roughness. The addition of BSA is effective for the improvement of traditional biosensing assays such as enzyme-linked immunosorbent assay (ELISA). More recent biosensors utilize other set-ups, such as surface plasmon resonance (SPR),[15] quartz crystal microbalance (QCM),[16] field-effect transistors (FETs)[17] and electrochemistry. These sensitive detection modalities are affected by the thicknesses and uniformity of the interface. Therefore, the modification of polymers, including glycopolymers, is advantageous for the sensitivity of the biosensors owing to their bioinertness, also minimizing possible protein denaturation.

9.1.3 Glycopolymer–Inorganic Conjugates

Different optical and electrical effects of substrates are utilized in modern biosensing. The substances used are precious metals and semiconductors. The optical and electrical effects occur near the interface at the nanometer level. Hence precise control of the polymer modification is important in biosensors.

The modification of inorganic substances has been extensively studied. Gold substrates are applicable in various optical and electrochemical

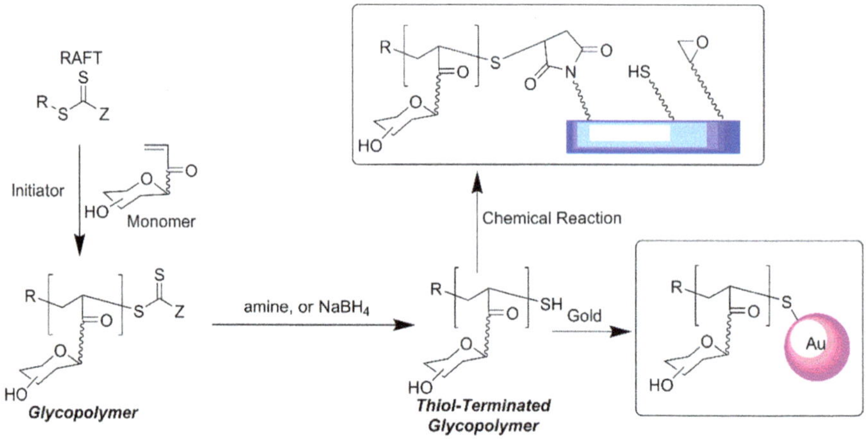

Scheme 9.1 Preparation of glycopolymer conjugates *via* RAFT living radical polymerization.

biosensing devices. These substrates are straightforwardly modified *via* the Au–S bond, which is utilized for the formation of self-assembled monolayers (SAMs),[18] and thiol-terminated polymers modify the gold substrate and the gold particles. Thiol-terminated glycopolymers are easily prepared, *e.g.* by reversible addition–fragmentation chain transfer (RAFT) polymerization (Scheme 9.1),[19] because *via* the RAFT method the polymer terminus presents di- or trithio groups that can be converted to thiol. Thiol-terminated poly(ethylene glycol) (PEG) has also been reported for modification of gold substrates, gold nanoparticles and quantum dots.[20] In addition to Au–S bond formation, thiol-terminated polymers are able to undergo various reactions such as thiol–maleimide coupling, disulfide bond formation and thioether bond formation.[19] We have reported glass substrate modification by maleimide coupling of glycopolymer *via* RAFT polymerization.[20] Thiol-terminated polymers permit the modification of various materials with various thiol reactivity.

The modification of silicon and metal oxides produces hydroxyl groups on the oxidized layer, which can be modified by alkoxysilanes from silane coupling reagents. The functional group-immobilized or hydrophobized surfaces were subjected to modification for biosensing and sugar arrays.[21] The hydrophobized substrates were applicable for the coating of glycopolymers based on hydrophobic interactions, as glycopolymers have an amphiphilic structure due to the hydrophobic polymer backbone and the hydrophilic sugar side chains.[22]

Similarly to glycopolymer coating with hydrophobic interactions, the self-assembling properties of polymers are advantageous in the preparation of glycopolymer–inorganic materials. For example, polymer–polymer interactions with polyelectrolytes are utilized for layer-by-layer (LBL) assembly.

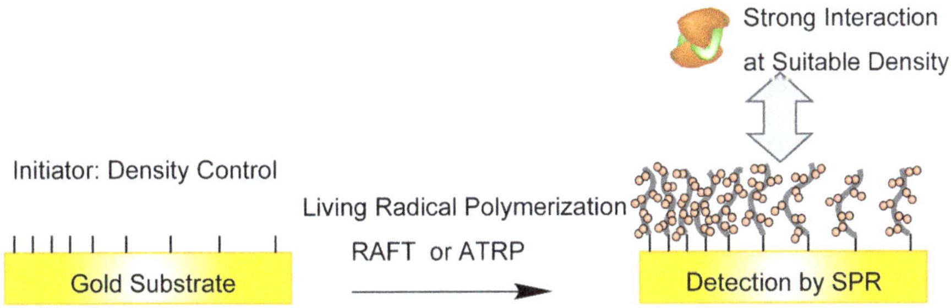

Figure 9.2 Control of glycopolymer density by surface-initiated living radical polymerization.

Saccharide-carrying vinyl monomers were copolymerized with acrylic acid to provide anionic glycopolymers, which were self-assembled with polycations such as polyethylenimine (PEI) by LBL deposition.[23]

Polymer brush formation is an important method for anchoring glycopolymers to surfaces. Polymer brushes can be prepared by 'grafting from' and 'grafting to' methods.[24] Especially the 'grafting from' method of surface-initiated living radical polymerization is a useful means of preparing polymer brushes. In this approach, a densely packed glycopolymer thin layer can be formed and the density of the glycopolymer layers is adjustable by the amount of initiator used (Figure 9.2). Stenzel *et al.* reported glycopolymer brush formation by using an immobilized RAFT reagent on a silicon substrate.[25] Xu and co-workers reported glycopolymer brush synthesis by the surface-initiated atom-transfer radical polymerization (SI-ATRP), where lactose-carrying glycopolymers formed the glycopolymer brush with various density and thickness.[26] They showed that the glycopolymer density and thickness greatly affected the interactions with lectin and suitable glycopolymer densities induced strong interactions with lectin. The 'grafting to' method involves polymer immobilization with a suitable functional group. As mentioned above, glycopolymers obtained *via* RAFT polymerization are easily converted to thiol-terminated polymers, which are utilized for glycopolymer brush formation by the 'grafting to' method *via* Au–S bond formation.[27] Glycopolymers with functional groups such as trimethoxysilane are utilized in the 'grafting to' method.[28] It is difficult to control the polymer density at the interface in the 'grafting to' method.

A novel method for surface modification involves dopamine, which recently has attracted much attention not only for its bioactivity but also its surface adhesiveness to various substrates. Dopamine and polydopamine have been reported to modify various substrates, including metals, glass and polymers.[29] Dopamine permits modification in the absence of an oxide layer and is preferred to silane coupling reagents as a modifier. Muñoz-Bonilla *et al.* reported that glycopolymers were immobilized on magnetite using dopamine methacrylate.[30]

9.2 Biosensing with Glycoconjugate Materials

9.2.1 Biosensing on Glycopolymer-Modified Gold Substrates

Glycopolymer-modified substrates can be used to detect specific proteins. Glycopolymer-modified gold substrates are utilized for biosensing with SPR, QCM and electrochemistry. We modified gold substrates with α-Man- and β-Gal-carrying glycopolymers with a thiol terminus *via* RAFT polymerization and demonstrated lectin detection [(α-Man for concanavalin A (ConA) and β-Gal for peanut agglutinin (PNA)] with SPR.[31] The glycopolymer [poly (α-Man-*co*-acrylamide)]-modified gold substrate showed the α-Man-specific lectin binding of ConA. The affinity of the glycopolymer layer was much stronger ($K_a \approx 10^7\ M^{-1}$) than that of monomeric saccharides. The non-specific protein binding of BSA and fibrinogen was inhibited by the hydrophilic glycopolymer layer. We compared the non-specific adsorption of proteins on the glycopolymer layers with that on a glycolipid monolayer. The amount of fibrinogen bound to the monolayer at 500 nM was larger than that of bound lectin because the bioinert properties of the glycolipid monolayer were not effective enough (Figure 9.3). The ratio of specific adsorption to non-specific adsorption of glycopolymer-coated gold substrate (S/N) was >10. The glycopolymer-coated gold substrate was utilized for biosensing of lectins. The glycopolymer layer could detect a specific lectin at the nanomolar level and the glycopolymer layer did not adsorb any non-specific lectin such as wheat germ agglutinin (WGA) (a GlcNAc-specific lectin).

Glycopolymer-modified gold substrates have been utilized for biosensing by other groups. Uzawa *et al.* reported biosensing of Shiga toxins (Stxs) using glycopolymer with a globobiose (Gb2)-modified gold substrate, which was prepared by the LBL technique.[32] Stxs showed a strong affinity to the glycopolymer layer and the affinity of the toxins was dependent on the kind of toxin and the sugar content. They synthesized two glycopolymers with lower (13%)

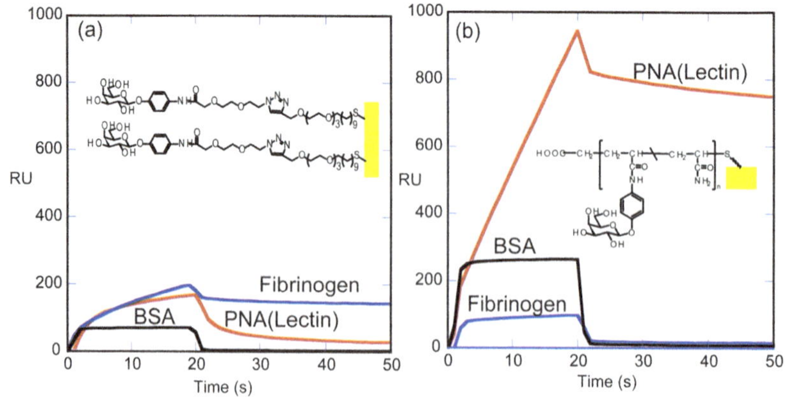

Figure 9.3 SPR diagrams of saccharide-immobilized interfaces of (a) glycomonolayer and (b) glycopolymer layer.

and higher (43%) Gb2. Stxs showed high affinities to the glycopolymer with a higher sugar content due to the greater multivalent effects. The affinity of the glycopolymer layer (43%) with Stx-1 reached 1.8×10^9 M^{-1}, which was comparable to or stronger than the affinity between antigen and antibody. They also reported the discrimination of Stx-1 and Stx-2 by the addition of another glycopolymer inhibitor. They succeeded in rapid and highly sensitive Stxs detection (10 ng mL^{-1}, 30 min). The same group also reported the detection of a toxic protein of ricin with lactose.[33]

Kiessling and co-workers reported the SPR analysis of glycopolymers with selectin and cells.[34] They immobilized the glycopolymer with a carboxyl group terminus on the carboxymethyldextran chip of a gold substrate. Maynard's group reported the immobilization of a glycopolymer on a gold substrate.[35] They synthesized a glycopolymer with a biotin terminus *via* ATRP and immobilized it on the gold substrate by biotin–streptavidin interaction. The glycopolymer-immobilized surface was utilized for the detection of proteins.

9.2.2 Biosensing With Glycopolymer-Coated Gold Nanoparticles

Glycopolymer-modified gold nanoparticles are optically and electrically active, and are utilized for biosensing. The chromogenic effect of gold nanoparticle is based on surface plasmons and is stable in various environments compared with an organic chromogenic group. The electrical activity is utilized for detection using voltammetry.

We prepared glycopolymer [poly(α-Man-*co*-acrylamide)]-modified gold nanoparticles using the same method as in Section 9.2.1 *via* RAFT polymerization.[27,36] The glycopolymer-modified gold nanoparticles showed a pink color with a plasmon peak around 530 nm. The color of the gold nanoparticles varied according to the distance between them due to the plasmon coupling. Since lectins have multiple binding sites, they could be a crosslinker of the glycopolymer-modified gold nanoparticles. ConA was added to the solution of the poly[α-Man-*co*-acrylamide]-modified gold nanoparticles, which induced the dramatic red shift in the UV spectra from 530 to 590 nm and the color changed from pink to blue. The dissociation constants from the UV shift were in the order of 10^{-8} M, indicating the strong interaction due to the multivalent effect. Sugar recognition protein (ConA) and bacteria (*Escherichia coli* ORN178) were detected by color change. The sensitivity of the color change was dependent on the amount of sugar in the glycopolymers, and glycopolymer with a high sugar content (100%) showed high sensitivity to lectin due to the multivalency effect.

The chromogenic effect of the glycopolymer-modified gold nanoparticle was applied in a biosensor using immune chromatography (lateral flow assay) (Figure 9.4).[37] Immune chromatography provides a simple and portable biosensor without any electrical device. We prepared an immune chromatography device with glycopolymer-modified gold nanoparticles

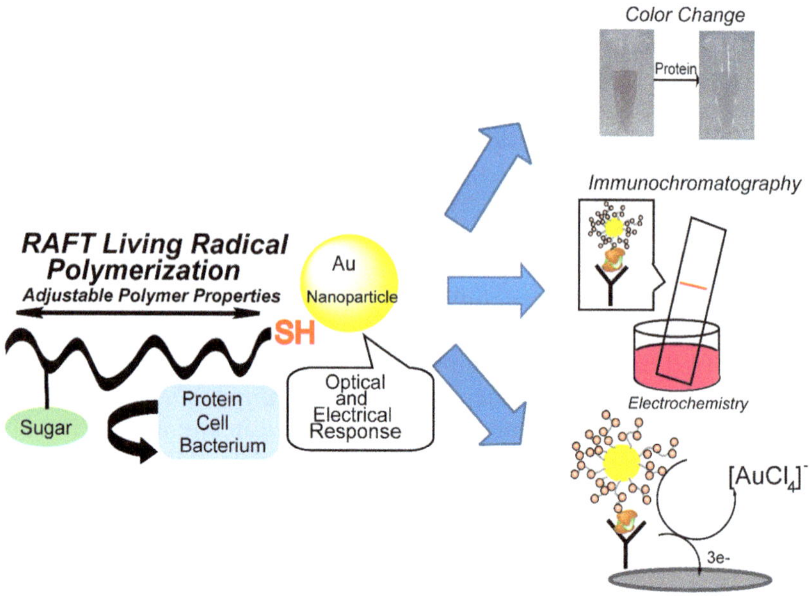

Figure 9.4 Biosensing of glycopolymer-modified gold nanoparticles.

and anti-lectin (ConA) antibody. While the nanoparticles with higher sugar contents showed high sensitivity in the colorimetric assay, the sensitivity of the glycopolymer-modified gold nanoparticles was low in the glycopolymer with a high sugar content (100%). The sensitivity is determined not only by the interaction with lectin but also by the aggregation properties and non-specific lectin on the chromatography paper. The glycopolymer with a 100% sugar content [poly(α-Man)] has less dispersion stability in aqueous solution due to the stiff polymer structure. The affinity of the glycopolymer–nanoparticles to lectin and the dispersion properties were a trade-off. In our case, the glycopolymer with a 66% sugar content showed the highest sensitivity in the detection of lectin by immune chromatography and lectin (ConA) was detectable at concentrations as low as 60 nM.

Gold nanoparticles are also utilized for detection with redox signals.[38] Since gold nanoparticles are subject to oxidation, the sugar recognition protein of ConA was detectable by voltammetry.[39] The lectin was labeled with glycopolymer–gold nanoparticles and captured by antibody, where the oxidation signal from Au(0) to Au^{3+} was monitored. The amount of protein correlates with the gold nanoparticle oxidation signal and the detection limit was around 100 pM.

Another important property of gold nanoparticles is the exhibition of local surface plasmon resonance (LSPR).[40] Whereas an SPR device with a gold substrate needs special apparatus such as a laser and a goniometer, LSPR can be measured with a regular spectrometer to accomplish label-free sensing. Kitano and co-workers reported LSPR measurements of a gold particle-immobilized substrate.[41] They immobilized gold nanoparticles on amino-terminal glass substrates and glycopolymers with mannose or lactose were immobilized on the gold nanoparticles. The substrate showed an LSPR peak at around 520 nm, The LSPR peak of the substrate increased with addition of lectins (ConA–Man or RCA_{120}–lactose) and decreased with addition of competitor due to the specific interaction. The peak did not decrease with the addition of other proteins such as BSA. The glycopolymer-modified gold nanoparticles also increased the sensitivity of the SPR signal due to the plasmon coupling.[31] Narain and co-workers reported the immobilization of biotinylated glycopolymer-modified gold nanoparticles on a gold substrate and biosensing with SPR measurement.[42]

9.3 Glyco-π-Conjugated Polymers for Biosensing

The most important property of glycopolymers is the exhibition of multivalent effects for molecular recognition. In the previous sections, glycopolymers were regarded as having a molecular recognition function and the conjugation of inorganic materials with glycopolymers was described, because the incorporation of optical or electrical activity was necessary for biosensing. On the other hand, π-conjugated polymers have been reported to show fluorescence, chromogenic effects and electrical activity, which were utilized in biosensors and chemosensors. Thus glyco-π-conjugated polymers represent novel biosensing polymers.

A facile and convenient method for the preparation of π-conjugated polymers is photopolymerization of polydiacetylene (PDA) lipids (Figure 9.5 (a)).[43] Various groups have reported sensing with PDA prepared, for example, with a commercially available PDA lipid such as 10,12-pentacosadiynoic acid. PDA lipids are easily polymerized by UV irradiation and can form micelle, liposome and Langmuir–Blodgett (LB) layers. Charych's group reported saccharide-modified PDAs.[44] They synthesized sialic acid-, lactose- and glucose-substituted PDA lipids and prepared an LB membrane on a substrate and also liposomes. Sialic acid-substituted PDA showed a UV peak around 500–700 nm, visible with the naked eye. The addition of an influenza virus decreased the absorbance to around 600 nm and showed a blue shift. Influenza viruses could be detected from 10 hemagglutinin units due to the specific and strong interaction with hemagglutinin. A natural glycolipid mixture with PDA also provided a facile method to prepare glyco-conjugated polymers. Pan and Charych reported a PDA liposome with a ganglioside mixture.[45] Ganglioside of GM1 was mixed with a diacetylene lipid to form liposome. Similarly to the PDA with sialic acid, GM1 mixed PDA liposome showed a blue shift in the UV spectrum on addition of cholera toxin. The detection limit of cholera toxin was 100 $\mu g\ mL^{-1}$.

Nishimura and co-workers reported proteome analyses with PDA. They prepared PDA with an aminooxy group or saccharide. They monitored the glycoprotein interactions colorimetrically and the saccharide of the glycoprotein was separated and analyzed with PDA liposome.[46] They also prepared

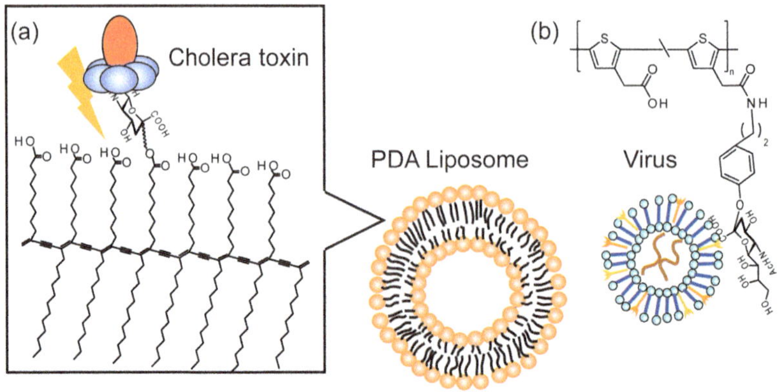

Figure 9.5 Glycopolymer-π-conjugated polymers for biosensing.

a glycolipid of PDA, which was utilized for the colorimetric sensing of galactosyltransferase activity.[47]

Other π-conjugated polymers have also been investigated for biosensing. Baek *et al.* reported the syntheses of saccharide (sialic acid and Man)-modified polythiophene and biosensing with glycopolythiophene (Figure 9.5 (b)).[48] They prepared polythiophene carrying saccharides with a sugar content of around 40%. The polythiophene was water soluble due to the side groups of saccharides and carboxylic acid and showed a broad UV peak around 400 nm. In the case of polythiophene with sialic acid, the addition of influenza viruses (influenza virus A and B) induced a red shift of around 10 nm, although sialic acid recognition of WGA did not induce a peak shift. Similarly, in the case of polythiophene with mannose, the addition of *E. coli* induced a red shift of around 20–30 nm. The addition of mannose recognition lectin (ConA) quickly induced aggregation of the polymer.

Disney *et al.* reported poly(*p*-phenyleneethynylene) with mannose, which exhibited a green color with a peak around 500 nm.[49] The addition of ConA quickly induced aggregation of the polymer and the fluorescence decreased owing to this aggregation. ConA was detected by the measuring decrease in fluorescence and the detection limit of ConA was 0.1 μM, whereas a polymer with galactose did not show any change. Poly(*p*-phenyleneethynylene) with Man also attached to *E. coli* and the bacteria were made visible with fluorescence. Similarly, the syntheses of various representative π-conjugated polymers with sugars such as poly(*p*-phenylene),[50] polyphenylenevinylene[51,52] and polyphenylacetylene[53] have been reported. For biosensing, the fluorescence change is important for detecting the target molecules. The π-conjugated polymers easily aggregate by crosslinking with lectins, viruses and bacteria, compared with low molecular weight fluorophores. The fluorescence of the conjugated polymers was quenched by aggregation and the π-conjugated glycopolymers showed a larger fluorescence change due to the high molecular weight. Most of the π-conjugated glycopolymers detect proteins by quenching, but π-conjugated glycopolymers that which detect protein by emission have also been investigated.[54]

9.4 Conclusion

In this chapter, biosensing with glycopolymers has been described. The glycopolymer biosensors were produced with two materials, inorganic material conjugates and π-conjugated polymers. Various techniques have been described for the conjugation of glycopolymer–inorganic materials. In the glycopolymer–inorganic conjugates, various biosensing experiments with existing devices have been reported and many groups have studied biosensing with SPR. Notably, the detection of AB_5 toxins showed very high sensitivity. In other sensing methods, the interaction between the target protein and saccharide was amplified by multivalency of the glycopolymer and the glycopolymer sensing showed modest sensitivity except for redox signal measurements.

Practical biosensing using immune chromatography has also been reported. In π-conjugated polymers, biosensing with observations with the naked eye was achieved. Polydiacetylene lipids with sugars were prepared that showed a color change with addition of the target protein and pathogens. Other π-conjugated polymers showed quenching of fluorescence on addition of proteins. The various examples indicate that glycopolymers are useful materials for specific biosensing and that the fabrication of novel glycopolymers is important for the development of new biosensors.

References

1. R. A. Dwek, *Chem. Rev.*, 1996, **96**, 683–720.
2. F. Santoyo-Gonzalez and F. Hernadez-Mateo, *Chem. Soc. Rev.*, 2009, **38**, 3449–3462.
3. Y. Miura, *Polym. J.*, 2012, **44**, 679–689.
4. M. Mammen, S.-K. Choi and G. M. Whitesides, *Angew. Chem. Int. Ed.*, 1998, **37**, 2754–2794.
5. G. M. Kuziemko, M. Stroh and R. C. Stevens, *Biochemistry*, 1996, **35**, 6375–6384.
6. H. Nakajima, N. Kiyokawa, Y. U. Katagiri, T. Tatuchi, T. Suzuki, T. Sekino, K. Mimori, T. Ebata, M. Saito, H. Nakao, T. Takeda and J. Fujimoto, *J. Biol. Chem.*, 2001, **276**, 42915–42922.
7. Y. Miura, Y. Sasao, H. Dohi, Y. Nishida and K. Kobayashi, *Anal. Biochem.*, 2002, **310**, 27–35.
8. M. Waldmann, R. Jirmann, K. Hoelscher, M. Wienke, F. C. Niemeyer, D. Rehders and B. Meyer, *J. Am. Chem. Soc.*, 2014, **136**, 783–788.
9. J. M. Rini, *Annu. Rev. Biophys. Biomol. Struct.*, 1995, **24**, 551–577.
10. J. J. Gray, *Curr. Opin. Struct. Biol.*, 2004, **14**, 110–115.
11. K. L. Prime and G. M. Whitesides, *Science*, 1991, **252**, 1164–1167.

12. Y. Miura, H. Sato, T. Ikeda, H. Sugimura, O. Takai and K. Kobayashi, *Biomacromolecules*, 2004, **5**, 1708–1713.
13. Y. L. Jeyachandran, J. A. Mielczarski, E. Mielczarski and B. Rai, *J. Colloid Interface Sci.*, 2010, **341**, 136–142.
14. R. F. Vogt Jr., D. L. Philipps, L. O. Henderson, W. Whitefield and F. W. Spierto, *J. Immunol. Methods*, 1987, **101**, 43–50.
15. J. Homola, *Chem. Rev.*, 2008, **108**, 462–493.
16. Y. Ebara and Y. Okahata, *Langmuir*, 1993, **9**, 574–576.
17. T. Sakata and Y. Miyahara, *Angew. Chem. Int. Ed.*, 2006, **45**, 2225–2228.
18. J. C. Love, L. A. Estroff, J. K. Kriebel, R. G. Nuzzo and G. M. Whitesides, *Chem. Rev.*, 2005, **105**, 1103–1169.
19. P. J. Roth, C. Boyer, A. B. Lowe and T. P. Davis, *Macromol. Rapid Commun.*, 2011, **32**, 1123–1143.
20. H. Otsuka, Y. Akiyama, Y. Nagasaki and K. Kataoka, *J. Am. Chem. Soc.*, 2001, **123**, 8226–8230.
21. H. Seto, M. Takara, C. Yamashita, T. Murakami, T. Hasegawa, Y. Hoshino and Y. Miura, *ACS Appl. Mater. Interfaces*, 2012, **4**, 5125–5133.
22. K.-S. Ko, F. A. Jaipuri and N. L. Pohl, *J. Am. Chem. Soc.*, 2005, **127**, 13162–13163.
23. C. Boyer, A. Bousquet, J. Rohdolo, M. R. Whittaker, M. H. Stenzel and T. P. Davis, *Macromolecules*, 2010, **43**, 3775–3784.
24. O. Azzaroni, *J. Polym. Sci. Part A: Polym. Chem.*, 2012, **50**, 3225–3258.
25. M. H. Stenzel, L. Zhang and W. T. S. Huck, *Macromol. Rapid Commun.*, 2006, **27**, 1121–1126.
26. X.-L. Meng, Y. Fang, L.-S. Wan, X.-J. Huang and Z.-K. Xu, *Langmuir*, 2012, **28**, 13616–13623.
27. M. Toyoshima and Y. Miura, *J. Polym. Sci. Part A: Polym. Chem.*, 2009, **47**, 1412–1421.
28. H. Seto, Y. Ogata, T. Murakami, Y. Hoshino and Y. Miura, *ACS Appl. Mater. Interfaces*, 2012, **4**, 411–417.
29. H. Lee, S. M. Dellatore, W. M. Miller and P. B. Messersmith, *Science*, 2007, **318**, 426–430.
30. A. Muñoz-Bonilla, G. Marcelo, C. Casado, F. J. Teran and M. Fernández-García, *J. Polym. Sci. Part A: Polym. Chem.*, 2012, **50**, 5087–5096.
31. M. Toyoshima, T. Oura, T. Fukuda, E. Matsumoto and Y. Miura, *Polym. J.*, 2010, **42**, 172–178.
32. H. Uzawa, H. Ito, P. Neri, H. Mori and Y. Nishida, *ChemBioChem*, 2007, **8**, 2117–2124.
33. T. Nagatsuka, H. Uzawa, K. Sato, S. Kondo, M. Izumi, K. Yokoyama, I. Ohsawa, Y. Seto, P. Neri, H. Mori, Y. Nishida, M. Saito and E. Tamiya, *ACS Appl. Mater. Interfaces*, 2013, **5**, 4173–4180.
34. J. E. Gestwicki, C. W. Cairo, D. A. Mann, R. M. Owen and L. L. Kiessling, *Anal. Biochem.*, 2002, **305**, 149–155.
35. V. Vazquez-Dorbatt and H. D. Maynard, *Biomacromolecules*, 2006, **7**, 2297–2302.
36. M. Takara, M. Toyoshima, H. Seto, Y. Hoshino and Y. Miura, *Polym. Chem.*, 2014, **5**, 931–939.

37. J. W. Davis, M. Demetriades and J. M. Bowen, *J. Immunol. Methods*, 1984, **68**, 157–165.
38. K. Idegami, M. Chikae, K. Kerman, N. Nagatani, T. Yuhi, T. Endo and E. Tamiya, *Electroanalysis*, 2008, **20**, 14–21.
39. J. Ishii, M. Chikae, M. Toyoshima, Y. Ukita, Y. Miura and Y. Takamura, *Electrochem. Commun.*, 2011, **13**, 830–833.
40. A. J. Haes and R. P. V. Duyne, *J. Am. Chem. Soc.*, 2002, **124**, 10596–10604.
41. (a) K. Mizukami, H. Takakura, T. Matsunaga and H. Kitano, *Colloids Surf., B*, 2008, **66**, 110–118; (b) S. Morokoshi, K. Ohhori, K. Mizukami and H. Kitano, *Langmuir*, 2004, **20**, 8897–8902.
42. X. Jiang, A. Housni, G. Gody, P. Boullanger, M.-T. Charreyre, T. Delair and R. Narain, *Bioconjugate Chem.*, 2010, **21**, 521–530.
43. M. Reppy and B. Pindzola, *Chem. Commun.*, 2007, 4317–4338.
44. (a) D. H. Charych, J. O. Nagy, W. Spevak and M. D. Bednarski, *Science*, 1993, **261**, 585–588; (b) D. Charych, Q. Cheng, A. Reichertl, G. Kuziemko, M. Stroh, J. O. Nagyl, W. Spevakl and R. C. Stevens, *Chem. Biol.*, 1996, **3**, 113–120.
45. J. J. Pan and D. Charych, *Langmuir*, 1997, **13**, 1365–1367.
46. K. Niikura, R. Kamitani, M. Kurogochi, R. Uematsu, Y. Shinohara, H. Nakagawa, K. Deguchi, K. Monde, H. Kondo and S.-I. Nishimura, *Chem. Eur. J.*, 2005, **11**, 3825–3834.
47. N. Nagahori, K. Niikura, R. Sadamoto, M. Taniguchi, A. Yamagishi, K. Monde and S.-I. Nishimura, *Adv. Synth. Catal.*, 2003, **345**, 729–734.
48. M. G. Baek, R. C. Stevens and D. H. Charych, *Bioconjugate Chem.*, 2000, **11**, 777–788.
49. M. D. Disney, J. Zheng, T. M. Swager and P. H. Seeberger, *J. Am. Chem. Soc.*, 2004, **126**, 13343–13346.
50. C. Xue, S. P. Jog, P. Murthy and H. Liu, *Biomacromolecules*, 2006, **7**, 2470–2474.
51. I.-B. Kim, J. N. Wilson and U. H. F. Bunz, *Chem. Commun.*, 2005, 1273–1275.
52. A. Takasu, K. Iso, T. Dohmae and T. Hirabayashi, *Biomacromolecules*, 2006, **7**, 411–414.
53. I. Otsuka, T. Hongo, H. Nakade, A. Narumi, R. Sakai, T. Satoh, H. Kaga and T. Kakuchi, *Macromolecules*, 2007, **40**, 8930–8937.
54. J. Shi, L. Cai, K.-Y. Pu and B. Liu, *Chem. Asian J.*, 2010, **5**, 301–308.

Subject Index